PARTITIONS

Optimality and Clustering

Vol. II: Multi-Parameter

SERIES ON APPLIED MATHEMATICS

Editor-in-Chief: Zhong-ci Shi

---

*Published**

Vol. 7 The Splitting Extrapolation Method
by C. B. Liem, T. Lü and T. M. Shih

Vol. 8 Quaternary Codes
by Z.-X. Wan

Vol. 9 Finite Element Methods for Integrodifferential Equations
by C. M. Chen and T. M. Shih

Vol. 10 Statistical Quality Control — A Loss Minimization Approach
by D. Trietsch

Vol. 11 The Mathematical Theory of Nonblocking Switching Networks
by F. K. Hwang

Vol. 12 Combinatorial Group Testing and Its Applications (2nd Edition)
by D.-Z. Du and F. K. Hwang

Vol. 13 Inverse Problems for Electrical Networks
by E. B. Curtis and J. A. Morrow

Vol. 14 Combinatorial and Global Optimization
eds. P. M. Pardalos, A. Migdalas and R. E. Burkard

Vol. 15 The Mathematical Theory of Nonblocking Switching Networks
(2nd Edition)
by F. K. Hwang

Vol. 16 Ordinary Differential Equations with Applications
by S. B. Hsu

Vol. 17 Block Designs: Analysis, Combinatorics and Applications
by D. Raghavarao and L. V. Padgett

Vol. 18 Pooling Designs and Nonadaptive Group Testing — Important Tools
for DNA Sequencing
by D.-Z. Du and F. K. Hwang

Vol. 19 Partitions: Optimality and Clustering
Vol. I: Single-parameter
by F. K. Hwang and U. G. Rothblum

Vol. 20 Partitions: Optimality and Clustering
Vol. II: Multi-Parameter
by F. K. Hwang, U. G. Rothblum and H. B. Chen

*For the complete list of the volumes in this series, please visit
http://www.worldscibooks.com/series/sam_series.shtml

Series on
Applied Mathematics
Volume 20

PARTITIONS

Optimality and Clustering
Vol. II: Multi-Parameter

Frank K Hwang
National Chiao-Tung University, Taiwan

Uriel G Rothblum
Technion, Israel

Hong-Bin Chen
Academia Sinica, Taiwan

NEW JERSEY • LONDON • SINGAPORE • BEIJING • SHANGHAI • HONG KONG • TAIPEI • CHENNAI

Published by

World Scientific Publishing Co. Pte. Ltd.
5 Toh Tuck Link, Singapore 596224
USA office: 27 Warren Street, Suite 401-402, Hackensack, NJ 07601
UK office: 57 Shelton Street, Covent Garden, London WC2H 9HE

Library of Congress Cataloging-in-Publication Data
Hwang, Frank.
Partitions : optimality and clustering (v. 1. Single-parameter) / by Frank K. Hwang & Uriel G. Rothblum.
p. cm. -- (Series on applied mathematics ; v. 19)
Includes bibliographical references and index.
ISBN-13 978-981-270-812-0 (hardcover : alk. paper)
ISBN-10 981-270-812-X (hardcover : alk. paper)
1. Partitions (Mathematics). I. Title. II. Rothblum, Uriel G.
QA165 .H83 2011
512.7'3--dc22
2010038315

British Library Cataloguing-in-Publication Data
A catalogue record for this book is available from the British Library.

Partitions: Optimality and Clustering
Vol. II: Multi-Parameter
Series on Applied Mathematics — Vol. 20
ISBN 978-981-4412-34-6

Printed in Singapore.

Dedicated to our wives –

Aileen, Naomi and Ada

Contents

Preface

The partition problem considered in the book is to partition a set A of points in $\mathbb{R}^d$ (actually, A can have identical elements, so we abuse the term "set" here) into several parts to maximize a given objective function. Suppose A contains n elements. Then we can index the elements $A^1, \cdots, A^n$ so as to distinguish one from the other. However, in $\mathbb{R}^1$, A^i's are just numbers which can be ordered. Thus it is natural (and also advantageous) to index A^i's according to this order, i.e.,

$$i < i' \Rightarrow A^i \leq A^{i'}.$$

Another crucial choice was made in Vol. I, which is to do the partition on the index-set $\mathcal{N} = \{1, 2, \cdots, n\}$ (*index-partition*), rather than on $A = \{A^1 \leq A^2 \leq \cdots \leq A^n\}$. Note that A is a geometric set while $\mathcal{N}$ is just a combinatorial set. By replacing A with $\mathcal{N}$, information on the distances between every pair of adjacent A^i's is lost; in particular, information on the zero-distances which implies two identical points is lost. Luckily, most of the important partition properties are order-dependent and little need for more details on distances, except the zero-distances.

For $\mathbb{R}^d$, $d \geq 2$, there is no overwhelming natural order to index the points (while coordinates index is an index, its ordering ignores distances and has very little impact on our partition theory). Therefore, random indices from the set $\mathcal{N}$ are matched to the set A, just to provide names, but no implication of any sort. Consequently, for the $d \geq 2$ case, the partition is on A, called a *vector-partition*, not on $\mathcal{N}$ (but we sometimes still write $i \in \pi_j$ as short for $A^i \in \pi_j$).

One consequence of dealing with the partition of A is that we can no longer ignore the possibility that A^i's are not necessarily all distinct. We deal with this new problem by introducing an equivalence relation among partitions: two partitions are equivalent if they differ only in interchanging identical vectors. We can still use indices to distinguish elements in A, the important things are that equivalent elements are recognized and no order is attached to the indices.

We say that A is in general position, or A is generic, if any set of $d+1$ vectors in A is linearly independent (see Vol. I, p. 85). The new complications in partitioning A actually go beyond the nondistinctness of A but also when A is distinct but not generic. It turns out that for most results on partitions, whether A is generic not only affects the proofs, but may also affect the results. So new methods have to be developed to deal with these new situations.

We also make another generalization in this volume by allowing parts to be empty. Note that a p-partition problem allowing empty parts is the same as a bounded-size partition problem not allowing empty parts with bounds $(1, p)$, a problem we skipped in Vol. I.

Here a chapter dealing with a generalization in a different direction is included. This time, the 1-dim is preserved, but the partitioned elements are divided into t types, each having its own restrictions on size and shape. This problem is originated from the "system assembly" problem where each part π_j requires a certain number n_{uj} (or a bound) of type u (see Chapter 1 of Vol. I or Chapter 6 of this volume).

Finally, this volume gives a detailed analysis of the application problems raised in Section 1.4 of Vol. I by using the results reported in the two volumes.

I (Frank) deeply lament the untimely death of Uri after having laid down the foundation of the book. This huge loss of his wisdom and warmth hit me both academically and personally. I also thank Hongbin for kindly joining us to complete Volume II and thank Maydan (Modi), Guy and Roni, the three charming sons of Uri and Naomi, for helping us to search through Uri's files for material needed for this volume.

Chapter 1

Bounded-Shape Sum-Partition Problems: Polyhedral Approach

In this chapter, we consider partition problems where the elements to be partitioned are associated with multiple, say d, parameters; so, each element $i = 1, \ldots, n$ is associated with a vector A^i of dimension d and $A = (A^1, \ldots, A^n) \in \mathbb{R}^{d \times n}$ (A will denote both the $d \times n$ matrix and the set of n $d \times 1$ vectors in A). In the sum-partition problem, the task is to maximize an objective function $F(.)$ over a family of p-partitions Π where the objective value of a p-partition $\pi = (\pi_1, \ldots, \pi_p)$ is given by $F(\pi) = f(A_\pi)$, with $A_\pi = [\sum_{i \in \pi_1} A^i, \ldots, \sum_{i \in \pi_p} A^i] \in \mathbb{R}^{d \times p}$ and $f(.)$ a real-valued function on $\mathbb{R}^{d \times p}$ (or a relevant subset thereof). Here, as in the case of the one-dimensional parameter space where each element is associated with a single parameter ($d = 1$), we use the approach of considering the optimization problem over the partition polytope P_A^Π defined as the convex hull of $\{A_\pi : \pi \in \Pi\}$. When the function f is guaranteed to obtain a maximum over P_A^Π at a vertex, enumerating the vertices of P_A^Π and selecting a partition corresponding to a vertex that maximizes f will produce a solution to the partition problem.

As noted in the Preface, we allow partition-parts to be empty in general. Of course, for single/bounded/constrained-shape families, we can control empty parts through specifying the shapes, but not for single/bounded/constrained-size families. What we will do is to classify the size families into two types: those allowing empty parts are called *relaxed size families* and those not simply size families. Note that optimizing over a relaxed size family can lead to a better solution because it can choose from a larger set of options.

Throughout this chapter we consider partition problems with the structure described in the above paragraph. In particular, unless stated explicitly otherwise, we assume that data d, n and a $d \times n$ matrix A are given and fixed. As usual, bounds that are expressed in terms of "$O(\cdot)$" or "$\Omega(\cdot)$" refer to varying variable n with d and p fixed (recall that $O(\cdot)$ and $\Omega(\cdot)$ denote the order of complexity of an upper bound and a lower bound, respectively).

1.1 Linear Objective: Solution by LP

In the current section, we assume that the function $f(.)$ is linear and show how linear programming can be used to solve the corresponding partition problem. Recall that for 1-dim problems, Chapter 5 of Vol. I provided a linear-inequality representation of the partition polytope through the function θ_*^Π. However, in the multi-parameter case, the A^i's cannot be meaningfully ordered and there is generally no natural definition for the multi-parameter version of θ_*^Π that is useful for obtaining a linear-inequality representation of the partition polytope. The approach followed in the current section does not depend on such a representation—instead, we obtain a representation of the problem as a projection of an optimization problem over a (generalized transportation) polytope which has a simple linear inequality representation. The resulting algorithm is polynomial in the number of partitioned vectors, their dimension and the number of parts of the partitions. Following Hwang, Onn and Rothblum [54], the approach builds on the ideas outlined in Vol. I (Lemma 3.5.7 of and Algorithm PLP of Section 3.5 of projecting linear programming solvability to the image of linear operators); here, we explicitly use the special structure of the source polytope being a generalized transportation-polytope. The development also yields a polynomial test for vertices of bounded-shape partition polytopes.

Throughout this section, we use superscripts to denote columns of matrices, subscripts for rows and double indices for elements, e.g., U^j, U_t and U_t^j. The vector of 1's of appropriate dimension is denoted by e. Also, for positive integer j, let e^j denote the standard j-unit vector in the corresponding Euclidean space (usually, either $\mathbb{R}^p$ or $\mathbb{R}^n$). For matrices U and V of common dimension, say $d \times p$, the *inner product* of U and V is defined as the scalar $\langle U, V\rangle \equiv \sum_{t=1}^d \sum_{j=1}^p U_t^j V_t^j$. For matrices U, V and W of dimension $d \times p$, $d \times q$ and $q \times p$, respectively, we have that

$$\langle U, VW\rangle = \langle V^T U, W\rangle = \langle UW^T, V\rangle . \tag{1.1.1}$$

Note that for every p-partition π,

$$A_\pi = \left[\sum_{i\in\pi_1} A^i, \ldots, \sum_{i\in\pi_p} A^i\right] = \sum_{j=1}^{p}\left(\sum_{i\in\pi_j} A^i\right)(e^j)^T = \sum_{j=1}^{p}\sum_{i\in\pi_j} A^i(e^j)^T. \tag{1.1.2}$$

Let I be the $n \times n$ identity matrix. We shall consider the columns of I as vectors associated with the elements that are to be partitioned and for each p-partition π, we use the notation I_π in the standard form; applying (1.1.2) with I replacing A, we have

$$(I_\pi)_t^j \equiv \begin{cases} 1 & \text{if } t \in \pi_j \text{ and} \\ 0 & \text{otherwise} \end{cases} = \left[\sum_{j=1}^{p}\sum_{t\in\pi_j} e^t(e^j)^T\right]_t^j, \tag{1.1.3}$$

where e^t and e^j denote the corresponding unit vectors in $\mathbb{R}^n$ and $\mathbb{R}^p$, respectively. For a set of partitions Π, we refer to $P_I^\Pi = \text{conv}\{I_\pi : \pi \in \Pi\}$ as a *generalized transportation-polytope* (*associated with* Π) and to the corresponding partition problem as a *generalized transportation problem.* We observe that classical transportation-polytopes are obtained when Π consists of a single shape (see the forthcoming Corollary 1.1.4). The special attention given to this class of partition problems is justified by the following properties that we are about to establish:

(i) The correspondence $\pi \to I_\pi$ is a bijection; further, none of the matrices I_π are expressible as a convex combination of others (the forthcoming Lemma 1.1.1 and Corollary 1.1.2).
(ii) Explicit representing systems of linear inequalities are available for bounded-shape generalized transportation-polytopes (the forthcoming Theorem 1.1.3).
(iii) Constrained-shape partition problems are reducible to generalized transportation problems (the forthcoming Theorem 1.1.5 and Corollary 1.1.6).

We next verify these properties and use them to develop efficient computational methods for solving the (linear) partition problem.

Lemma 1.1.1. *For each p-partition π, the matrix I_π cannot be expressed as a convex combination of matrices $I_{\pi'}$ for p-partitions $\pi' \neq \pi$, in particular, the correspondence $\pi \to I_\pi$ mapping p-partitions into $n \times p$ matrices is one-to-one.*

Proof. For a given p-partition π, the rows of I_π are the transpose of unit vectors in $\mathbb{R}^p$ with $(I_\pi)_i = e^j$ if and only if $A^i \in \pi_j$. As no unit vector in $\mathbb{R}^p$ is expressible as a convex combination of other unit vectors in $\mathbb{R}^p$, I_π cannot be expressed as a convex combination of matrices $I_{\pi'}$ for p-partitions $\pi' \neq \pi$. The conclusion that the correspondence $\pi \to I_\pi$ mapping partitions into $n \times p$ matrices is one-to-one follows immediately. □

Corollary 1.1.2. *Let Π be a set of p-partitions. Then the correspondence $\pi \to I_\pi$ is a bijection of Π onto the vertices of P_I^Π; in particular, the vertices of P_I^Π are precisely the matrices $\{I_\pi : \pi \in \Pi\}$. Further, the inverse correspondence of vertices of P_I^Π onto the partitions of Π has vertex v corresponding to the partition π with $\pi_j = \{t : v_t^j = 1\}$ for $j = 1, \ldots, p$.*

Proof. Let V be the set of vertices of P_I^Π. As P_I^Π is the convex hull of $\{I_\pi : \pi \in \Pi\}$, $V \subseteq \{I_\pi : \pi \in \Pi\}$ (Proposition 3.1.1 of Vol. I, part (i)). Also, as P_I^Π is a polytope, it is the convex hull of its vertices (Proposition 3.1.1 of Vol. I, part (a)). Now, for a partition $\pi \in \Pi$, $I_\pi \in P_I^\Pi$; hence, I_π is in the convex hull of $V \equiv \{I_\pi : \pi \in \Pi$ and I_π is a vertex of $P_I^\Pi\}$. But, Lemma 1.1.1 implies that this is possible only if $I_\pi \in V$. Lemma 1.1.1 further assures that no vertex of P_I^Π corresponds to more than one single partition. Finally, the constructive representation of the inverse correspondence of vertices of P_I^Π onto the partitions of Π is immediate from (1.1.3). □

An important feature of the bijection $\pi \to I_\pi$ of Lemma 1.1.1 and Corollary 1.1.2 is the fact that it is constructive in both directions; namely, it is easy to determine I_π from π and, vice versa, π from I_π.

We next provide explicit linear-inequality representation for bounded-shape generalized transportation-polytopes.

Theorem 1.1.3. *Let L and U be positive integer p-vectors satisfying $L \leq U$ and $\sum_{j=1}^p L_j \leq n \leq \sum_{j=1}^p U_j$, and let $\Pi \equiv \Pi^{(L,U)}$. Then P_I^Π is the solution set of the linear system:*

$$\begin{aligned} &(a)\ X_t^j \geq 0 \ \text{ for } \ t = 1, \ldots, n \ \text{ and } \ j = 1, \ldots, p, \\ &(b)\ \sum_{j=1}^p X_t^j = 1 \ \text{ for } \ t = 1, \ldots, n, \\ &(c)\ L_j \leq \sum_{t=1}^n X_t^j \leq U_j \ \text{ for } \ j = 1, \ldots, p. \end{aligned} \tag{1.1.4}$$

Proof. Let K be the solution set of (1.1.4). Trivially, $I_\pi \in K$ for each $\pi \in \Pi$, implying that the convex hull of these matrices, namely P_I^Π, is contained in K. Next, standard results (that rely on the fact that the constraint matrix of the inequality system (1.1.4) is totally unimodular) assure that the vertices of K are integer solutions of (1.1.4) (cf., Schrijver [85]); as integer solutions of (1.1.4) correspond to p-partitions in Π, that is, have representation as I_π for some $\pi \in \Pi$, each vertex of K is in P_I^Π. As K is the convex hull of its vertices (Proposition 3.1.1 of Vol. I, part (a)), we conclude that $K \subseteq \text{conv}\ P_I^\Pi = P_I^\Pi$. □

When $L = U$, Theorem 1.1.3 specializes to the following inequality description of classical transportation-polytopes.

Corollary 1.1.4. *Let $n_1, \ldots, n_p$ be positive integers with $\sum_{j=1}^p n_j = n$ and let $\Pi \equiv \Pi^{(n_1,\ldots,n_p)}$. Then P_I^Π is the solution set of the linear system:*

$$\begin{aligned}
&(a)\ X_t^j \geq 0 \ \textit{ for }\ t = 1,\ldots,n \ \textit{ and }\ j = 1,\ldots,p,\\
&(b)\ \sum_{j=1}^p X_t^j = 1 \ \textit{ for }\ t = 1,\ldots,n, \qquad (1.1.5)\\
&(c)\ \sum_{t=1}^n X_t^j = n_j \ \textit{ for }\ j = 1,\ldots,p.
\end{aligned}$$

Theorem 1.1.5. *Let $A \in \mathbb{R}^{d\times n}$ and let Π be a set of p-partitions. Then the partition polytope P_A^Π is the image of the generalized transportation-polytope P_I^Π under the linear function mapping $X \in P_I^\Pi \subseteq \mathbb{R}^{n\times p}$ into $AX \in \mathbb{R}^{d\times p}$. In particular, for every p-partition π, $A_\pi = AI_\pi$.*

Proof. Observe from (1.1.3) and (1.1.2) that for every p-partition π

$$AI_\pi = \sum_{j=1}^p \sum_{t\in\pi_j} Ae^t(e^j)^T = \sum_{j=1}^p \sum_{t\in\pi_j} A^t(e^j)^T = A_\pi. \qquad (1.1.6)$$

As $P_I^\Pi = \text{conv}\{I_\pi : \pi \in \Pi\}$, a standard argument about convex hulls of sets combines with (1.1.6) to show that

$$\begin{aligned}
\{AX : X \in P_I^\Pi\} &= \{AX : X \in \text{conv}\{I_\pi : \pi \in \Pi\}\}\\
&= \text{conv}\{AI_\pi : \pi \in \Pi\} = \text{conv}\{A_\pi : \pi \in \Pi\} = P_A^\Pi.
\end{aligned}$$ □

The representation of partition polytopes as a projection of generalized transportation-polytopes derived in Theorem 1.1.5 yields the following test for vertices of constrained-shape partition polytopes.

Corollary 1.1.6. *Let $A \in \mathbb{R}^{d\times n}$, let Π be a set of p-partitions and let $V \in P_A^\Pi$. Then V is a vertex of P_A^Π if and only if for every representation $V = \frac{1}{2}A(X' + X'')$ with $X', X'' \in P_I^\Pi$, we have that $AX' = AX''$.* □

Using the linear-inequality representation of bounded-shape generalized transportation-polytopes available from Theorem 1.1.3, we get an efficient test for vertices of bounded-shape partition polytopes.

Testing If a Vector (say A_π) is a Vertex of a given Bounded-Shape Partition Polytope

Let $A \in \mathbb{R}^{d\times n}$, L and U be positive integer p-vectors satisfying $L \le U$ and $\sum_{j=1}^{p} L_j \le n \le \sum_{j=1}^{p} U_j$ and $\Pi \equiv \Pi^{(L,U)}$. Also, let $V \in P_A^\Pi$. For each $i = 1, \ldots, d$ and $j = 1, \ldots, p$, let $E^{i,j}$ be (i,j)-unit matrix in $\mathbb{R}^{d\times p}$, that is, the matrix with all of its coordinates are zero except for the (i,j)-coordinate which is 1. Corollary 1.1.6 implies that V is a vertex of P_A^Π if and only if for each $i = 1, \ldots, d$ and $j = 1, \ldots, p$, the maximum of $\langle E^{i,j}, A(X' - X'')\rangle = \langle A^T E^{i,j}, X' - X''\rangle$ over $\{(X', X'') \in P_I^\Pi \times P_I^\Pi : AX' + AX'' = 2V\}$ is zero. Using the representation of P_I^Π from Theorem 1.1.3, these conditions can be checked by solving dp linear programming problems, each with $2np$ variables (corresponding to X' and X'') and $2(n+p)+dp$ constraints ($n+p$ constraints corresponding to $X' \in P_I^\Pi$, $n+p$ constraints corresponding to $X'' \in P_I^\Pi$ and dp constraints corresponding to $AX' + AX'' = 2V$). These linear programs are solvable in polynomial time in d, p and n. □

The above method extends to constrained-shape partition polytopes, provided the linear optimization problem over $\{(X', X'') \in P_I^\Pi \times P_I^\Pi : AX' + AX'' = 2V\}$ is solvable efficiently. An alternative method for testing if the vector associated with a partition is a vertex of a corresponding bounded-shape partition polytopes appears in Section 2.3.

Theorem 1.1.5 yields the following result which shows that partition problems over Π may be lifted to optimization problems over generalized transportation-polytopes.

Corollary 1.1.7. *Let $A \in \mathbb{R}^{d\times n}$, Π be a set of p-partitions, $f : \mathbb{R}^{d\times p} \to \mathbb{R}$ and $F : \Pi \to \mathbb{R}$ with $F(\pi) = f(A_\pi)$ for each $\pi \in \Pi$. Consider the function $h : P_I^\Pi \to \mathbb{R}$ with $h(X) = f(AX)$ for each $X \in P_I^\Pi$. If π is a p-partition such that I_π maximizes h over $\{I_{\pi'} : \pi' \in \Pi\}$, then π maximizes $F(\cdot)$ over Π; further if f is linear with $f(X) = \langle C, X\rangle$ for every $X \in \mathbb{R}^{d\times p}$, then*

$$h(Z) = \langle C, AZ\rangle = \langle A^T C, Z\rangle \quad \text{for all } Z \in P_I^\Pi \tag{1.1.7}$$

and a partition π with I_π maximizing h over $P_I^\Pi \supseteq \{I_{\pi'} : \pi' \in \Pi\}$ exists.

Proof. Let π be a partition in Π where I_π maximizes f over $\{I_{\pi'} : \pi' \in$

$\Pi\}$. By Theorem 1.1.5, for every partition $\pi' \in \Pi$, we have $I_{\pi'} \in P_I^\Pi$ and

$$F(\pi') = f(A_{\pi'}) = f(AI_{\pi'}) = h(I_{\pi'}) \leq h(I_\pi) = F(\pi).$$

Thus, π maximizes F over Π. Next, if f is linear with $f(X) = \langle C, X \rangle$ for all $X \in \mathbb{R}^{d\times p}$, (1.1.7) is immediate from (1.1.1). Further, in this case, the boundedness of P_I^Π assures that h attains a maximum over P_I^Π at a vertex and Corollary 1.1.2 assures that such a vertex has a representation I_π for some $\pi \in \Pi$. □

Solution of Bounded-Shape Partition Problems with Linear Objective Function

Let $A \in \mathbb{R}^{d\times n}$, $C \in \mathbb{R}^{d\times p}$, L and U be positive integer vectors satisfying $L \leq U$ and $\sum_{j=1}^p L_j \leq n \leq \sum_{j=1}^p U_j$, and let $F : \Pi^{(L,U)} \to \mathbb{R}$ with $F(\pi) = \langle C, A_\pi \rangle$ for each $\pi \in \Pi^{(L,U)}$. The goal is to maximize $F(.)$ over $\Pi \equiv \Pi^{(L,U)}$. Let $h : P_I^\Pi \to \mathbb{R}$ with $h(X) = \langle C, AX \rangle = \langle A^T C, X \rangle$ for each $X \in P_I^\Pi$. In view of Corollary 1.1.7 and Corollary 1.1.2, the problem of maximizing F over $\Pi^{(L,U)}$ reduces to finding a vertex of P_I^Π that maximizes the linear objective $h(X)$ with coefficients $\{(A^T C)_t^j : t = 1, \ldots, n$ and $j = 1, \ldots, p\}$ over P_I^Π; with $a \equiv \max\{\lg |A_i^t| : A_i^t \neq 0\}$ and $c \equiv \max\{\lg |C_i^j| : C_i^j \neq 0\}$, these coefficients are bounded by ke^{a+c}, where k is a constant and e is the base of natural logarithm, and are computable in time $O(npd(a + c))$ (we ignore the availability of sophisticated fast algorithms for multiplying matrices); using the explicit representation of P_I^Π provided in Theorem 1.1.3, the partition problem reduces to a minimum cost flow problem. Figure 1.1.1 shows the minimum cost flow network corresponding to the bounded-shape partition problem with the given parameters. Standard results show that an optimal solution can be found in strongly polynomial time (see Ahuja, Magnanti and Orlin [1]). □

The above solution method applies (in fact, in a simplified form) to single-shape partition problems (with linear objective), and consequently to constrained-shape problems where the set of shapes is tractable. Thus, for a set of shapes Π that are polynomial in the parameters d, n and p, we get a polynomial solution method by solving $|\Pi|$ linear programs, each having $L = U$.

We next show two examples of partition problems that can be converted to problems having linear objective and thereby solved by the above

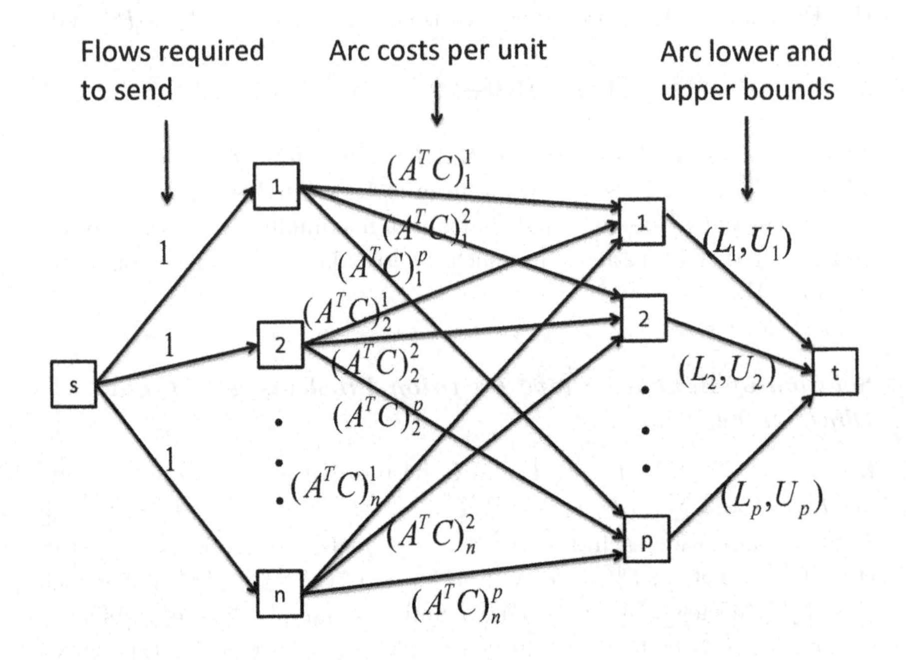

Figure 1.1.1 Network for the bounded-shape partition problem with linear objective function.

method.

Example 1.1.1. Consider the objective function determined in terms of given vectors $t_1, \ldots, t_p$ and an arbitrary (distance) function $D(\cdot, \cdot)$ on $\mathbb{R}^d \times \mathbb{R}^d$; specifically, the objective function associates with a partition π the value

$$F(\pi) = \sum_{j=1}^{p} \sum_{i \in \pi_j} D(A^i, t_j).$$

For $i = 1, \ldots, n$, consider the p-vector D^i having $(D^i)_j = D(A^i, t_j)$ for $j = 1, \ldots, p$. Let D be the $p \times n$ matrix $(D^1, \ldots, D^n)$. It then follows that for every partition π, $D_\pi = (\sum_{i \in \pi_1} D^i, \ldots, \sum_{i \in \pi_p} D^i)$ and

$$F(\pi) = \sum_{j=1}^{p} \sum_{i \in \pi_j} (e^j)^T D^i = \sum_{j=1}^{p} (e^j)^T (D_\pi)^j,$$

where $(D_\pi)^j$ is the j-th term of D_π. This representation of the objective function is linear as $F(\pi) = f(D_\pi)$ with $f : \mathbb{R}^{p \times p} \to \mathbb{R}$ having $f(X) =$

$\langle I, X\rangle$ (that is, f sums all the diagonal coordinates of its $p \times p$ matrix-variable). Consequently, the solution methods of bounded-shape problems with linear objective function apply. $\square$

Example 1.1.2. Consider a bounded-shape partition problem where the objective function is the sum of a linear function of the type considered throughout the current section and a linear combination of the part-sizes. Formally, for some matrix $C \in \mathbb{R}^{d\times p}$ and $c = (c_1, c_2, \cdots, c_p) \in \mathbb{R}^p$, we have that for each (feasible) partition π, $F(\pi) = \langle C, A_\pi\rangle + \sum_{j=1}^p c_j|\pi_j|$. This objective function can be cast as a regular linear one by appending a first coordinate 1 to each of the columns of A. Let $\widetilde{A}$ be the resulting $(d+1) \times n$ matrix, that is, $\widetilde{A} = \binom{e^T}{A}$ where $e = (1, \dots, 1)^T \in \mathbb{R}^n$. We then have that for each partition π, the first row of $\widetilde{A}_\pi$ is the shape of π and $F(\pi) = \langle \binom{c}{C}, \widetilde{A}_\pi\rangle$. $\square$

We next show that, with linear objective and special structure, a solution to single-shape partition problems can be obtained explicitly without addressing the linear programming problem over P_I^Π.

Theorem 1.1.8. *Let $A \in \mathbb{R}^{d\times n}$, $C \in \mathbb{R}^{d\times p}$, $n_1, \dots, n_p$ be positive integers with $\sum_{j=1}^p n_j = n$, and $F : \Pi^{(n_1,\dots,n_p)} \to \mathbb{R}$ with $F(\pi) = \langle C, A_\pi\rangle$ for each $\pi \in \Pi^{(n_1,\dots,n_p)}$. If A and C satisfy*

$$(A^t - A^{t+1})^T(C^j - C^{j+1}) \geq 0 \quad \textit{for} \quad 1 \leq t < n \quad \textit{and} \quad 1 \leq j < p, \tag{1.1.8}$$

then the p-partition π with $\pi_j = \left\{\sum_{u=1}^{j-1} n_u + 1, \dots, \sum_{u=1}^{j} n_u\right\}$ for $j = 1, \dots, p$ maximizes $F(.)$ over $\Pi^{(n_1,\dots,n_p)}$; further, if the inequality of (1.1.8) holds strictly, then π is a unique maximizer.

Proof. Corollaries 1.1.4 and 1.1.7 imply that the problem of maximizing $F(.)$ over $\Pi^{(n_1,\dots,n_p)}$ reduces to the problem of finding an optimal vertex over the solution set of (1.1.5) with respect to the linear function with coefficient matrix $W = A^TC$; we refer to the latter as the derived generalized transportation problem. As $W_t^j = (A^TC)_t^j = (A^t)^TC^j$ for $t = 1, \dots, n$ and $j = 1, \dots, p$, (1.1.8) asserts that W satisfies the (supermodularity) condition

$$W_t^j + W_{t+1}^{j+1} \geq W_{t+1}^j + W_t^{j+1} \quad \text{for} \quad 1 \leq t < n \quad \text{and} \quad 1 \leq j < p. \tag{1.1.9}$$

It is well known (e.g., Hoffman [47, 48], Dietrich [26] or Shamir and Dietrich [86]) that under this condition, the derived generalized transportation problem admits a greedy optimal solution obtained by the following algorithm: set all entries of the matrix $X = (X_t^j)$ to zero and sweep the entries in the order:

$$X_1^1, \ldots, X_n^1, X_1^2, \ldots, X_n^2, \ldots, X_1^p, \ldots, X_n^p,$$

increasing each X_t^j, in turn, as much as possible subject to the constraints $\sum_{u=1}^p X_t^u \leq 1$ and $\sum_{v=1}^n X_v^j \leq n_j$. The resulting outcome of this procedure is obviously the 0-1 matrix with

$$X_t^j = \begin{cases} 1 & \text{if } t \in \left\{\sum_{w=1}^{j-1} n_w + 1, \ldots, \sum_{w=1}^{j} n_w\right\} \\ 0 & \text{otherwise,} \end{cases}$$

corresponding to the partition π with $\pi_j = \left\{\sum_{w=1}^{j-1} n_w + 1, \ldots, \sum_{w=1}^{j} n_w\right\}$ for $j = 1, \ldots, p$. The optimality of this partition now follows from Corollary 1.1.7. Finally, if the inequalities of (1.1.8) hold strictly, so do the inequalities of (1.1.9); in this case, the outcome of the greedy procedure is known to be the unique optimal solution for the generalized transportation problem, immediately implying that the constructed partition is the only optimal partition. □

Recall from standard arguments that conditions (1.1.8) and (1.1.9) are equivalent, respectively, to (the seemingly stronger assertions):

$$\left(A^t - A^{t'}\right)^T \left(C^j - C^{j'}\right) \geq 0 \text{ for } 1 \leq t < t' \leq n \text{ and } 1 \leq j < j' \leq p, \tag{1.1.10}$$

and

$$W_t^j + W_{t'}^{j'} \geq W_{t'}^j + W_t^{j'} \text{ for } 1 \leq t < t' \leq n \text{ and } 1 \leq j < j' \leq p. \tag{1.1.11}$$

We now analyze the time complexity of checking whether a fixed matrix W (implying permutations of A and C are fixed) satisfies (1.1.11). Since there are $\binom{n}{2}\binom{p}{2}$ choices of t, t', j, j', it takes $O((np)^2)$ time for a fixed W. There are $n!p!$ permutations of A and C, and therefore $n!p!$ different W's; thus the total checking time is inefficient $O(n!p!(np)^2)$. In general, a permutation of A and C such that the corresponding W satisfying (1.1.11) may not exist. However, when $d = 1$ such a permutation does always exist and can be easily identified. The next result (which was also derived earlier in Theorem 2.1.1 of Vol. I and its following paragraph) provides an answer.

Corollary 1.1.9. *Assume that*

$$A^1 \leq \cdots \leq A^n \text{ and } C^1 \leq \cdots \leq C^p. \tag{1.1.12}$$

Then the p-partition π with $\pi_j = \left\{\sum_{u=1}^{j-1} n_u + 1, \ldots, \sum_{u=1}^{j} n_u\right\}$ for $j = 1, \ldots, p$ maximizes $F(.)$ over $\Pi^{(n_1,\ldots,n_p)}$; further, if the inequalities of (1.1.12) hold strictly, then π is the only maximizer.

Proof. Evidently, weak and strict inequalities in (1.1.12) imply the corresponding inequalities in (1.1.8), and the conclusions about π follow directly from Theorem 1.1.8. □

For $d = 1$, one can always sort A and C into nondecreasing sequences, thereby yielding an explicit solution to the partition problem. But for $d \geq 2$, one can hardly expect (1.1.12) to hold. We now introduce a new greedy algorithm which needs not verify (1.1.11) for every permutation of A and C.

For easy presentation, $\{W_t^j : 1 \leq j \leq p; 1 \leq t \leq n\}$ is called a W-set, any ordered W-set is called a W-sequence denoted by w and any n-subsequence of w with all t's distinct and with n_j occurrences of superscript j (of W) in the subsequence is called a *partition sequence* and denoted by w_π. By interpreting $W_t^j \in w_\pi$ as A^t goes to $\pi_j \in \pi$, w_π yields a partition π with shape $(n_1, \cdots, n_p)$. We also denote the k-th term in w_π by $w_\pi(k)$.

The precise algorithm goes as follows:

(1) Construct W. Compute W_t^j for $1 \leq j \leq p$ and $1 \leq t \leq n$ to obtain the W-set.
(2) Construct w. Sort elements in the W-set into a nonincreasing sequence. Let w^* denote the set of all such sequences by considering all permutations of equal elements. Order w in w^* arbitrarily.
(3) Select the first w in w^*.
(4) Construct w_π. Initially, set $N_j = 0$ and $\pi_j = \emptyset$ for $j = 1, \cdots, p$. Construct $w_\pi(k)$ (from w) in order of $k = 1, 2, \cdots$. Suppose we are constructing $w_\pi(k)$. Select the first term in w to be $w_\pi(k)$, say, $w_\pi(k) = W_t^j$. Delete all $W_t^{j'}$'s from w. Add 1 to N_j and delete all $W_{t'}^j$'s with $N_j = n_j$ from w. Set $k = k + 1$.
(5) Delete w from w^*. If w^* is empty, stop. If not and $k = n$, go to (3). If not and $k < n$, go to (4).

Example 1.1.3. Let $A = \begin{pmatrix} 3 & 3 & 7 \\ 5 & 2 & 1 \end{pmatrix}$ and $C = \begin{pmatrix} 2 & 1 \\ 2 & 2 \end{pmatrix}$ with shape $(1, 2)$. Then we have $W = \begin{pmatrix} W_1^1 & W_1^2 \\ W_2^1 & W_2^2 \\ W_3^1 & W_3^2 \end{pmatrix} = \begin{pmatrix} 16 & 13 \\ 10 & 7 \\ 16 & 9 \end{pmatrix}$. Sort the W_t^j's into nonincreasing sequences: $w^* = \{w^1 = (W_3^1, W_1^1, W_1^2, W_2^1, W_3^2, W_2^2), w^2 = (W_1^1, W_3^1, W_1^2, W_2^1, W_3^2, W_2^2)\}$. Then the corresponding output $w_\pi^1 = (W_3^1, W_1^2, W_2^2) = (16, 13, 7)$ and $w_\pi^2 = (W_1^1, W_3^2, W_2^2) = (16, 9, 7)$. □

We next analyze the time complexity of this greedy algorithm.
In step (1). Since each multiplication of A^t and C^j takes $O(d)$ time, the total time is $O(dpn)$.
In step (2). A sorting algorithm on pn distinct elements can be easily modified into one allowing nondistinctness (in fact the distinct case requires the largest number of paired comparisons). It is well-known that a $O(np\log(np))$ sorting algorithm exists.
In step (4). Suppose W contains q distinct values, which partition W into q equivalent classes $e_1, \cdots, e_q$. Then there are in general

$$\sharp = \prod_{i=1}^{q} |e_i|$$

w-sequences. For each w, each k, generating $w_\pi(k)$ takes $O(n) + O(p) = O(n)$ time. So the total time to generate w_π is $O(\sharp n^2)$.

Summing up, the total time needed by the greedy algorithm is $O(\sharp n^2)$ time, which makes the algorithm ineffective unless $\sharp$ is a small number; for example $\sharp = 1$ when W is a real set (no equal members). But $\sharp$ can be an exponential number if no additional constraint is added. For example, when $q = n/2$ and each equivalent class has 2 members, then $\sharp = 2^{np/2}$. Even though the number of generated partition sequences w_π's can be exponential, it is much smaller than the number $n!p!$ of all permutations.

Let w_i denote the updated sequence w in the $(i-1)$-th round of the greedy algorithm and $w_1 = w$. The following result shows that given a fixed partition sequence w_π, it suffices to check for each $i = 1, \cdots, n$ such that

$$W_t^j + W_{t'}^{j'} \geq W_{t'}^j + W_t^{j'}, \quad \text{where } W_t^j = w_\pi(i), \text{ for any } W_{t'}^{j'}, W_{t'}^j, W_t^{j'} \in w_i. \tag{1.1.13}$$

Theorem 1.1.10. *Assume that $w_\pi(1) \geq \cdots \geq w_\pi(n)$ is the partition sequence w_π output by the above greedy algorithm. Suppose that (1.1.13) holds for $i = 1, \cdots, n$. Then the p-partition π corresponding to w_π maximizes $F(\cdot)$ over $\Pi^{(n_1,\dots,n_p)}$; further, if the inequality of (1.1.13) holds strictly, then π is the only maximizer.*

Proof. Let π be the output of the above algorithm and let $w_\pi(1) = W_{t_1}^{j_1}, \cdots, w_\pi(n) = W_{t_n}^{j_n}$ be the corresponding sequence. It suffices to show that π maximizes $F(.)$ over $\Pi^{(n_1,\cdots,n_p)}$. Suppose to the contrary that there exists some $\pi' \in \Pi^{(n_1,\cdots,n_p)}$ such that $F(\pi) < F(\pi')$. Among all such counter-examples, choose the one σ (let $X_{t'_1}^{j'_1} \geq \cdots \geq X_{t'_n}^{j'_n}$ be the corresponding sequence of σ) with the largest k satisfying $(j'_1, t'_1) = (j_1, t_1), \cdots, (j'_k, t'_k) = (j_k, t_k)$. (Note that we use X here to denote the entries corresponding to the partition σ; the value X_t^j is actually the value W_t^j in the coefficient matrix W.)

Consider the two sub-partitions π^1 and σ^1 obtained from the subsequences $W_{t_{k+1}}^{j_{k+1}}, \cdots, W_{t_n}^{j_n}$ and $X_{t'_{k+1}}^{j'_{k+1}}, \cdots, X_{t'_n}^{j'_n}$, respectively. Obviously, π^1 and σ^1 have the same shape and elements. We may assume $t_{k+1} \in \sigma_{j'}^1$ for some $j' \neq j_{k+1}$ for otherwise, $(j_{k+1}, t_{k+1}) = (j'_\ell, t'_\ell)$ for some $\ell > k+1$, we can rearrange the corresponding sequence of σ to obtain another sequence Y with the largest $k+1$ terms coincident with the largest $k+1$ terms of the corresponding sequence of π; thus a contradiction. On the other hand, since π^1 and σ^1 have the same shape, there exists some $t' \neq t_{k+1}$ such that $t' \in \sigma_{j_{k+1}}^1$. Consider the moment of putting t_{k+1} into $\pi_{j_{k+1}}$ in step (4) of the algorithm. At this moment, all $W_{t_{k+1}}^{j_{k+1}}, \cdots, W_{t_n}^{j_n}$ and $W_{t'_{k+1}}^{j'_{k+1}}, \cdots, W_{t'_n}^{j'_n}$ must belong to the updated sequence w_{k+1}. According to the algorithm, $W_{t_{k+1}}^{j_{k+1}}$ is the largest one among them; consequently, $W_{t_{k+1}}^{j_{k+1}} \geq \max\{W_{t_{k+1}}^{j'}, W_{t'}^{j_{k+1}}\}$ as $\{W_{t_{k+1}}^{j'}, W_{t'}^{j_{k+1}}\} = \{X_{t_{k+1}}^{j'}, X_{t'}^{j_{k+1}}\} \subseteq \{X_{t'_{k+1}}^{j'_{k+1}}, \cdots, X_{t'_n}^{j'_n}\} = \{W_{t'_{k+1}}^{j'_{k+1}}, \cdots, W_{t'_n}^{j'_n}\}$. By hypothesis this implies $W_{t_{k+1}}^{j_{k+1}} + W_{t'}^{j'} \geq W_{t_{k+1}}^{j'} + W_{t'}^{j_{k+1}}$. Let γ be a partition obtained from σ by interchanging t_{k+1} and t'. Obviously, γ has the same shape with σ and has a corresponding sequence (after rearrangement) with the largest $k+1$ terms coincident with the largest $k+1$ terms of the corresponding sequence of π. In addition, $F(\gamma) = F(\sigma) + W_{t_{k+1}}^{j_{k+1}} + W_{t'}^{j'} - W_{t_{k+1}}^{j'} - W_{t'}^{j_{k+1}} \geq F(\sigma)$. This contradicts the assumption that σ is the one among counter-examples with its corresponding sequence having the largest k satisfying $(j'_1, t'_1) = (j_1, t_1), \cdots, (j'_k, t'_k) = (j_k, t_k)$. Therefore, π is an optimal partition maximizing $F(\cdot)$.

If the inequality in (1.1.13) holds strictly, then $F(\gamma) = F(\sigma) + W_{t_k}^{j_k} + W_{t_{k'}}^{j'} - W_{t_k}^{j'} + W_{t_{k'}}^{j_k} > F(\sigma)$, a contradiction to the assumption that σ is optimal. This completes the proof. □

Example 1.1.4. Consider the case in Example 1.1.3. It is easy to verify that the resulting partition sequence w_π^1 satisfies (1.1.13) while w_π^2 does not; therefore, the partition $(\{3\}\{1,2\})$ corresponding to w_π^1 is optimal while the other ({1}{2, 3}) corresponding to w_π^2 is not. □

Note that (1.1.13) implies if

$$\max\{W_t^j, W_{t'}^{j'}\} = \max\{W_t^{j'}, W_{t'}^j\}, \tag{1.1.14}$$

then

$$\min\{W_t^j, W_{t'}^{j'}\} = \min\{W_t^{j'}, W_{t'}^j\} \tag{1.1.15}$$

since both maximum terms in (1.1.14) must satisfy (1.1.13), which forces (1.1.15).

We now analyze the time complexity of verifying (1.1.13) over all $i = 1, \cdots, n$. For a fixed i, it takes $O(np)$ time since there are at most $O(np)$ choices of j', t'. Thus, the total time complexity is $O(n^2 p)$ for a fixed partition sequence w_π.

1.2 Enumerating Vertices of the Partition Polytopes and Corresponding Partitions Using Edge-Directions

Corollary 3.3.7 of Vol. I assures that when the function $f(\cdot)$ is (edge-)quasi-convex on a partition polytope, it attains a maximum at a vertex of that polytope. When $d > 1$, it is usually difficult to recognize (edge-)quasi-convex functions and most applications are for functions $f(\cdot)$ that are just quasi-convex. As all vertices are matrices that correspond to partitions, it follows that partitions corresponding to vertices that maximize $f(\cdot)$ over the partition polytope are optimal. A natural solution method is then to enumerate the vertices of the partition polytope along with the corresponding partitions. An optimal partition can then be found by evaluating the function $f(\cdot)$ at each vertex and picking a partition corresponding to the best vertex.

With Γ as a set of shapes, the number of matrices in the set $\{A_\pi : \pi \text{ has shape in } \Gamma\}$ is typically exponential in n, even for fixed d, p. Therefore, although the dimension of the corresponding partition polytope is bounded by dp, this polytope can potentially have exponentially many vertices and facets. But, in the current section and in Section 2.2, we shall demonstrate that, in fact, bounded- and constrained-shape partition polytopes are exceptionally well behaved.

The approach we follow in the current section is to generate a complete list of vertices of a bounded-shape partition polytope by the projection technique developed in Section 3.5 of Vol. I. Recall that the approach requires the availability of a complete list of edge-directions and solvability of linear programs over the corresponding polytope. Section 1.1 demonstrated how linear objective functions can be maximized over bounded-shape partition polytopes (specializing the projection method of Algorithm PLP of Section 3.5 of Vol. I). So, it remains to develop a method for listing a complete set of edge-directions for bounded-shape partition polytopes. Here again, we build on the projection technique developed in Section 3.5 of Vol. I. In particular, we use results of Section 3.7 of Vol. I to obtain lists of edge-directions of network polytopes and show how these can be projected. The derivation of the list of edge-directions simplifies when considering single-size partition problems.

Let L and U be positive integer p-vectors satisfying $L \leq U$ and $\sum_{j=1}^{p} L_j \leq n \leq \sum_{j=1}^{p} U_j$. We recall the notation $P_A^{(L,U)}$ and $P_I^{(L,U)}$ for the corresponding bounded-shape partition polytopes. In particular, Theorem 1.1.3 provides a linear inequality representation for $P_I^{(L,U)}$ through (1.1.4). Adding variables X_0^j for $j = 1, \ldots, p$, replacing (1.1.4c) by the constraints $X_0^j = \sum_{t=1}^{n} X_t^j$ and $L_j \leq X_0^j \leq U_j$ for $j = 1, \ldots, p$ and adding the constraint $\sum_{j=1}^{p} X_0^j = n$, (1.1.4) is converted to the equivalent linear system:

$$
\begin{aligned}
&(a)\ \sum_{j=1}^{p} X_0^j = n,\\
&(b)\ X_0^j = \sum_{t=1}^{n} X_t^j \quad \text{for} \quad j = 1, \ldots, p,\\
&(c)\ \sum_{j=1}^{p} X_t^j = 1 \quad \text{for} \quad t = 1, \ldots, n,\\
&(d)\ L_j \leq X_0^j \leq U_j \quad \text{for} \quad j = 1, \ldots, p, \quad \text{and}\\
&(e)\ 0 \leq X_t^j \leq 1 \quad \text{for} \quad t = 1, \ldots, n \ \text{and} \ j = 1, \ldots, p;
\end{aligned}
\tag{1.2.1}
$$

(the upper bound in (1.2.1e) is obviously superfluous, as it is implied by the lower bound of (1.2.1e) and (1.2.1c)). Now, (1.2.1) corresponds to a network flow problem with integer lower and upper bounds on the arc-flows (see Figure 1.2.1).

We next consider the (orthogonal) projection which maps $\{X_t^j : t = 0, \ldots, n \text{ and } j = 1, \ldots, p\}$ into $\{X_t^j : t = 1, \ldots, n \text{ and } j = 1, \ldots, p\}$ — it maps the network polytope (the solution set of (1.2.1)) onto P_I^{Π} (the solution set of (1.1.4)). As the projection is one-to-one, it maps vertices of the network polytope onto vertices of $P_I^{(L,U)}$. We will use underbars

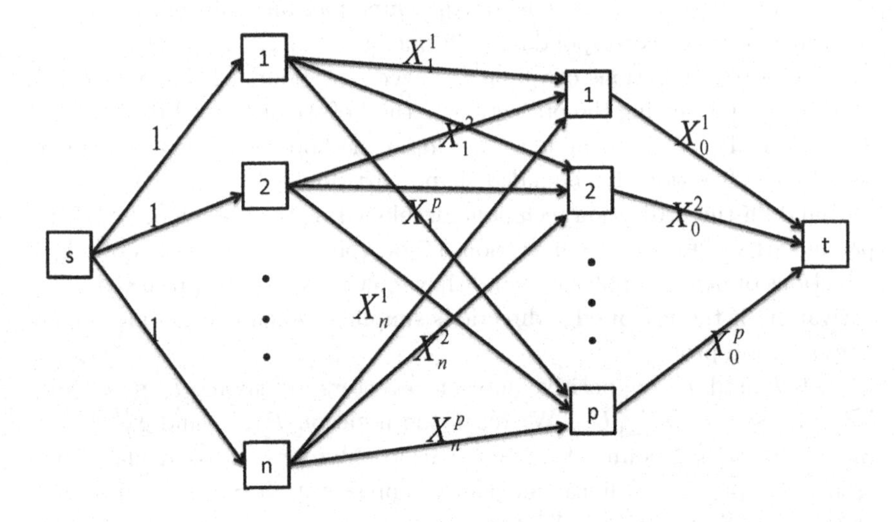

Figure 1.2.1 Network flow problem.

to denote the projection and overbars to denote its inverse correspondence (that maps $P_I^{(L,U)}$ onto the network polytope). In particular, Theorem 1.1.5 implies that $P_A^{(L,U)}$ is the image of the polytope corresponding to the above network (represented by (1.2.1)) under the linear transformation

$$Y \in \{X_t^j : t = 0, \ldots, n \text{ and } j = 1, \ldots, p\} \rightarrow A\underline{Y}. \tag{1.2.2}$$

We have seen that bounded-shape partition polytopes are projections of network polytopes. As the edge-directions of network polytopes are available (see Section 3.7 of Vol. I), the Projected Vertex Enumeration (PVE) Algorithm (of Section 3.5 of Vol. I) can be applied to enumerate the vertices of bounded-shape partition polytopes. The method is outlined below.

Enumerating Vertices of Bounded-Shape Partition Polytopes along with Corresponding Partitions Using Edge-Directions

1. Use the results of Section 3.7 in Vol. I to derive a list of vectors that contains a direction for each edge of the network polytope defined by (1.2.1); details about the execution of this step are described below.
2. Use the map described in (1.2.2) to calculate the image of each of the vectors in the list determined in step 1. By Corollary 3.5.6 of Vol. I,

the resulting list contains a direction for each edge of $P_A^{(L,U)}$.

3. Use the Vertex Enumeration (VE) Algorithm of Section 3.5 of Vol. I to enumerate all the vertices of $P_A^{(L,U)}$. The algorithm requires the solution of linear programs over the bounded-shape partition polytope $P_A^{(L,U)}$ and Section 1.1 demonstrates how this can be accomplished. Algorithm VE will then list the vertices of $P_A^{(L,U)}$ along with vertices of the network polytope that correspond to them under the transformation given in (1.2.2). Dropping coordinates (that correspond to edges containing 0), the vertices of the network polytope become vertices of $P_I^{(L,U)}$; the images of the latter under the transformation $X \to AX$ are then the vertices of $P_A^{(L,U)}$. Corollary 1.1.2 further shows that each vertex v of $P_I^{(L,U)}$ is of the form $v = I_\pi$ with $\pi \in \Pi^{(L,U)}$ easily determined from v. So, the output of Algorithm VE will list a set of partitions $\Pi^* \subseteq \Pi^{(L,U)}$ along with their images under the map $I_\pi \to AI_\pi = A_\pi$ such that $\{A_\pi : \pi \in \Pi^*\}$ is the set of vertices of $P_A^{(L,U)}$. □

Once the vertices of a bounded-shape partition polytope are enumerated along with corresponding partitions, a partition problem with objective function $F(\pi) = f(A_\pi)$ where $f(\cdot)$ is (edge-)quasi-convex can be solved by evaluating $f(.)$ at the vertices and selecting a partition corresponding to a vertex that maximizes $f(\cdot)$.

We next provide details about the execution of step 1 of the above algorithm for enumerating vertices and derive bounds on the number of cycles of the network polytope that generate edge directions. Results of Section 3.5 of Vol. I will then be used to derive bounds on the number of vertices of $P_A^{(L,U)}$ and a bound on the effort that is needed for generating them by the above method. We note that the dimension of the partition polytope is $(p-1)d$ (the -1 accounts for the fact that the row sum of each point in the partition polytope is $\sum_{i=1}^n A^i$). Hence, if the number of edge directions is $O(n^m)$, the results of Section 3.5 of Vol. I yield a bound of $O(n^{m[(p-1)d-1]})$ on the number of vertices of the corresponding partition polytope.

We shall use the common double subscripts to denote edges. Now, consider the network whose graph is demonstrated in Figure 1.2.2, with the lower and upper bounds $l_{(r,n+s)} = 0$ and $u_{(r,n+s)} = 1$ for $1 \le r \le n$ and $1 \le s \le p$, while $l_{(n+s,n+p+1)}$ and $u_{(n+s,n+p+1)}$ represent the bounds of the shape for $1 \le s \le p$. Finally, the node 0 has an outflow $b_r = 1$ for each $r = 1, \cdots, n$ and therefore total outflow n, and the node $n + p + 1$

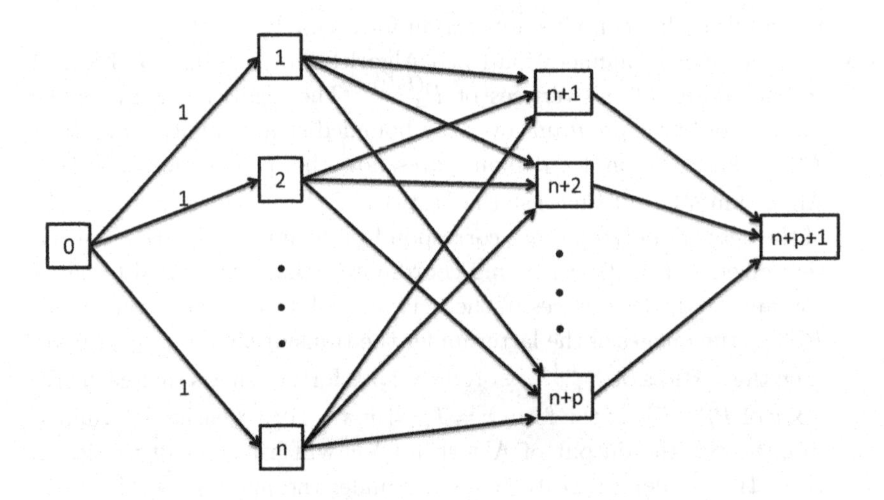

Figure 1.2.2 Graph of network.

has total inflow $b_{n+p+1} = n$. For brevity, we let $l_s \equiv l_{(n+s,n+p+1)}$ and $u_s \equiv u_{(n+s,n+p+1)}$ for $s = 1, \ldots, p$.

With b integral, a vertex of the corresponding network polyhedron is known to be integral (e.g., Schrijver [85]). In particular, for each vertex x, we have that for every $r = 1, \ldots, n$, $x_{(r,n+s)} = 1$ for exactly one $s \in \{1, \ldots, p\}$, and for each $s = 1, \ldots, p$, $x_{(n+s,n+p+1)}$ is the number of nodes $r \in \{1, \ldots, n\}$ with $x_{(r,n+s)} = 1$. Thus, a vertex corresponds to an assignment of nodes $1, \ldots, n$ to the p destinations $n+1, n+2, \ldots, n+p$, subject to requirements/capacity constraints on the number of nodes assigned to each destination; so, vertices correspond to *partitions* of $\{1, \ldots, n\}$ into parts, indexed by $1, \ldots, p$, subject to lower and upper bounds. In Figure 1.2.3, we illustrate the support of a vertex using the network representation of Figure 1.2.2. It is observed that for a vertex x, $n+p+1$ occurs in every arc in $\text{float}(x) = \{j : l_j < x_j < u_j\}$ (defined in p. 113, Vol. I).

We shall use the standard representation of cycles through sequences of nodes. Given a vertex x, a cycle z for which $x + \alpha z$ is feasible (that is, in the network polyhedron) has either of the following two representations:

(i) For some $2 \le k \le p$, there are sequences $r_1, \ldots, r_k$ and $s_1, \ldots, s_k$ of distinct elements from $\{1, \ldots, n\}$ and from $\{1, \ldots, p\}$, respectively, and z is represented by the sequence $r_1, n+s_1, r_2, n+s_2, \ldots, r_k, n+s_k$; this representation corresponds to a cyclic change where for $j = 1, \ldots, k$, r_j

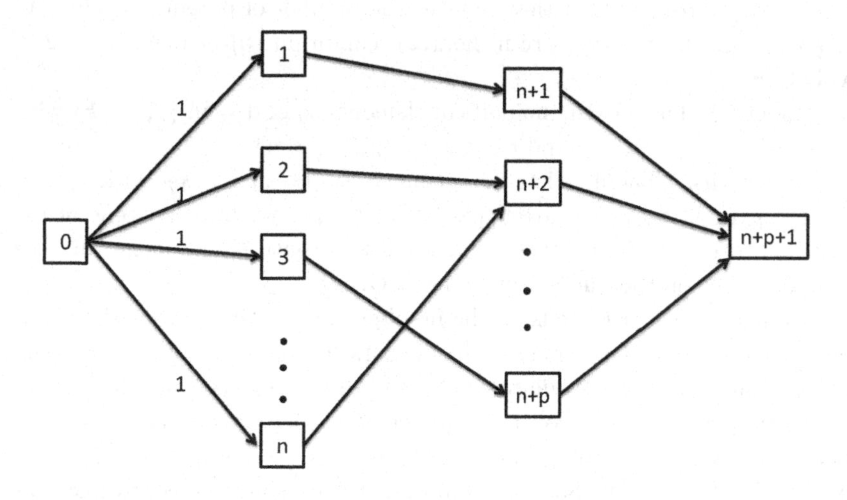

Figure 1.2.3 Support of a vertex.

is moved from part s_{j-1} to part s_j. The requirement that $x + \alpha z \in P$ for some $\alpha > 0$ imposes the constraint $x_{r_j,n+s_{j-1}} = 1$ for $j = 1, \ldots, k$ (with $s_0 = s_k$); thus, $s_1, \ldots, s_k$ are determined by $r_1, \ldots, r_k$ and the latter sequence characterizes z.

(ii) For some $1 \le k \le p$, there are sequences $r_2, \ldots, r_k$ and $s_1, \ldots, s_k$ of distinct elements from $\{1, \ldots, n\}$ and from $\{1, \ldots, p\}$, respectively, and z is represented by the sequence $n+p+1, n+s_1, r_2, n+s_2, \ldots, r_k, n+s_k$; this representation corresponds to a sequential change where for $j = 2, \ldots, k$, r_j is moved from part s_{j-1} to part s_j, with the assignment of part s_1 decreased by one element and the assignment of part s_k increased by one element. The requirement that $x + \alpha z \in P$ for some $\alpha > 0$ imposes the constraints $x_{r_j,n+s_{j-1}} = 1$ for $j = 2, \ldots, k$, $x_{n+s_1,n+p+1} > l_{s_1}$ and $x_{n+s_k,n+p+1} < u_{s_k}$; thus, $s_1, \ldots, s_{k-1}$ are determined by $r_2, \ldots, r_k$ and $r_2, \ldots, r_k$, s_k characterizes z.

With p fixed, the number of cycles z with the first representation is then bounded by $\sum_{k=1}^{p} \binom{n}{k}(k-1)! = O(n^p)$ (we counted for permutation-invariance of cycle-representation), and the number of cycles z with the second representation is bounded by $\sum_{k=1}^{p-1} \binom{n}{k} p = O(n^p)$. The total number of cycles is then $O(n^p)$.

Corollary 3.7.4 of Vol. I shows that such a cycle z is a direction of a 1-dim face that contains x, if and only if float$(x) \setminus$ supp(z) cannot be used to bisect z; a requirement that reduces the number of potential cycles. As $n+p+1$ occurs in every arc in float(z), condition (b) of Lemma 3.7.2 of Vol. I means that:

Under (i): There exists no pairs of elements s_u and s_v in $\{1,\ldots,k\}$ with $l_{s_u} < x_{n+s_u,n+p+1} < u_{s_u}$ and $l_{s_v} < x_{n+s_v,n+p+1} < u_{s_v}$.

Under (ii): There exists no element s_u in $\{1,\ldots,k\}$ with $l_{s_u} < x_{n+s_u,n+p+1} < u_{s_u}$. For arbitrary l_s's and u_s's, we have no general expression to count for the cycles satisfying the above requirement, and the bound we get on the number of cycles is $O(n^p)$.

We next consider bounds on the number of edge-directions without reference to a particular vertex x that occurs in the edge. In cases where no prior information is available for edges which are necessarily in float(x) for vertices x, the bound on the number of $\sim$-edge-directions is the number of cycles. The total number of cycles under either (i) or (ii) is $\sum_{k=2}^{p}\binom{n}{k}\binom{p}{k}k!(k-1)!$. So, the bound on the number of $\sim$-edge-directions is $2\sum_{k=2}^{p}\binom{n}{k}\binom{p}{k}k!(k-1)!$. The resulting bound on the number of vertices is then $O(n^{p[(p-1)d-1]})$.

Single-Size Problems

Single-size problems correspond to the case where $u_s = n+1$ and $l_s = 0$ for every s, that is, there are no upper bounds on part-capacities. In this case, every arc $(n+s, n+p+1)$ is in float(x) and the only case where float(x) cannot be used to bisect z is for cycles of the second form with $k = 2$. These cycles correspond to a single switch of one element from one part to another. The number of such cycles is then $n\binom{p}{2} = O(n)$. Consequently, the number of $\sim$-edge-directions for the corresponding partition polytopes is bounded by $n\binom{p}{2}$. The resulting bound on the number of vertices is then $O(n^{(p-1)d-1})$.

The bound on edge-directions also allows us to bound the number of facets.

Theorem 1.2.1. *For any $A \in \mathbb{R}^{d\times n}$ the number of facets of $P_A^{\Pi^p}$ is bounded by $O(n^{d^2p^3})$.*

Proof. By the well-known Upper Bound Theorem (e.g., McMullen [68]), the number of facets of any k-dim polytope with m vertices is $O(m^{\frac{k}{2}})$. Applying this to $P_A^{\Pi^p}$ with $k \le dp$ and $m = O(n^{d\binom{p}{2}})$ we get the asserted bound on the number of facets of $P_A^{\Pi^p}$. □

Theorem 1.2.2. *For any $A \in \mathbb{R}^{d \times n}$ and set $\Gamma \neq \emptyset$ of positive integer p-vectors with coordinate-sum n, the number of facets of P_A^Γ is bounded by $O(n^{d^2 p^3})$.*

Proof. The proof is the same as that of Theorem 1.2.1. □

We next derive a method for producing all facets of constrained-shape partition polytopes (by which we mean finding, for each facet F, a hyperplane $\{X \in \mathbb{R}^{d \times p} : \langle H, X \rangle = h\}$ supporting P_A^Γ at F).

Enumerating the Facets of a Constrained-Shape Partition Polytope Using Generic Partitions (along with Supporting Hyperplanes)

Let $A \in \mathbb{R}^{d \times n}$ and let Γ be a nonempty set of positive integer p-vectors with coordinate-sum n. Construct first the set V of vertices of P_A^Γ using the algorithm that was described earlier. Next compute the affine dimension a of $\text{aff}(P_A^\Gamma) = \text{aff}(V)$ and compute a (possibly empty) set S of $dp - a$ points that together with V form an affine basis of $\mathbb{R}^{d \times p}$. For each affine basis T of V, compute the hyperplane $\{X \in \mathbb{R}^{d \times p} : \langle H, X \rangle = h\}$ spanned by $S \cup T$. This hyperplane supports a facet of P_A^Γ if and only if all points in V lie on one of its closed half-spaces. Clearly, all facets of P_A^Γ are obtained in that way and the number of arithmetic operations and queries to the Γ-oracle is bounded by $O(n^{d^2 p^3})$. Obviously, we can enumerate all facets of a bounded-shape partition polytope by viewing the bounded-shape set as a constrained-shape set. □

1.3 Representation, Characterization and Enumeration of Vertices of Partition Polytopes: Distinct Partitioned Vectors

One may wonder about the degree of freedom in representing a vertex as the vector associated with a partition. It is shown in the current section that for bounded-shape problems with the A^i's nonzero and distinct, there is no degree of freedom at all—each vertex corresponds to a unique partition. This is accomplished by a characterization that allows one to test whether or not A_π associated with a partition π is a vertex of the corresponding bounded-shape polytope. The assumption that the A^i's are nonzero is relaxed at the end of the current section and the assumption that the A^i's are distinct is relaxed in the forthcoming Section 1.4. Under these relaxations, the only degree of freedom is the obvious one—permuting identical vectors and shifting zero vectors.

We start by considering the (nondegenerate) situations where the A^i's are nonzero and distinct. The following analysis of the vertices of a bounded-shape partition polytope relies on further examination of the corresponding partition polytope associated with the $n \times n$ identity matrix, examined in Section 1.1. The first result (Theorem 1.3.1) given by Hwang, Onn and Rothblum [53] characterizes vertices of bounded-shape partition polytopes in terms of unique representation, the second (Theorem 1.3.3) provides a characterization in terms of the feasibility of a linear system.

Theorem 1.3.1. *Let $A \in \mathbb{R}^{d\times n}$ have nonzero and distinct columns; let L and U be positive integer p-vectors satisfying $L \le U$ and $\sum_{j=1}^p L_j \le n \le \sum_{j=1}^p U_j$, and let $\pi \in \Pi^{(L,U)}$. Then the following are equivalent:*
(a) A_π is a vertex of $P_A^{(L,U)}$,
(b) $\{X \in P_I^{(L,U)} : AX = A_\pi\} = \{I_\pi\}$,
(c) $\{X \in P_I^{(L,U)} : AX = A_\pi$ and $\langle I_\pi, X\rangle < n\} = \emptyset$, and
(d) $\{X \in P_I^{(L,U)} : AX = A_\pi$ and $\langle I_\pi, X\rangle \le n-1\} = \emptyset$.

Proof. Recall from Theorem 1.1.5 that $A_\sigma = AI_\sigma$ for each $\sigma \in \Pi^{(L,U)}$.

$(a) \Rightarrow (b)$: Recall the linear inequality representation of $P_I^{(L,U)}$ in Theorem 1.1.3, given by

$$\begin{aligned}
&(a)\ \ X_t^j \ge 0 \ \text{ for } \ t = 1,\dots,n \ \text{ and } \ j = 1,\dots,p,\\
&(b)\ \ \sum_{j=1}^p X_t^j = 1 \ \text{ for } \ t = 1,\dots,n, \qquad\qquad (1.3.1)\\
&(c)\ \ L_j \le \sum_{t=1}^n X_t^j \le U_j \ \text{ for } \ j = 1,\dots,p.
\end{aligned}$$

Adding variables X_0^j for $j = 1,\dots,p$, replacing (1.3.1c) by the constraints $X_0^j = \sum_{t=1}^n X_t^j$ and $L_j \le X_0^j \le U_j$ for $j = 1,\dots,p$ and adding the constraint $\sum_{j=1}^p X_0^j = n$, the linear system (1.3.1) is then converted to the equivalent linear system:

$$(a)\ \sum_{j=1}^{p} X_0^j = n\,,$$

$$(b)\ X_0^j = \sum_{t=1}^{n} X_t^j \ \text{ for } \ j = 1, \ldots, p,$$

$$(c)\ \sum_{j=1}^{p} X_t^j = 1 \ \text{ for } \ t = 1, \ldots, n, \tag{1.3.2}$$

$$(d)\ L_j \le X_0^j \le U_j \ \text{ for } \ j = 1, \ldots, p, \ \text{ and}$$

$$(e)\ 0 \le X_t^j \le 1 \ \text{ for } \ t = 1, \ldots, n \ \text{ and } j = 1, \ldots, p\,.$$

The latter is a network flow problem with integer lower and upper bounds on some of the arc-flows (see Figure 1.2.1).

A *circuit* z is a nonzero normalized, minimal support solution of the (homogeneous) system

$$(a)\ \sum_{j=1}^{p} z_0^j = 0\,,$$

$$(b)\ z_0^j - \sum_{t=1}^{n} z_t^j = 0 \ \text{ for } \ j = 1, \ldots, p\,, \tag{1.3.3}$$

$$(c)\ \sum_{j=1}^{p} z_t^j = 0 \ \text{ for } \ t = 1, \ldots, n$$

where *normalized* means that $\max\{|x_t^j| : j = 1, \ldots, p \text{ and } t = 1, \ldots, n\} = 1$ and *minimal support* means that if $z' \neq 0$ satisfies (1.3.3), then the set of nonzero coordinates of z' is not strictly included in that of z. Standard results show that the coordinates of a circuit z are all $-1, 0$ and 1 and that each node of the network presented in Figure 1.2.1 appears in either 0 or 2 indices corresponding to nonzero coordinates of z. It follows that each column j and each row t of a circuit z has at most two nonzero elements which can take only the values -1 and 1. The former implies that each column of Az is the difference of two columns of A, indexed by distinct integers. As the columns of A are assumed to be nonzero and distinct, it follows that

$$\text{circuit } \ z \ \text{ satisfies } \ Az = 0 \ \text{ if and only if } \ z = 0\,. \tag{1.3.4}$$

As (1.3.3b) determines the z_0^j's of a circuit z from the remaining coordinates, we identify such a circuit with its projection $\underline{z} \in \mathbb{R}^{n \times p}$ (in particular, $z \neq 0$ if and only if $\underline{z} \neq 0$).

Suppose π is a partition in $\Pi^{(L,U)}$ with A_π a vertex of $P_A^{(L,U)}$ and $X \in P_I^{(L,U)}$ satisfies $AX = A_\pi$; we will show that $X = I_\pi$. As A_π is a vertex of $P_A^{(L,U)}$, A_π is the unique maximizer over $P_A^{(L,U)}$ of some linear functions; let such a linear function be represented by the matrix $C \in \mathbb{R}^{d\times p}$. Viewing X and I_π as solutions of the network flow problem, a standard result (e.g., Denardo [24, p. 99]) implies that $X - I_\pi$ can be decomposed into a sum $\sum_{s=1}^q \beta_s \underline{z}^s$ where for each $s = 1, \ldots, q$, z^s is a circuit of the network flow problem, β_s is a positive number and $I_\pi + \underline{z}^s$ is feasible, that is, $I_\pi + \underline{z}^s \in P_I^{(L,U)}$. By Theorem 1.1.5, for each $s = 1, \ldots, q$, $AI_\pi + A\underline{z}^s = A(I_\pi + \underline{z}^s) \in P_A^{(L,U)}$. Now, the unique optimality of $A_\pi = AI_\pi$ over $P_A^{(L,U)}$ under the linear function represented by C implies that for $s = 1, \ldots, q$, $\langle C, A_\pi\rangle \geq \langle C, AI_\pi + A\underline{z}^s\rangle = \langle C, A_\pi\rangle + \langle C, A\underline{z}^s\rangle$ with equality holding if and only if $AI_\pi + A\underline{z}^s = A_\pi$, that is, for each s, $\langle C, A\underline{z}^s\rangle \leq 0$ with equality holding if and only if $A\underline{z}^s = 0$, the latter being equivalent by (1.3.4) to the assertion $\underline{z}^s = 0$. As we are assuming that $AX = A_\pi$ and $X = I_\pi + \sum_{s=1}^q \beta_s \underline{z}^s$, we have that $\langle C, A_\pi\rangle = \langle C, AX\rangle = \langle C, AI_\pi + \sum_{s=1}^q \beta_s A\underline{z}^s\rangle = \langle C, A_\pi\rangle + \sum_{s=1}^q \beta_s \langle C, A\underline{z}^s\rangle \leq \langle C, A_\pi\rangle$; it follows that the inequalities in the above string hold as equalities. Thus, for $s = 1, \ldots, q$, $\langle C, A\underline{z}^t\rangle = 0$ and consequently $\underline{z}^s = 0$. Hence, $X = I_\pi + \sum_{s=1}^q \beta_s \underline{z}^s = I_\pi$.

$(b) \Rightarrow (c)$: This implication is trivial as $\langle I_\pi, I_\pi\rangle = n$.

$(c) \Rightarrow (d)$: This implication is trivial.

$(d) \Rightarrow (a)$: Assume that A_π is not a vertex of $P_A^{(L,U)}$ and we will show that (d) does not hold.

The vertices of $P_A^{(L,U)}$ are all in the set $\{A_\sigma : \sigma \in \Pi^{(L,U)}$ (Proposition 3.1.1 of Vol. I, part (i)). As A_π (like every other point in $P_A^{(L,U)}$) has a representation as a convex combination of vertices of $P_A^{(L,U)}$ all of which are vectors associated with partitions (Proposition 3.1.1 of Vol. I, parts (a) and (i)), there exist partitions $\sigma^1, \ldots, \sigma^q$ in $\Pi^{(L,U)}$, all distinct from π, and positive coefficients $\alpha_1, \ldots, \alpha_q$ that sum to 1 such that $A_\pi = \sum_{s=1}^q \alpha_s A_{\sigma^s}$. As $A_{\sigma^s} = AI_{\sigma^s}$ for $s = 1, \ldots, q$, we have that $A_\pi = \sum_{s=1}^q \alpha_s AI_{\sigma^s} = A(\sum_{s=1}^q \alpha_s I_{\sigma^s})$. As $P_I^{(L,U)}$ is convex, we have that $\sum_{s=1}^q \alpha_s I_{\sigma^s} \in P_I^{(L,U)}$. Noting that $\langle I_\pi, I_\sigma\rangle \leq n-1$ for each p-partition σ that is distinct from π, we conclude that $\langle I_\pi, \sum_{s=1}^q \alpha_s I_{\sigma^s}\rangle = \sum_{s=1}^q \alpha_s \langle I_\pi, I_{\sigma^s}\rangle \leq \sum_{s=1}^q \alpha_s (n-1) = n-1$. So, $X \equiv \sum_{s=1}^q \alpha_s I_{\sigma^s} \in P_I^{(L,U)}$ has $AX = A_\pi$ and $\langle I_\pi, X\rangle \leq n-1$, implying that $\{X \in P_I^{(L,U)} : AX = A_\pi$ and $\langle I_\pi, X\rangle \leq n-1\}$ is not empty, that is, (d) is false. □

The next corollary of Theorem 1.3.1 specializes the equivalence of conditions (a) and (b) of that theorem to vectors X associated with partitions, showing that such a vector is a vertex (of the corresponding bounded-shape partition polytope) only if it corresponds to a unique partition. This result was first established by Barnes, Hoffman and Rothblum [7].

Corollary 1.3.2. *Let $A \in \mathbb{R}^{d\times n}$ have nonzero and distinct columns, let L and U be positive integer p-vectors satisfying $L \leq U$ and $\sum_{j=1}^{p} L_j \leq n \leq \sum_{j=1}^{p} U_j$, and let $\pi \in \Pi^{(L,U)}$. If A_π is a vertex of $P_A^{(L,U)}$, then $A_\pi \neq A_\sigma$ for every partition $\sigma \in \Pi^{(L,U)} \setminus \{\pi\}$.*

Proof. Let $\sigma \in \Pi^{(L,U)} \setminus \{\pi\}$. Lemma 1.1.1 assures that $I_\sigma \neq I_\pi$, and by definition $I_\sigma \in P_I^{(L,U)}$. Thus, the implication $(a) \Rightarrow (b)$ of Theorem 1.3.1 implies that $AI_\sigma \neq A_\pi$. □

The next four examples demonstrate, respectively, that (with the columns of A nonzero and distinct):

(i) for a partition π, the necessary condition for A_π to be a vertex of $P^{(L,U)}$ stated in Corollary 1.3.2 is not sufficient, even for single-shape partition polytopes;
(ii) vectors in the interior of bounded-shape partition polytopes may have multiple representation in the form A_π for π in $\Pi^{(L,U)}$, in particular, they have multiple representation of the form AX with $X \in P_I^{(L,U)}$;
(iii) vertices of constrained-shape partition polytopes, which are not bounded-shape, may have multiple representation in the form AX for X in $P_I^{(L,U)}$ (the example has the columns of A nonzero and distinct);
(iv) unique representation of vectors in bounded-shape (in fact, even single-shape) partition polytopes is not sufficient for being a vertex, when the vector is not associated with a partition.

For positive indices i and j, let $\Pi^{(i,j)}$ be the partitions with shape (i,j) and $P^{(i,j)}$ be the corresponding partition polytopes.

Example 1.3.1. Let $d = 1, n = 3, A = (1,2,3), p = 2, n_1 = 1$ and $n_2 = 2$. There are three partitions in $\Pi^{(1,2)}$ and the associated matrices are $(1,5)$, $(2,4)$ and $(3,3)$. Here, $(2,4)$ has unique representation as A_π with $\pi = (\{2\},\{1,3\})$, but it is not a vertex of $P_A^{(1,2)}$ since $(2,4) = \frac{1}{2}(1,5) + \frac{1}{2}(3,3)$. □

Example 1.3.2. Let $d = 1$, $n = 3$, $A = (1,2,3)$, $p = 2$, $L_1 = L_2 = 1$ and $U_1 = U_2 = 2$. There are six partitions in $\Pi^{(L,U)}$ and the associated matrices are $(1,5)$, $(2,4)$, $(3,3)$, $(3,3)$, $(4,2)$ and $(5,1)$. Here, $(3,3)$ is not a vertex of $P_A^{(L,U)}$ and it has the representation A_π for two partitions of $\Pi^{(L,U)}$, namely, for $\pi = (\{1,2\},\{3\})$ and for $\pi = (\{3\},\{1,2\})$. □

Example 1.3.3. Let $d = 1$, $n = 4$, $A = (-2,-1,1,2)$ and $p = 2$. In this case, $\{A_\pi : \pi \in \Pi^{(1,3)}\} = \{A_\pi : \pi \in \Pi^{(3,1)}\} = \{(-2,2),(-1,1),(1,-1),(2,-2)\} = \{A_\pi : \pi \in \Pi^{(1,3)} \cup \Pi^{(3,1)}\}$. So, the vertices of the constrained-shape partition polytope corresponding to $\Pi^{(1,3)} \cup \Pi^{(3,1)}$ are $(-2,2)$ and $(2,-2)$ and each is realizable by two partitions. Of course, while $\Pi^{(1,3)} \cup \Pi^{(3,1)}$ is a set of partitions which is constrained-shape, it is not bounded-shape. □

Example 1.3.4. Consider the case where $A = I$. In this case, for every set of partitions Π, the partition polytope P_A^Π coincides with the corresponding transportation-polytope P_I^Π and the map $X \to AX = IX$ of P_I^Π onto P_A^Π asserted in Theorem 1.1.5 is the identity. In particular, every point in $P_A^\Pi = P_I^\Pi$ has a unique representation; but, when $n > p$, the $P_A^\Pi = P_I^\Pi$ is not a singleton and contains points which are not vertices. □

The next result provides a necessary and sufficient condition for a partition to be a vertex of bounded-shape partition polytopes when the columns of A are nonzero and distinct. The characterization is due to Barnes, Hoffman and Rothblum [7].

Theorem 1.3.3. *Let $A \in \mathbb{R}^{d \times n}$ have nonzero and distinct columns; let L and U be positive integer p-vectors satisfying $L \le U$ and $\sum_{j=1}^p L_j \le n \le \sum_{j=1}^p U_j$, and let $\pi \in \Pi^{(L,U)}$. Then the following are equivalent:*

(a) A_π is a vertex of $P_A^{(L,U)}$, and
(b) there exist a matrix $C \in \mathbb{R}^{d \times n}$ and vector $\alpha \in \mathbb{R}^p$ such that:

1) $(C^r - C^s)^T A^t > \alpha_s - \alpha_r$ for $r, s \in \{1,\ldots,p\}$ with $r \ne s$ and $t \in \pi_r$,
2) $\alpha_r \le 0$ if $|\pi_r| > L_r$, and
3) $\alpha_r \ge 0$ if $|\pi_r| < U_r$.

Proof. $(a) \Rightarrow (b)$: Suppose A_π is a vertex of $P_A^{(L,U)}$, then A_π is a unique maximizer over $P_A^{(L,U)}$ of some linear functions, say, one that is determined by the matrix $C \in \mathbb{R}^{d \times p}$. So,

$$\langle C, Y \rangle < \langle C, A_\pi \rangle \text{ for each } Y \in P_A^{(L,U)} \setminus \{A_\pi\}. \qquad (1.3.5)$$

By Theorem 1.1.5, $P_A^{(L,U)} = \{AX : X \in P_I^{(L,U)}\}$, and by Theorem 1.3.1, I_π is the only vector X in $P_I^{(L,U)}$ with $AX = A_\pi$. Hence

$$\langle A^T C, X\rangle = \langle C, AX\rangle < \langle C, AI_\pi\rangle = \langle A^T C, I_\pi\rangle \text{ for each } X \in P_I^{(L,U)} \backslash \{I_\pi\}, \quad (1.3.6)$$

where (1.1.1) is used for the two equalities in (1.3.6). So, I_π is the only solution of the linear program where the linear function defined through the matrix $A^T C \in \mathbb{R}^{n\times p}$ is maximized over $P_I^{(L,U)}$. Using the linear inequality representation of $P_I^{(L,U)}$ given in (1.1.4) (Theorem 1.1.3), the dual of this linear program has unrestricted variables $\beta_1, \ldots, \beta_n$, nonnegative variables $\gamma_1, \ldots, \gamma_p$ and $\delta_1, \ldots, \delta_p$, constraints

$$\beta_t + \delta_j - \gamma_j \geq (A^t)^T C^j \text{ for } j = 1, \ldots, p \text{ and } t = 1, \ldots, n,$$

and objective function

$$\min\left(\sum_{t=1}^n \beta_t - \sum_{j=1}^p \gamma_j L_j + \sum_{j=1}^p \delta_j U_j\right).$$

As I_π is the unique solution of the primal linear program, it follows from the Strong Complementarity Theorem (see Theorem A.2 of Appendix A of this chapter) that there exists a solution (β, γ, δ) of the dual program with

$$\begin{aligned}
&(a)\ \beta_t + \delta_j - \gamma_j = (A^t)^T C^j \text{ if } (I_\pi)_t^j > 0,\\
&(b)\ \beta_t + \delta_j - \gamma_j > (A^t)^T C^j \text{ if } (I_\pi)_t^j = 0,\\
&(c)\ \gamma_j = 0 \text{ if } \sum_{t=1}^n (I_\pi)_t^j > L_j,\\
&(d)\ \gamma_j > 0 \text{ if } \sum_{t=1}^n (I_\pi)_t^j = L_j,\\
&(e)\ \delta_j = 0 \text{ if } \sum_{t=1}^n (I_\pi)_t^j < U_j,\\
&(f)\ \delta_j > 0 \text{ if } \sum_{t=1}^n (I_\pi)_t^j = U_j.
\end{aligned} \quad (1.3.7)$$

Let $\alpha_j = \gamma_j - \delta_j$ for $j = 1, \ldots, p$. Also, consider $t = 1, \ldots, n$ and $r, s = 1, \ldots, p$ where $t \in \pi_r$ and $t \notin \pi_s$. As $(I_\pi)_t^r = 1$ and $(I_\pi)_t^s = 0$, we have that $\beta_t - \alpha_r = (A^t)^T C^r$ and $\beta_t - \alpha_s > (A^t)^T C^s$, implying that $\alpha_s - \alpha_r < (A^t)^T (C^r - C^s)$. Also, for $j = 1, \ldots, p$, $\sum_{t=1}^n (I_\pi)_t^j = |\pi_j|$; hence, we get from (1.3.7) that if $|\pi_j| < U_j$ then $\alpha_j = \gamma_j - \delta_j = \gamma_j - 0 \geq 0$, and if $|\pi_j| > L_j$, $\alpha_j = \gamma_j - \delta_j = -\delta_j \leq 0$.

$(b) \Rightarrow (a)$: Assume that π satisfies (b) with $C \in \mathbb{R}^{d \times p}$ and $\alpha \in \mathbb{R}^p$; we will show that A_π is a vertex of $P_A^{(L,U)}$. As $A_\pi \in P_A^{(L,U)}$, it suffices to show that the linear functional that is defined on $\mathbb{R}^{d \times p}$ by C is uniquely maximized over $P_A^{(L,U)}$ at A_π. Suppose $\sigma \in \Pi^{(L,U)} \setminus \{\pi\}$. For each $i = 1, \ldots, n$, let $r(\pi)_i$ and $r(\sigma)_i$ be the indices of the parts of π and σ, respectively, that contain i. Evidently, if $r(\pi)_i = r(\sigma)_i$, then trivially

$$\left[C^{r(\pi)_i} - C^{r(\sigma)_i}\right]^T A^i = 0 = \alpha_{r(\pi)_i} - \alpha_{r(\sigma)_i}. \tag{1.3.8}$$

Also, if $r(\pi)_i \neq r(\sigma)_i$, then $(b1)$ implies that

$$\left[C^{r(\pi)_i} - C^{r(\sigma)_i}\right]^T A^i > \alpha_{r(\sigma)_i} - \alpha_{r(\pi)_i}. \tag{1.3.9}$$

As $\pi \neq \sigma$, $r(\pi)_i \neq r(\sigma)_i$ for at least one $i \in \{1, \ldots, n\}$ and adding these inequalities over all i we get that

$$\sum_{i=1}^{n} \left[C^{r(\pi)_i} - C^{r(\sigma)_i}\right]^T A^i > \sum_{i=1}^{n} \left(\alpha_{r(\sigma)_i} - \alpha_{r(\pi)_i}\right), \tag{1.3.10}$$

and by arranging terms, we have that

$$\langle C, A_\pi \rangle - \langle C, A_\sigma \rangle = \sum_{j=1}^{p} (C^j)^T[(A_\pi)^j - (A_\sigma)^j] > \sum_{j=1}^{p} \alpha_j[|\sigma_j| - |\pi_j|]. \tag{1.3.11}$$

Next, observe from $(b3)$ and $(b2)$, respectively, that if $\alpha_j < 0$ then $|\pi_j| = U_j \geq |\sigma_j|$ and if $\alpha_j > 0$ then $|\pi_j| = L_j \leq |\sigma_j|$; thus, for each j, $\alpha_j[|\sigma_j| - |\pi_j|]$ is nonnegative. It follows from (1.3.11) that $\langle C, A_\pi \rangle - \langle C, A_\sigma \rangle > 0$ for each $\sigma \in \Pi^{(L,U)}$. So, A_π is the unique maximizer over $\{A_\sigma : \sigma \in \Pi^{(L,U)}\}$ of the linear function on $\mathbb{R}^{d \times p}$ determined by the matrix C; we can therefore conclude from part (m) of Proposition 3.1.1 in Vol. I that A_π is a vertex of $P_A^{(L,U)} = \text{conv}\{A_\sigma : \sigma \in \Pi^{(L,U)}\}$. □

The characterization of vertices of bounded-shape partition polytopes in Theorem 1.3.3 specializes to characterizations of vertices of single-shape and of single-size partition polytopes.

Corollary 1.3.4. *Let $A \in \mathbb{R}^{d \times n}$ have nonzero and distinct columns, let $(n_1, \cdots, n_p)$ be a nonnegative integer p-vector satisfying $\sum_{j=1}^{p} n_j = n$, and let $\pi \in \Pi^{(n_1, \cdots, n_p)}$. Then A_π is a vertex of $P_A^{(n_1, \cdots, n_p)}$ if and only if there exist a matrix $C \in \mathbb{R}^{d \times p}$ and a vector $\alpha \in \mathbb{R}^p$ such that $(C^r - C^s)^T A^t > \alpha_s - \alpha_r$ for $r, s \in \{1, \cdots, p\}$ with $r \neq s$ and $t \in \pi_r$.*

Proof. The corollary follows immediately from Theorem 1.3.3 with $L_j = U_j = n_j$ for each $j \in \{1, \cdots, p\}$. □

Corollary 1.3.5. *Let $A \in \mathbb{R}^{d\times n}$ have nonzero and distinct columns and let $\pi \in \Pi^p$. Then A_π is a vertex of P_A^p if and only if there exists a matrix $C \in \mathbb{R}^{d\times p}$ such that $(C^r - C^s)^T A^t > 0$ for $r, s \in \{1, \cdots, p\}$ with $r \neq s$ and $t \in \pi_r$.*

Proof. Let $L_j = 0$ and $U_j = n$ for each $j = 1, \cdots, n$. Then $P_A^{(L,U)} = P_A^p$. For sufficiency, assume that there exists $C \in \mathbb{R}^{d\times n}$ satisfying $(C^r - C^s)^T A^t > 0$ for $r, s \in \{1, \cdots, p\}$ with $r \neq s$ and $t \in \pi_r$. Then condition (b) of Theorem 1.3.3 holds with C and with $\alpha = 0$, implying A_π is a vertex of P_A^p.

Next assume that A_π is a vertex of P_A^p. By Theorem 1.3.3, there exist $C \in \mathbb{R}^{d\times n}$ and $\alpha \in \mathbb{R}^p$ such that condition $(b1)$ holds. Consider an arbitrary r and $t \in \pi_r$. It follows $|\pi_r| \geq 1 > L_r$ and $|\pi_s| \leq n - 1 < U_s$ for each $s \neq r$. Hence, by $(b2)$, $\alpha_r \leq 0$, and by $(b3)$, $\alpha_s \geq 0$. Therefore, $(C^r - C^s)^T A^t > \alpha_s - \alpha_r \geq 0$ for $r, s \in \{1, \cdots, p\}$ with $r \neq s$ and $t \in \pi_r$. □

For a vector $v \in \mathbb{R}^d$, recall (Vol. I, p. 85) that $\widetilde{v} \equiv \binom{1}{v} \in \mathbb{R}^{d+1}$; and for a matrix A with d rows, $\widetilde{A}$ is the matrix obtained from A by replacing each column v by $\widetilde{v}$ (i.e., a row of 1's is added at the top of A). If $A \in \mathbb{R}^{d\times n}$ and π is a p-partition with shape $(n_1, \cdots, n_p)$, then $\widetilde{A}_\pi = \binom{n_1, \cdots, n_p}{A_\pi}$.

Corollaries 1.3.4 and 1.3.5 are next combined to establish a characterization of partitions associated with vertices of a single-shape partition polytope as partitions associated with vertices of a single-size partition polytope, with the underlying matrix A replaced by $\widetilde{A}$. Let $V[P_A]$ denote the vertex set of the polytope on A. Rothblum proved

Theorem 1.3.6. *Let $A \in \mathbb{R}^{d\times n}$ have nonzero and distinct columns, let $(n_1, \cdots, n_p)$ be a nonnegative integer p-vector satisfying $\sum_{j=1}^p n_j = n$, and let $\pi \in \Pi^{(n_1,\cdots,n_p)}$. Then $A_\pi \in V[P_A^{(n_1,\cdots,n_p)}]$ if and only if $\widetilde{A}_\pi \in V[P_{\widetilde{A}}^p]$.*

Proof. Assume that A_π is a vertex of $P_A^{(n_1,\cdots,n_p)}$ and let $C \in \mathbb{R}^{d\times p}$ and $\alpha \in \mathbb{R}^p$ be as in the conclusion of Corollary 1.3.4. With $C_\alpha \equiv \binom{\alpha}{C} \in \mathbb{R}^{(d+1)\times p}$, it follows that

$$(C_\alpha^r - C_\alpha^s)^T \widetilde{A}^t = [(C^r)^T A^t + \alpha_r] - [(C^s)^T A^t + \alpha_s] > 0$$

for $r, s \in \{1, \cdots, p\}$ with $r \neq s$ and $t \in \pi_r$. Hence, by Corollary 1.3.5, $\widetilde{A}_\pi$ is a vertex of $P_{\widetilde{A}}^p$.

Next assume that $\widetilde{A}_\pi$ is a vertex of $P_{\widetilde{A}}^p$. By Corollary 1.3.5, there exists $C_\alpha \in \mathbb{R}^{(d+1)\times p}$ such that $(C_\alpha^r - C_\alpha^s)^T \widetilde{A}^t > 0$ for $r, s \in \{1, \cdots, p\}$ with $r \neq s$

and $t \in \pi_r$. Designating the first row of C_α as α^T and its remaining d rows as C, it then follows that

$$[(C^r)^T A^t + \alpha_r] - [(C^s)^T A^t + \alpha_s] = (C^r_\alpha - C^s_\alpha)^T \widetilde{A}^t > 0$$

for $r, s \in \{1, \cdots, p\}$ with $r \neq s$ and $t \in \pi_r$. Hence, by Corollary 1.3.4, A_π is a vertex of $P_A^{(n_1,\cdots,n_p)}$. □

This characterization allows us to enumerate vertices of a single-shape partition polytope through enumerating vertices of the corresponding single-size partition polytope with $\widetilde{A}$ replacing A (see Section 1.2, enumeration of vertices of single-size partition polytopes). This observation is recorded in the next corollary.

Corollary 1.3.7. *Let $A \in \mathbb{R}^{d\times n}$ have nonzero and distinct columns, let $(n_1, \cdots, n_p)$ be a nonnegative integer p-vector satisfying $\sum_{j=1}^p n_j = n$. Then $|V[P_A^{(n_1,\cdots,n_p)}]| \leq |V[P_{\widetilde{A}}^p]| \leq O(n^{(d+1)(p-1)-1})$, and a strongly polynomial algorithm using $O(n^{(d+1)(p-1)-1})$ arithmetic operations for enumerating all vertices of $P_A^{(n_1,\cdots,n_p)}$ is available.*

Clearly, Theorem 1.3.6 also yields an algorithm of counting the vertices of a bounded-shape partition polytope by simply adding up the number of vertices of each single-shape partition polytope where the shape satisfies the given bound.

Testing if a Vector A_π is a Vertex of the Bounded-Shape Partition Polytope When the Columns of A are Nonzero and Distinct

Theorem 1.3.3 identifies a system of linear inequalities (including strict ones) whose feasibility characterizes whether or not the vector A_π associated with partition π is a vertex of the bounded-shape partition polytope $P_A^{(L,U)}$. The system is homogenous in its variables—the matrix C and the vector α; hence, its feasibility is equivalent to the feasibility of the following inhomogenous system with weak inequalities

$$\begin{aligned} (C^r - C^s)^T A^t \geq 1 + \alpha_s - \alpha_r \quad &\text{for } r, s \in \{1, \ldots, p\} \text{ with } r \neq s \text{ and } t \in \pi_r, \\ \alpha_r \leq 0 \quad &\text{if } |\pi_r| > L_r, \text{ and} \\ \alpha_r \geq 0 \quad &\text{if } |\pi_r| < U_r. \end{aligned} \tag{1.3.12}$$

(Note that the strict inequality in Theorem 1.3.3($b1$) can be changed to weak inequality by adding an $\epsilon > 0$ to the right-hand side. Multiplying both sides by $\frac{1}{\epsilon}$ yields the first inequality of (1.3.12)). Thus, we have a test for checking whether or not a vector associated with a partition is a vertex of a corresponding bounded-shape partition polytope by checking feasibility of a system of linear inequalities with $(d+1)p$ variables, $(p^2-p)n$ inequalities and at most p nonnegativity/nonpositivity constraints. In particular, this test is polynomial in p, d and n. □

An alternative test with polynomial complexity for being a vertex is available from Corollary 1.1.6; while this test has to solve dp linear programs and its total efficiency is apparently slightly worse than that of the test described above, it is applicable without the restriction about the columns of A.

The forthcoming Theorems 1.3.8 and 1.3.10 extend, respectively, parts of Theorems 1.3.1 and 1.3.3 to the case where A has a zero vector while its columns are still assumed distinct. A generalization of these results to the case where A is allowed to have repeated columns, see Section 1.4. These results were established by Hwang, Onn and Rothblum [1998].

Theorem 1.3.8. *Let $A \in \mathbb{R}^{d\times n}$ have distinct columns, let L and U be positive integer p-vectors satisfying $L \le U$ and $\sum_{j=1}^{p} L_j \le n \le \sum_{j=1}^{p} U_j$, and let $\pi \in \Pi^{(L,U)}$. Suppose $A^{t_1} = 0$ and $t_1 \in \pi_{r_1}$, and consider the subset Δ of $\{1,\dots,p\}$ given by*

$$\Delta = \begin{cases} \{s \in \{1,\dots,p\} \setminus \{r_1\} : |\pi_s| < U_s\} & \text{if } |\pi_{r_1}| > L_{r_1} \\ \emptyset & \text{if } |\pi_{r_1}| = L_{r_1}. \end{cases}$$

For each $s \in \Delta$, let σ^s be the partition with

$$\sigma_j^s = \begin{cases} \pi_j & \text{if } j \in \{1,\dots,p\} \setminus \{r_1, s\} \\ \pi_{r_1} \setminus \{t_1\} & \text{if } j = r_1, \text{ and} \\ \pi_s \cup \{t_1\} & \text{if } j = s\,. \end{cases}$$

Then the following are equivalent:

a) A_π is a vertex of $P_A^{(L,U)}$,

b) $\{X \in P_I^{(L,U)} : AX = A_\pi\} = conv(\{I_\pi\} \cup \{I_{\sigma^s} : s \in \Delta\})$.

Proof. $(a) \Rightarrow (b)$: Assume that A_π is a vertex of $P_A^{(L,U)}$ and $Y \in P_I^{(L,U)}$ satisfies $AY = A_\pi$. Consider the network constructed in the proof of Theorem 1.3.1. The arguments of the proof of that theorem show that $Y \in P_I^{(L,U)}$ has a decomposition $Y = I_\pi + \sum_{u=1}^{q} \beta_u \underline{z}^u$ where for each $u = 1,\dots,q$, z^u is a circuit of the network flow problem, β_u is a positive number and $I_\pi + \underline{z}^u \in P_I^{(L,U)}$. As in the proof of the necessity part in Theorem 1.3.1, the existence of a linear function attaining a unique maximum at A_π implies that $A\underline{z}^u = 0$ for each $u = 1,\dots,q$. Now, if $\Delta = \emptyset$, we still have that (1.3.4) holds and $A\underline{z} = 0$ implies $\underline{z} = 0$. The remainder of the proof of the implication $(a) \Rightarrow (b)$ in Theorem 1.3.1 can then be used to establish (b) (which coincides with condition (b) of Theorem 1.3.1 when $\Delta = \emptyset$). Next, assume that $\Delta \neq \emptyset$. In this case, (1.3.4) needs not hold and $Az = 0$ is possible—it will happen only for a circuit that shifts the zero vector from $j = r_1$ to some $s \in \Delta$; for such circuit $\underline{z}$ we have that $z = I_{\sigma^s} - I_\pi$ for some $s \in \Delta$. It follows that each $\underline{z}^u$ has a representation as $I_{\tau^u} - I_\pi$ for some τ^u in $\{\sigma^s : s \in \Delta\}$ and

$$Y = I_\pi + \sum_{u=1}^{q} \beta_u \underline{z}^u = \left(1 - \sum_{u=1}^{q} \beta_u\right) I_\pi + \sum_{u=1}^{q} \beta_u I_{\tau^u}.$$

Considering the (t_1, r_1) coordinate of this equation, and noting that $Y_{t_1}^{r_1} \geq 0$ (from (1.1.4a), $(I_\pi)_{t_1}^{r_1} = 1$ and $(I_{\sigma^s})_{t_1}^{r_1} = 0$ for $s \in \Delta$, one can conclude that

$$0 \leq (Y)_{t_1}^{r_1} = \left(1 - \sum_{u=1}^{q} \beta_u\right) (I_\pi)_{t_1}^{r_1} + \sum_{u=1}^{q} \beta_u \left(I_{\tau^u}\right)_{t_1}^{r_1} = 1 - \sum_{u=1}^{q} \beta_u,$$

implying that $\sum_{u=1}^{q} \beta_u \leq 1$; thus, $Y \in \text{conv}(\{I_\pi\} \cup \{I_{\sigma^s} : s \in \Delta\}$. We established that $\{Y \in P_I^{(L,U)} : AY = A_\pi\} \subseteq \text{conv}(\{I_\pi\} \cup \{I_{\sigma^s} : s \in \Delta\})$. The inverse inclusion is trivial as $AI_{\sigma^s} = A_\pi$ for each $s \in \Delta$.

$(b) \Rightarrow (a)$: Assume that (b) is satisfied and A_π is not a vertex of $P_A^{(L,U)}$. The vertices of $P_A^{(L,U)}$ are all in the set $\{A_\sigma : \sigma \in \Pi^{(L,U)}\}$. As A_π (like every other point in $P_A^{(L,U)}$) has a representation as a convex combination of vertices of $P_A^{(L,U)}$ (Proposition 3.1.1 of Vol. I, parts (i) and (a)), there exist partitions $\tau^1, \ldots, \tau^q$ in $\Pi^{(L,U)}$, all with $A_{\tau^u} \neq A_\pi$, and positive coefficients $\alpha_1, \ldots, \alpha_q$ that sum to 1 such that $A_\pi = \sum_{u=1}^{q} \alpha_u A_{\tau^u}$. As $A_{\tau^u} = AI_{\tau^u}$ for $u = 1, \ldots, q$, we have that $A_\pi = \sum_{u=1}^{q} \alpha_u AI_{\tau^u} = A(\sum_{u=1}^{q} \alpha_u I_{\tau^u})$. As $P_I^{(L,U)}$ is convex, $Y \equiv \sum_{u=1}^{q} \alpha_u I_{\tau^u} \in P_I^{(L,U)}$. We conclude from the asserted condition (b), that Y is in $\text{conv}(\{I_\pi\} \cup \{I_{\sigma^s} : s \in \Delta\})$. Now, for $U \in \{I_\pi\} \cup \{I_{\sigma^s} : s \in \Delta\}$, $U_t^j = 0$ for pair of indices (t, j) with either $t \neq \{t_1\}$ and $t \notin \pi_j$, or $t = t_1$ and $j \notin \{r_1\} \cup \Delta$; as this conclusion extends to the convex hull of matrices satisfying it, we conclude that for such pairs (t, j), $Y_t^j = 0$. As $Y \equiv \sum_{u=1}^{q} \alpha_u I_{\tau^u}$ with positive α_u's and I_{τ^u} in $P_I^{(L,U)}$, we reach the same conclusion for each I_{τ^u}, but this implies that each τ^u is in $\{\pi\} \cup \{\sigma^s : s \in \Delta\}$ and $A_{\tau^u} = A_\pi$, a contradiction which proves that A_π must be a vertex of $P_A^{(L,U)}$. □

Similar to the derivation of Corollary 1.3.2, we obtain a necessary condition for being a vertex by restricting condition (b) of Theorem 1.3.8 to vectors X associated with partitions.

Corollary 1.3.9. *Let $A, L, U, \pi, t_1, r_1, \Delta$ and σ^s for $s \in \Delta$ be as in Theorem 1.3.8. If A_π is a vertex of $P_A^{(L,U)}$ and $\pi \neq \sigma \in \Pi^{(L,U)} \setminus \{\sigma^s : s \in \Delta\}$, then $A_\pi \neq A_\sigma$.*

Proof. Suppose $\sigma \in \Pi^{(L,U)} \setminus \{\sigma^s : s \in \Delta\}$. Lemma 1.1.1 assures that I_σ is not in the convex hull of $\{I_{\sigma^s} : s \in \Delta\}$, and by definition $I_\sigma \in P_I^{(L,U)}$. Thus, Theorem 1.3.8 implies that $AI_\sigma \neq A_\pi$. □

Theorem 1.3.10. *Let A, L, U, π, t_1, r_1 and Δ be as in Theorem 1.3.8. Then the following are equivalent:*

(a) A_π is a vertex of $P_A^{(L,U)}$, and
(b) there exists a matrix $C \in \mathbb{R}^{d\times n}$ and vector $\alpha \in \mathbb{R}^p$ such that:

1.1) $(C^r - C^s)^T A^t > \alpha_s - \alpha_r$ for $r, s \in \{1, \ldots, p\}$, $t \in \pi_r$ with $r \neq s$ and either $t \neq t_1$ or $t = t_1$ (which assures $r = r_1$) and $s \notin \Delta$,
1.2) $0 \geq \alpha_s - \alpha_{r_1}$ if $s \in \Delta$.
2) $\alpha_r \leq 0$ if $|\pi_r| > L_r$, and
3) $\alpha_r \geq 0$ if $|\pi_r| < U_r$.

Proof. Throughout, let σ^s for $s \in \Delta$ be defined as in Theorem 1.3.8.

$(a) \Rightarrow (b)$: If $\Delta = \emptyset$, *1.2)* is vacuous, (b) coincides with condition (b) of Theorem 1.3.3, and the arguments proving $(a) \Rightarrow (b)$ in the proof of Theorem 1.3.3 establish the validity of (b). Next assume that $\Delta \neq \emptyset$ and A_π is a vertex of $P_A^{(L,U)}$. As in the proof of the implication $(a) \Rightarrow (b)$ in Theorem 1.3.3, it follows that there exists a matrix $C \in \mathbb{R}^{d\times p}$ that satisfies (1.3.5); consequently, using the implication $(a) \Rightarrow (b)$ in Theorem 1.3.8 we get the following modification of (1.3.6):

$$\begin{aligned}\langle A^T C, X\rangle &= \langle C, AX\rangle < \langle C, AI_\pi\rangle \\ &= \langle A^T C, I_\pi\rangle \quad \forall\, X \in P_I^\Pi \setminus \mathrm{conv}(\{I_\pi\} \cup \{I_{\sigma^s} : s \in \Delta\}).\end{aligned}$$

As

$$\langle A^T C, I_{\sigma^s}\rangle = \langle C, AI_{\sigma^s}\rangle = \langle C, A_{\sigma^s}\rangle = \langle C, A_\pi\rangle \;\; \forall s \in \Delta,$$

we conclude that the vectors in $\mathrm{conv}(\{I_\pi\} \cup \{I_{\sigma^s} : s \in \Delta\})$ are precisely the optimal solutions of the linear program where the (linear) objective function defined through the matrix $A^T C \in \mathbb{R}^{n\times n}$ is to be maximized over $P_I^{(L,U)}$. In particular, for all pairs (t, j), except for (t_1, r_1) and (t_1, s) with $s \in \Delta$, an optimal solution Y must have $Y_t^r = (I_\pi)_t^r$ (since this holds for the I_{σ^s}'s where $s \in \Delta$). Consider the dual of the above linear program using considered linear inequality representation of $P_I^{(L,U)}$ in (1.1.4) (as was done in the proof of Theorem 1.3.3); we then get from the strong complementarity theorem (again, see Theorem A.2 of the Appendix of this chapter) that there is an optimal solution that satisfies (1.3.7), except that (1.3.7a) holds with $\geq$ replacing $=$ for $(t, j) = (t_1, r_1)$ and (1.3.7b) holds as a weak inequality for (t, s) with $s \in \Delta$. The verification of (b) now follows from these inequalities and the arguments of the proof of Theorem 1.3.3.

$(b) \Rightarrow (a)$: Assume that $C \in \mathbb{R}^{d\times p}$ and $\alpha \in \mathbb{R}^p$ satisfy the conditions of (b). Let $\sigma \in \Pi^{(L,U)} \setminus \{\pi\}$ and let $r(\pi)_i$ and $r(\sigma)_i$ be defined as in the sufficiency proof of Theorem 1.3.3. We then have that (1.3.8) holds trivially for each i satisfying $r(\pi)_i = r(\sigma)_i$. Also, for i with $(\pi)_i \neq r(\sigma)_i$, (1.3.9) holds as long as $i \neq t_1$ or $i = t_1$ and $r(\sigma)_i \notin \Delta$; when $i = t_1$ and $r(\sigma)_i \in \Delta$, (1.3.9) holds as a weak inequality. We conclude that unless σ and π coincide in the assignment of all elements except for t_1, (1.3.10) and (1.3.11) will hold (strictly). The remaining partitions σ are all in $\{\sigma^s : s \in \Delta\}$ and must satisfy $A_\sigma = A_\pi$; in particular, $\langle C, A_\pi\rangle = \langle C, A_\sigma\rangle$. It follows that $\langle C, A_\pi\rangle > \langle C, A_\sigma\rangle$ for every partition $\sigma \in \Pi^{(L,U)}$ with $A_\sigma \neq A_\pi$. So, A_π is the unique maximizer over $\{A_\sigma : \sigma \in \Pi^{(L,U)}\}$ of the linear function on $\mathbb{R}^{d\times p}$ determined by the matrix C; we can therefore conclude from part (m) of Proposition 3.1.1 of Vol. I that A_π is a vertex of $P_A^{(L,U)} = \text{conv}\{A_\sigma : \sigma \in \Pi^{(L,U)}\}$.

□

Testing if a Vector A_π is a Vertex of the Bounded-Shape Partition Polytope When the Columns of A are Distinct, but Contain the Zero Vector

Recall the text for a vector A_π associated with a partition π to be a vertex of the bounded-shape partition polytope $P_A^{(L,U)}$ when A's columns are nonzero and distinct. The test relied on testing feasibility of the system of linear inequalities identified in Theorem 1.3.3, using its homogeneity to convert the strict inequalities into weak inequalities (see (1.3.12)). We apply the same approach to the system identified in part (b) of Theorem 1.3.10. This system coincides with the one identified in Theorem 1.3.3 except that the inequalities corresponding to t with $A^t = 0$, r with $t \in \pi_r$ and $s \in \Delta$ (as defined in Theorem 1.3.8) are weak; consequently, in converting the system to a system with weak inequalities, we get (1.3.12) except that 1 is omitted in the inequality corresponding to such t, r and s. The dimensions of the linear system remain unchanged, and the resulting test is polynomial in p,

d and n. □

Mean-Partition Problems

Section 7.3 of Vol. I introduced the 1-dim mean-partition problem. We will continue the study here for multivariate partitions. The multivariate partition problem was first studied by Chang, Hwang and Rothblum [21] (Brieden and Gritzmann [11] studied a slightly different version). A *mean-partition* π can be represented by a $d \times p$ matrix

$$\overline{A}_\pi = (\overline{A}_{\pi_1}, \cdots, \overline{A}_{\pi_p}) \equiv \left(\frac{\sum_{i \in \pi_1} A^i}{|\pi_1|}, \cdots, \frac{\sum_{i \in \pi_p} A^i}{|\pi_p|} \right).$$

From now on, we simply write $\overline{A}_j$, instead of $\overline{A}_{\pi_j}$, as the mean of π_j. Chang, Hwang and Rothblum [21] noted

Lemma 1.3.11. *Consider a single shape $(n_1, \cdots, n_p)$. The mean-partition problem with the objective function $F(\pi) = g(\overline{A}_\pi)$ coincides with the corresponding sum-partition problem with the objective function $F(\pi) = f(A_\pi)$ where*

$$f(x^1, \cdots, x^p) = g(\frac{x^1}{n_1}, \cdots, \frac{x^p}{n_p})$$

for $x^i \in \mathbb{R}^d$.

Let $\overline{P}^\Pi$ denote the *mean-partition polytope* for the family Π. Lemma 1.3.11 implies an isomorphism between $\overline{P}^{(n_1,\cdots,n_p)}$ and $P^{(n_1,\cdots,n_p)}$. More specifically, let $D(n_1, \cdots, n_p)$ denote the diagonal matrix whose diagonal entries are $n_1, \cdots, n_p$. Then $\overline{P}^{(n_1,\cdots,n_p)} = \text{conv}\{D(n_1, \cdots, n_p)^T A_\pi : \pi \in P^{(n_1,\cdots,n_p)}\} = \text{conv}\{D(n_1, \cdots, n_p)^T x : x \in \overline{P}^{(n_1,\cdots,n_p)}\}$. Hence the linear transformation

$$X = (x^1, \cdots, x^p) \longrightarrow D(n_1, \cdots, n_p)^T X = (\frac{x^1}{n_1}, \cdots, \frac{x^p}{n_p}).$$

Since the linear transformation preserves vertices, any bound on the numbers of vertices in $P^{(n_1,\cdots,n_p)}$ is a bound on the number of vertices in $\overline{P}^{(n_1,\cdots,n_p)}$ and any algorithm for enumerating vertices in $P^{(n_1,\cdots,n_p)}$ is one for $\overline{P}^{(n_1,\cdots,n_p)}$.

Can this transformation be extended to other families of partitions? Brieden and Gritzmann [12] gave a counter-example for the bounded-shape family.

Example 1.3.5. Let $d = 1, p = 2, n = 4, A^1 = -15, A^2 = -3, A^3 = 3, A^4 = 15, L_j = 1, U_j = 3$ for $j = 1, 2$. Then $P^{(L,U)}$ has two vertices $(-18, 18)$ and $(18, -18)$, corresponding to the two partition $\pi^1 = (\{1,2\},\{3,4\})$ and $\pi^2 = (\{3,4\},\{1,2\})$, respectively. On the other hand, $\overline{P}^{(L,U)}$ has vertices $(-9, 9)$ and $(9, -9)$ corresponding to π^1 and π^2, but also $(-15, 5)$ and $(5, -15)$ corresponding to $\pi^3 = (\{1\},\{2,3,4\})$ and $\pi^4 = (\{2,3,4\},\{1\})$, and $(15, -5)$ and $(-5, 15)$ corresponding to $\pi^5 = (\{4\},\{1,2,3\})$ and $\pi^6 = (\{1,2,3\},\{4\})$. □

Brieden and Gritzmann [12] also gave an interesting geometric characterization of vertices of $\overline{P}^{(n_1,\cdots,n_p)}$. First, we introduce some geometric terms. A dissection $Q = (Q_1, \cdots, Q_p)$ of $\mathbb{R}^d$ is called a *power diagram* if for some p-vectors $(a_1, \cdots, a_p)$ and $(b_1, \cdots, b_p)$ where a_j's are in $\mathbb{R}^d$ and b_j's are real,

$$Q_j = \{x \in \mathbb{R}^d : \|x - a_j\|^2 - b_j \leq \|x - a_k\|^2 - b_k \text{ for all } k \neq j\}$$

for $j = 1, \cdots, p$. When a_j is the mean of π_j, then the power diagram is *centroidal*. Note that the well-known Voronoi diagram is a centroidal diagram with $b_j = 0$ for $j = 1, \cdots, p$. A partition $\pi = (\pi_1, \cdots, \pi_p)$ is said to admit a (centroidal) power diagram if there exists a (centroidal) power diagram $Q = (Q_1, \cdots, Q_p)$ such that $\text{conv}\pi_j \subset Q_j$ for $j = 1, \cdots, p$.

Brieden and Gritzmann [12] proved

Theorem 1.3.12. *Consider a single-shape $(n_1, \cdots, n_p)$. Suppose $A^i \in A$ are all distinct and $\overline{A}_j/n_j$ are all distinct over j. Then a partition π is a vertex of $\overline{P}^{(n_1,\cdots,n_p)}$ if and only if π admits a power diagram.*

Proof. By Corollary 1.3.4, a partition π is a vertex of $\overline{P}^{(n_1,\cdots,n_p)}$ if and only if there exist some matrix $C \in \mathbb{R}^{d\times p}$ and a vector $\alpha \in \mathbb{R}^p$ such that $(C^r - C^s)^T A^t > \alpha_s - \alpha_r$ for $r, s \in \{1, \cdots, p\}$ with $r \neq s$ and $t \in \pi_r$. Define

$$Q_j = \{x \in \mathbb{R}^d : (C^k - C^j)^T x \leq \alpha_j - \alpha_k \text{ for all } k \neq j\}$$

for $j = 1, \cdots, p$. Obviously, $\text{conv}\pi_j \subset Q_j$. Next we want to show that Q is indeed a power diagram. Let $a_j = C^j$ and $b_j = \|a_j\|^2 + 2\alpha_j$ for $j = 1, \cdots, p$.

Observe the following equivalences:

$$\begin{aligned}\|x-a_j\|^2-b_j &\le \|x-a_k\|^2-b_k\\ \|x\|^2-2a_j^Tx+\|a_j\|^2-b_j &\le \|x\|^2-2a_k^Tx+\|a_k\|^2-b_k\\ 2(a_k-a_j)^Tx &\le (\|a_k\|^2-b_k)-(\|a_j\|^2-b_j)\\ (C^k-C^j)^Tx &\le \alpha_j-\alpha_k.\end{aligned}$$

Thus, Q is a power diagram.

Conversely, given a power diagram Q one can easily obtain a corresponding matrix $C \in \mathbb{R}^{d\times p}$ and a vector $\alpha \in \mathbb{R}^p$ such that $(C^r - C^s)^T A^t > \alpha_s - \alpha_r$ for $r, s \in \{1, \cdots, p\}$ with $r \neq s$ and $t \in \pi_r$. □

Using Lemma 1.2.1 of Vol. I, we can easily extend the "only if" part of Theorem 1.3.12 to larger families.

Corollary 1.3.13. *Suppose $A^i \in A$ are all distinct. Let Γ denote a set of shapes. If a partition π corresponds to a vertex of $\overline{P}^\Gamma$, then π admits a power diagram.*

However, unlike Theorem 1.3.12, the inverse of Corollary 1.3.13 does not hold. The following is a counterexample, which modifies a counterexample of Brieden and Gritzmann [12] by eliminating identical A^i's, to illustrate this point.

Example 1.3.6. Let $d = 1, p = 2, n = 4, A^1 = -6, A^2 = -0.6, A^3 = 0.6, A^4 = 6, L_j = 1$ and $U_j = 3$ for $j = 1, 2$. Then $\overline{P}^{(L,U)}$ has four vertices $(-6, 2), (2, -6), (-2, 6), (6, -2)$ corresponding to the four partitions $\pi^1 = (\{1\}, \{2, 3, 4\}), \pi^2 = (\{2, 3, 4\}, \{1\}), \pi^3 = (\{1, 2, 3\}, \{4\})$ and $\pi^1 = (\{4\}, \{1, 2, 3\})$, respectively. Note that $\pi^5 = (\{1, 2\}, \{3, 4\})$ corresponds to an internal point $(-3.3, 3.3)$. But π^5 clearly admits a power diagram corresponding to the dissection $((-\infty, 0], [0, \infty))$. □

Let $\varphi : \mathbb{R}^{d\times n} \to [0, \infty)$ denote an ellipsoidal function defined on $z = (z_1, z_2, \cdots, z_p) \in \mathbb{R}^{d\times n}$ by

$$\varphi(z) = \varphi^{(n_1,\cdots,n_p)}(z) = \sum_{j=1}^{p} n_j \|z_j\|^2.$$

Brieden and Glitzmann showed that by adding the extra condition on π that it is a local maximizer of φ, the conclusion in Theorem 1.3.12 and Corollary 1.3.13 can be strengthened to admission of centroid power diagram. Further, this strengthening can be extended to bounded-shape polytopes if π is a vertex and also a local maximizer of

$$\varphi'(z) = \varphi^{(|\pi_1|,\cdots,|\pi_p|)}(z) = \sum_{j=1}^{p} |\pi_j| \|z_j\|^2.$$

Finally, by Lemma 1.3.11, these results concerning single-shape mean-partition polytopes also hold for single-shape sum-partition polytopes.

1.4 Representation, Characterization and Enumeration of Vertices of Partition Polytopes: General Case

Theorems 1.3.1 and 1.3.3 consider the case when the underlying matrix A has nonzero and distinct columns, and establish characterizations of partitions whose associated vectors are vertices of a bounded-shape partition polytope; these characterizations were stated in terms of unique representation and feasibility of systems of linear inequalities. In particular, we showed that vertices are associated with unique representing partitions. These results were extended to situations where A's columns are distinct but may contain the zero vector. In particular, it was shown in Theorem 1.3.8 that in such cases the only degree of freedom in multiple representation of vertices is in allowing for shifts of the zero vector. Similarly, in Theorem 1.3.10 we derive a characterization of partitions associated with vertices in terms of feasibility of a system of linear inequalities.

When A has repeated columns, any redistribution of indices whose corresponding vectors coincide will not affect the vector associated with the partition; thus, there is another source of nonuniqueness in the representation of vertices of bounded-shape partition polytopes, in addition to shifts of a zero vector. In the current section, following Hwang, Onn and Rothblum [52] we show that such redistributions and shifts are the only source of multiple representation of vertices. We also derive a characterization of partitions associated with vertices of corresponding bounded-shape partition polytopes through feasibility of a system of linear inequalities. But, the result does not yield an efficient test as it concerns circuits rather than individual inequalities.

A few additional definitions are needed before exploring general bounded-shape partition polytopes. Let $\bar{n}$ be the number of nonzero distinct columns of A. We will consider matrices with $\bar{n}+1$ rows or $\bar{n}+1$ columns, where these rows/columns are indexed by $0, 1, \ldots, \bar{n}$. Further, when a matrix has $\bar{n}+1$ rows indexed by $0, 1, \ldots, \bar{n}$, we use *underlin-*

ing to denote the submatrix obtained by truncating the 0-row, so, if $B \in \mathbb{R}^{(\bar{n}+1)\times n}$, then $\underline{B} \in \mathbb{R}^{\bar{n}\times n}$.

Given a $d \times n$ matrix A, let $\ddot{A}$ be the $d \times \bar{n}$ submatrix of A obtained by deleting zero and multiple columns that appear in A; for uniqueness assume that the first of any group of repeated columns of A is preserved while the others are deleted, and the order of $\ddot{A}$'s columns is induced from A. Also, let $\dot{A}$ be the $d \times (1 + \bar{n})$ matrix obtained from $\ddot{A}$ by adding the zero vector as the 0-column. Of course, $\ddot{A}$ has no zero vector but $\dot{A}$ does. Finally, let $J \in \mathbb{R}^{(1+\bar{n})\times n}$ be the $\{0,1\}$-matrix with $J^t = e^s$ (the s-unit vector in $\mathbb{R}^{\bar{n}+1}$) if $A^t = \dot{A}^s (\neq 0)$ and $J^t = e^0$ (the 0-unit vector in $\mathbb{R}^{(\bar{n}+1)}$) if $A^t = 00$, and $\bar{J}$ be obtained from J by deleting the first row (all zeros). For example, if

$$A = \begin{pmatrix} 1\,0\,3\,3\,1 \\ 2\,0\,4\,4\,2 \\ 1\,0\,0\,0\,1 \end{pmatrix}$$

then

$$\ddot{A} = \begin{pmatrix} 1\,3 \\ 2\,4 \\ 1\,0 \end{pmatrix}, \qquad \dot{A} = \begin{pmatrix} 0\,1\,3 \\ 0\,2\,4 \\ 0\,1\,0 \end{pmatrix},$$

$$J = \begin{pmatrix} 0\,0\,0\,0\,0 \\ 1\,0\,0\,0\,1 \\ 0\,0\,1\,1\,0 \end{pmatrix} \quad \text{and} \quad \underline{J} = \begin{pmatrix} 1\,0\,0\,0\,1 \\ 0\,0\,1\,1\,0 \end{pmatrix}.$$

The above definitions imply that for each matrix Y with $1 + \bar{n}$ rows, we have that

$$\dot{A}Y = \ddot{A}\underline{Y}, \tag{1.4.1}$$

in particular,

$$\dot{A}J = A = \ddot{A}\underline{J}. \tag{1.4.2}$$

When A has no zero vector, the forthcoming development can be carried out without the use of $\dot{A}$, but solely with the use of $\ddot{A}$. In particular, when A's columns are nonzero and distinct, $\ddot{A} = A$ and $\underline{J} = I \in \mathbb{R}^{n\times n}$.

We will use $J \in \mathbb{R}^{(1+\bar{n})\times n}$ and $\underline{J} \in \mathbb{R}^{\bar{n}\times n}$, in addition to $A \in \mathbb{R}^{d\times n}$, as data-matrices. In particular, for a p-partition π, $JI_\pi = J_\pi$, $\underline{J}I_\pi = \underline{J}_\pi$, $AI_\pi = A_\pi$ and, by (1.4.2),

$$A_\pi = AI_\pi = (\dot{A}J)I_\pi = \dot{A}(JI_\pi) = \dot{A}J_\pi \tag{1.4.3}$$

and

$$A_\pi = AI_\pi = (\ddot{A}\underline{J})I_\pi = \ddot{A}(\underline{J}I_\pi) = \ddot{A}\underline{J}_\pi. \tag{1.4.4}$$

Note that for a partition π, the nonzero elements of J_π record the number of copies of each vector that are assigned to each part of π; in particular, two partitions π and σ have $J_\pi = J_\sigma$ if and only if they are equivalent.

The next lemma shows that for a set of partitions Π, P_I^Π, P_J^Π, $P_{\underline{J}}^\Pi$ and P_A^Π form a sequence of polytopes where each is a projection of its predecessors; further, the composite projection of P_I^Π onto P_A^Π is given by $X \to AX$. The decompositions we are about to establish are demonstrated as follows.

$$\begin{array}{ccc} P_I^\Pi & & P_A^\Pi \\ \downarrow & & \uparrow \\ P_J^\Pi & \longrightarrow & P_{\underline{J}}^\Pi \end{array}$$

We shall refer to P_I^Π as the *generalized transportation-polytope corresponding to* Π.

Lemma 1.4.1. *Let $A \in \mathbb{R}^{d\times n}$ and let Π be a set of partitions. Then:*

(a) $P_J^\Pi = \{JX : X \in P_I^\Pi\} \subseteq \mathbb{R}^{n\times p}$,
(b) $P_{\underline{J}}^\Pi = \{\underline{Y} : Y \in P_J^\Pi\} \subseteq \mathbb{R}^{(1+\bar{n})\times p}$,
(c) $P_A^\Pi = \{\ddot{A}Z : Z \in P_{\underline{J}}^\Pi\} \subseteq \mathbb{R}^{\bar{n}\times p}$.

Further, the composite projections of P_I^Π onto $P\underline{J}^\Pi$, of P_I^Π onto P_A^Π and of P_J^Π onto P_A^Π are given, respectively, by $X \to \underline{J}X$, $X \to AX$ and $Y \to \dot{A}Y$.

Proof. A standard argument about convex hulls shows that

$$\begin{aligned} \{JX : X \in P_I^\Pi\} &= \{JX : X \in \text{conv}\{I_\pi : \pi \in \Pi\}\} \\ &= \text{conv}\{JI_\pi : \pi \in \Pi\} = \text{conv}\{J_\pi : \pi \in \Pi\} = P_J^\Pi, \end{aligned} \tag{1.4.5}$$

proving (a). The same argument combined with (1.4.4) shows that

$$\begin{aligned} \{\ddot{A}Z : Z \in P_{\underline{J}}^\Pi\} &= \{\ddot{A}Z : Z \in \text{conv}\{\underline{J}_\pi : \pi \in \Pi\}\} \\ &= \text{conv}\{\ddot{A}\underline{J}_\pi : \pi \in \Pi\} = \text{conv}\{A_\pi : \pi \in \Pi\} = P_A^\Pi, \end{aligned} \tag{1.4.6}$$

proving (c). Next, to establish (b), observe that the projection mapping $Y \in \mathbb{R}^{(1+\bar{n})\times p}$ into $\underline{Y} \in \mathbb{R}^{\bar{n}\times p}$ by eliminating the 0-row is a linear operator;

this operator is representable by a matrix, say $E \in \mathbb{R}^{\bar{n}\times(1+\bar{n})}$ with $EY = \underline{Y}$ for each $Y \in \mathbb{R}^{(+\bar{n})\times p}$. As in (1.4.6), we then get that

$$\begin{aligned}\{\underline{Y} : Y \in P_J^\Pi\} &= \{EY : Y \in P_J^\Pi\} \\ &= \{EY : Y \in \text{conv}\{J_\pi : \pi \in \Pi\}\} \qquad (1.4.7) \\ &= \text{conv}\{EJ_\pi : \pi \in \Pi\} = \text{conv}\{\underline{J}_\pi : \pi \in \Pi\} = P_{\underline{J}}^\Pi .\end{aligned}$$

Finally, the composite projections of P_I^Π onto $P_{\underline{J}}^\Pi$, of P_I^Π onto P_A^Π and of P_J^Π onto P_A^Π are given, respectively, by $X \to \underline{JX} = \underline{J}X$, $X \to \ddot{A}(\underline{JX}) = \ddot{A}(\underline{J}X) = (\ddot{A}\underline{J})X = AX$ (here we use (1.4.2)), and $Y \to \ddot{A}\underline{Y} = \dot{A}Y$ (here we use (1.4.1)). □

The composite transformation of the three transformations studied in Lemma 1.4.1 maps $X \in P_I^\Pi \subseteq \mathbb{R}^{n\times p}$ into $AX \in P_A^\Pi \subseteq \mathbb{R}^{d\times p}$. Lemma 1.4.1 implies that this map is onto (a fact that is demonstrated in Theorem 1.1.5 without relying on the assumption that the columns of A are nonzero and distinct).

We next obtain an explicit representation, through a system of linear inequalities, for bounded-shape partition polytopes when the data matrix is J. The result resembles Theorem 1.1.3 (this concerns the case where the data matrix is I).

Theorem 1.4.2. *Let L and U be positive integer p-vectors satisfying $L \le U$ and $\sum_{j=1}^p L_j \le n \le \sum_{j=1}^p U_j$. Then $P_J^{(L,U)}$ is the solution set of the linear system*

$$\begin{aligned}&(a)\ Y_s^j \ge 0 \ \textit{ for } \ s = 0,1,\ldots,\bar{n} \ \textit{ and } \ j = 1,\ldots,p, \\ &(b)\ \sum_{j=1}^p Y_s^j = (Je)_s \ \textit{ for } \ s = 0,1,\ldots,\bar{n}, \qquad (1.4.8) \\ &(c)\ L_j \le \sum_{s=0}^{\bar{n}} Y_s^j \le U_j \ \textit{ for } \ j = 1,\ldots,p,\end{aligned}$$

with e as the vector $(1,\ldots,1)^T \in \mathbb{R}^n$.

Proof. Trivially, each of the matrices J_π for $\pi \in \Pi^{(L,U)}$ satisfies (1.4.8), hence, the convex hull of these matrices, namely $P_J^{(L,U)}$, is contained in the solution set of (1.4.8) which we denote by K.

Adding variables W^j for $j = 1,\ldots,p$, replacing (1.4.8c) by the constraints $W^j = \sum_{s=0}^{\bar{n}} Y_s^j$ and $L_j \le W^j \le U_j$ for $j = 1,\ldots,p$ and adding the constraint $\sum_{j=1}^p W^j = \sum_{s=0}^{\bar{n}} (Je)_s = n$, the linear system (1.4.8) is expanded to a network flow problem with integer lower and upper bounds on arc flows (see Figure 1.4.1 below). By an argument similar to the one used in the proof of Theorem 1.1.3, we conclude that all vertices in K are integer matrices.

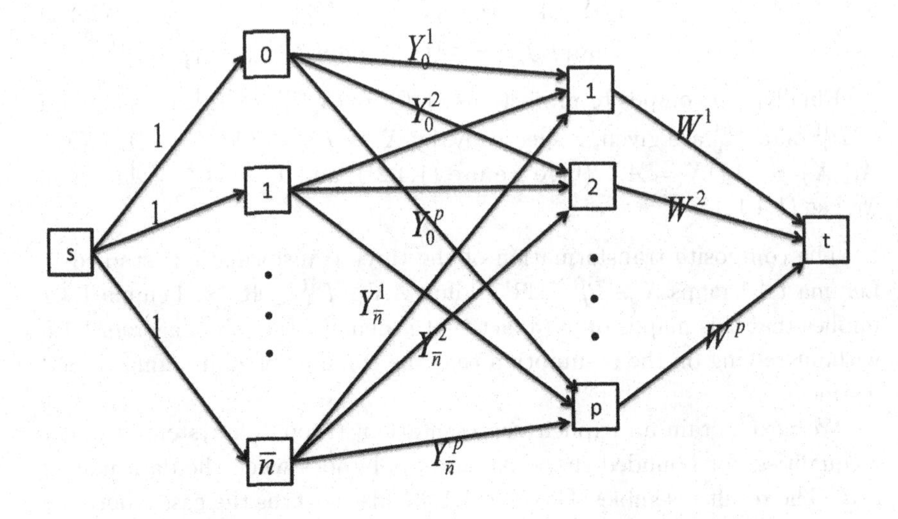

Figure 1.4.1 Network flow problem.

Let Y^* be a vertex of K; it then follows from the above paragraph that Y^* is an integer matrix. For each $s = 0, 1, \ldots, \bar{n}$, let $H_s \equiv \{t \in \{1, \ldots, n\} : A^t = \dot{A}^s\}$; these sets partition $\mathcal{N}$ and for each s, $\sum_{j=1}^{p} (Y^*)_s^j = (Je)_s = |H_s|$. The latter together with the integrality of Y^* implies that for $s = 0, 1, \ldots, \bar{n}$, there exists a (not necessarily unique) partition of H_s into set $\sigma_{s1}, \ldots, \sigma_{sp}$ such that $|\sigma_{sj}| = (Y^*)_s^j$. For $j = 1, \ldots, p$, let $\pi_j \equiv \cup_{s=0}^{\bar{n}} \sigma_{sj}$. It follows that $\pi_1, \ldots, \pi_p$ partition $\mathcal{N}$; also, for $j = 1, \ldots, p$,

$$|\pi_j| = \sum_{s=0}^{\bar{n}} |\sigma_{sj}| = \sum_{s=0}^{\bar{n}} (Y^*)_s^j,$$

so (1.4.8c) implies that $L_j \leq |\pi_j| \leq U_j$. Thus, $\pi \in \Pi^{(L,U)}$. Next, for $s = 0, 1, \ldots, \bar{n}$ and $j = 1, \ldots, p$,

$$\sum_{t \in H_s} (I_\pi)_t^j = |\sigma_{sj}| = (Y^*)_s^j.$$

Hence,

$$(J_\pi)_s^j = (JI_\pi)_s^j = \sum_{t \in H_s} (I_\pi)_t^j = (Y^*)_s^j,$$

implying that $Y^* = J_\pi \in P_J^{(L,U)}$. So, each vertex of K is in $P_J^{(L,U)}$. As K is the convex hull of its vertices, (Proposition 3.1.1 of Vol. I, part (a)), K is contained in the convex hull of the J_π's, that is, in $P_J^{(L,U)}$. □

The explicit representation of $P_J^{(L,U)}$ in Theorem 1.4.2 does not generally extend to $P_{\underline{J}}^{(L,U)}$; specifically, when zero vectors exist, they have to be accounted in consideration of the lower and upper bounds on the cardinality of the parts. But, part (b) of Lemma 1.4.1 shows that $P_{\underline{J}}^{(L,U)}$ is a projection of $P_J^{(L,U)}$. When A has no zero vector, $(Je)_0 = 0$ and all solutions of (1.4.8) have $Y_0^j = 0$ for $j = 1, \ldots, p$; by eliminating these variables, we get from (1.4.8) a characterization of $P_{\underline{J}}^{(L,U)}$.

The next theorem provides three necessary and sufficient conditions for partitions being vertices in bounded-shape partition polytopes, without the assumption that A's columns are nonzero and distinct. The three conditions look cumbersome and repetitive, but, they have distinct uses. Condition (b) extends the uniqueness of the representation of vertices when A's columns are nonzero and distinct (condition (b) of Theorem 1.3.1) while condition (d) explains the potential degrees of freedom in multiple representations of extreme points of the partition polytopes (see the discussions following the forthcoming Corollary 1.4.4). Conditions (c) and (d) concern the polytopes $P_{\underline{J}}^{(L,U)}$ and $P_I^{(L,U)}$ which have explicit representations of linear inequalities (Theorems 1.4.2 and 1.1.3) and are therefore useful for computable tests; in fact, the representation of $P_J^{(L,U)}$ is used to establish the (most difficult) implication $(a) \Rightarrow (b)$.

Theorem 1.4.3. *Let $A \in \mathbb{R}^{d \times n}$, let L and U be positive integer p-vectors satisfying $L \leq U$ and $\sum_{j=1}^p L_j \leq n \leq \sum_{j=1}^p U_j$, and let $\pi \in \Pi^{(L,U)}$. Then the following are equivalent:*

(a) A_π *is a vertex of* $P_A^{(L,U)}$,

(b) $\{Y \in P_J^{(L,U)} : \dot{A}Y = A_\pi\} = \{Y \in P_J^{(L,U)} : \underline{Y} = \underline{J}_\pi\}$ *and* $\underline{J}_\pi$ *is a vertex of* $P_{\underline{J}}^{(L,U)}$,

(c) $\{Z \in P_{\underline{J}}^{(L,U)} : \ddot{A}Z = A_\pi\} = \{\underline{J}_\pi\}$ *and* $\underline{J}_\pi$ *is a vertex of* $P_{\underline{J}}^{(L,U)}$, *and*

(d) $\{X \in P_I^{(L,U)} : AX = A_\pi\} = \{X \in P_I^{(L,U)} : \underline{J}X = \underline{J}_\pi\}$ *and* $\underline{J}_\pi$ *is a vertex of* $P_{\underline{J}}^{(L,U)}$.

Proof. $(a) \Rightarrow (b)$: Suppose A_π is a vertex of $P_A^{(L,U)}$. To see that $\{Y \in P_J^{(L,U)} : \underline{Y} = \underline{J}_\pi\} \subseteq \{Y \in P_J^{(L,U)} : \dot{A}Y = A_\pi\}$, let $Y \in P_J^{(L,U)}$ satisfy $\underline{Y} = \underline{J}_\pi$. It then follows from (1.4.1) and (1.4.4) that $\dot{A}Y = \ddot{A}\underline{Y} = \ddot{A}\underline{J}_\pi = A_\pi$.

To see the reverse inclusion, let $Y \in P_J^{(L,U)}$ satisfy $\dot{A}Y = A_\pi$. Consider the network flow (with integer lower and upper bounds on arc-flows) expansion of the linear system (1.4.8) as described in the proof of Theorem 1.4.2; see Figure 1.4.1. Theorem 1.4.2 and its proof show that $P_J^{(L,U)}$ is in one-to-one correspondence with the solution-set of this network flow problem with vertices mapped onto vertices. Thus, we may and will identify vectors in $P_J^{(L,U)}$, that is, solutions of (1.4.8), with their expansion to solutions of the network flow problem, in particular, this is the case for Y and J_π.

A *circuit* z is a nonzero, normalized, minimum support solution of the (homogeneous) system

$$(a)\ \sum_{j=1}^{p} w^j = 0\,,$$

$$(b)\ w^j - \sum_{s=0}^{\bar{n}} z_s^j = 0 \ \text{ for } \ j = 1, \ldots, p\,, \tag{1.4.9}$$

$$(c)\ \sum_{j=1}^{p} z_s^j = 0 \ \text{ for } \ s = 0, 1, \ldots, \bar{n}\,,$$

where *normalized* means that $\|z\|_\infty = 1$ (with $\| \ \|_\infty$ denoting the 1_∞ norm) and *minimal support* means that the set of nonzero variables of no solution of (1.4.9) is strictly contained in that of z. For a circuit (z, w), z determines w (by (1.4.9b)); hence, we can and will identify such a circuit with z. Now, standard results show that the coordinates of a circuit z are all $-1, 0$ and 1 and that each node of the network presented in Figure 1.4.1 appears in either 0 or 2 indices corresponding to nonzero coordinates of z. Also, each column of a circuit z has at most two nonzero elements which can take only the values -1 and 1; as $\dot{A}^0 = 0$ and $\dot{A}^1, \ldots, \dot{A}^{\bar{n}}$ are nonzero and distinct, it follows that

$$\text{circuit } \ z \ \text{ satisfies } \ \dot{A}z = 0 \ \text{ if and only if } \ \underline{z} = 0\,. \tag{1.4.10}$$

As A_π is a vertex of $P_A^{(L,U)}$, A_π is the unique maximizer over $P_A^{(L,U)}$ of some linear function; let the coefficients of such a linear function be represented by the matrix $C \in \mathbb{R}^{d \times p}$. Also, as Y and J_π are solutions of the network flow problem, a standard result about network flows (e.g., Denardo [24, p. 99]) implies that $Y - J_\pi$ can be decomposed into a sum $\sum_{t=1}^{q} \beta_t z^t$ where for each $t = 1, \ldots, q$, z^t is a circuit of the network flow problem, β_t is a positive number and $J_\pi + z^t$ is a feasible solution of the network flow problem, that is, $J_\pi + z^t \in P_J^{(L,U)}$, and by Lemma 1.4.1, $\dot{A}J_\pi + \dot{A}z^t = \dot{A}(J_\pi + z^t) \in P_A^{(L,U)}$.

Now, (1.4.3) and the unique optimality of A_π over $P_J^{(L,U)}$ under the linear function represented by C implies that for $t = 1, \ldots, q$,

$$\langle C, A_\pi \rangle \geq \langle C, \dot{A}J_\pi + \dot{A}z^t \rangle = \langle C, A_\pi \rangle + \langle C, \dot{A}z^t \rangle$$

with equality holding if and only if $\dot{A}J_\pi + \dot{A}z^t = A_\pi$, that is, $\langle C, \dot{A}z^t \rangle \leq 0$ with equality holding if and only if $\dot{A}z^t = 0$, and by (1.4.10) this is the case if and only if $\underline{z}^t = 0$. As we are assuming that $\dot{A}Y = A_\pi$ and $Y = J_\pi + \sum_{t=1}^q \beta_t z^t$, we have that $\langle C, A_\pi \rangle = \langle C, \dot{A}Y \rangle = \langle C, \dot{A}J_\pi + \sum_{t=1}^q \beta_t \dot{A}z^t \rangle = \langle C, A_\pi \rangle + \sum_{t=1}^q \beta_t \langle C, \dot{A}z^t \rangle \leq \langle C, A_\pi \rangle$; it follows that the inequality in the above string holds as equality. Thus, for $t = 1, \ldots, q$, $\langle C, \dot{A}z^t \rangle = 0$ and consequently $\underline{z}^t = 0$. Thus, $\underline{Y} = \underline{J}_\pi + \sum_{t=1}^q \beta_t \underline{z}^t = \underline{J}_\pi$.

It remains to show that $\underline{J}_\pi$ is a vertex of $P_{\underline{J}}^{(L,U)}$. Indeed, assume that $\underline{J}_\pi$ has a representation $\underline{J}_\pi = \beta Z^1 + (1 - \beta)Z^2$ where $0 < \beta < 1$ and $Z^1, Z^2 \in P_{\underline{J}}^{(L,U)}$, and we will show that $Z^1 = Z^2 = \underline{J}_\pi$. By (1.4.4), $A_\pi = \ddot{A}\underline{J}_\pi$, and by Lemma 1.4.1, $\ddot{A}Z^1$ and $\ddot{A}Z^2$ are in $P_A^{(L,U)}$. As A_π is assumed to be a vertex of $P_A^{(L,U)}$ and $A_\pi = \ddot{A}\underline{J}_\pi = \beta\ddot{A}Z^1 + (1-\beta)\ddot{A}Z^2$, we conclude that $A_\pi = \ddot{A}Z^1 = \ddot{A}Z^2$. By part (b) of Lemma 1.4.1 there exist matrices Y^1 and Y^2 in $P_J^{(L,U)}$ with $\underline{Y}^1 = Z^1$ and $\underline{Y}^2 = Z^2$, and (1.4.1) implies that $\dot{A}Y^1 = \ddot{A}\underline{Y}^1 = \ddot{A}Z^1 = A_\pi$ and $\dot{A}Y^2 = \ddot{A}\underline{Y}^2 = \ddot{A}Z^2 = A_\pi$. Hence, the established conclusion $\{Y \in P_J^{(L,U)} : \dot{A}Y = A_\pi\} = \{Y \in P_J^{(L,U)} : \underline{Y} = \underline{J}_\pi\}$ implies that $Z^1 = \underline{Y}^1 = \underline{J}_\pi$ and $Z^2 = \underline{Y}^2 = \underline{J}_\pi$.

(*b*) $\Rightarrow$ (*c*): It suffices to show that if $\{Y \in P_J^{(L,U)} : \dot{A}Y = A_\pi\} = \{Y \in P_J^{(L,U)} : \underline{Y} = \underline{J}_\pi\}$, then $\{Z \in P_{\underline{J}}^{(L,U)} : \ddot{A}Z = A_\pi\} = \{\underline{J}_\pi\}$. So, assume that the first equality holds. By (1.4.4), $\ddot{A}\underline{J}_\pi = A_\pi$; as $J_\pi \in P_J^{(L,U)}$, it follows that $\underline{J}_\pi \in \{Z \in P_{\underline{J}}^{(L,U)} : \ddot{A}Z = A_\pi\}$. To establish the reverse inclusion, let $Z \in P_{\underline{J}}^{(L,U)}$ satisfy $\ddot{A}Z = A_\pi$. It then follows from part (b) of Lemma 1.4.1 that $Z = \underline{Y}$ for some $Y \in P_J^{(L,U)}$; for such Y we have from (1.4.1) that $\dot{A}Y = \ddot{A}\underline{Y} = \ddot{A}Z = A_\pi$ and, by assumption, this implies that $\underline{Y} = \underline{J}_\pi$, that is, $Z = \underline{Y} = \underline{J}_\pi$.

(*c*) $\Rightarrow$ (*d*): From (1.4.1) and (1.4.4), if $X \in P_I^{(L,U)}$ and $\underline{J}X = \underline{J}_\pi$, then $AX = (\ddot{A}\underline{J})X = \ddot{A}(\underline{J}X) = \ddot{A}\underline{J}_\pi = A_\pi$. So, $\{X \in P_I^{(L,U)} : \underline{J}X = \underline{J}_\pi\} \subseteq \{X \in P_I^{(L,U)} : AX = A_\pi\}$. Thus, it suffices to show that if $\{X \in P_I^{(L,U)} : \underline{J}X = \underline{J}_\pi\} \subset \{X \in P_I^{(L,U)} : AX = A_\pi\}$ then there exists a matrix $Z \in P_{\underline{J}}^{(L,U)}$ with $\ddot{A}Z = A_\pi$ and $Z \neq \underline{J}_\pi$. So, suppose that $X \in P_I^{(L,U)}$ satisfies $AX = A_\pi$ and $\underline{J}X \neq \underline{J}_\pi$, and let $Z \equiv \underline{J}X$. Then $Z = \underline{J}X \neq \underline{J}_\pi$ and by Lemma 1.4.1 and (1.4.2), respectively, $Z = \underline{J}X \in P_{\underline{J}}^{(L,U)}$ and $\ddot{A}Z = \ddot{A}(\underline{J}X) = (\ddot{A}\underline{J})X = AX = A_\pi$.

$(d) \Rightarrow (a)$: Suppose condition (d) holds. We will assume that A_π is not a vertex of $P_A^{(L,U)}$ and establish a contradiction. As each vertex of $P_A^{(L,U)}$ is in the set $\{A_\pi : \pi \in \Pi\}$ and A_π (like every other vector in $X \in P_A^{(L,U)}$) has a representation as a convex combination of vertices of $P_A^{(L,U)}$, there exist partitions $\pi^1, \ldots, \pi^q$ in $\Pi^{(L,U)}$ and positive coefficients $\alpha_1, \ldots, \alpha_q$ which sum to 1 such that $A_\pi = \sum_{t=1}^q \alpha_t A_{\pi^t}$ and each A_{π^t} is a vertex of $P_A^{(L,U)}$. As $A_{\pi^t} = AI_{\pi^t}$ for $t = 1, \ldots, q$, implying that $A_\pi = \sum_{t=1}^q \alpha_t A_{\pi^t} = \sum_{t=1}^q \alpha_t AI_{\pi^t} = A\left(\sum_{t=1}^q \alpha_t I_{\pi^t}\right)$. As $P_I^{(L,U)}$ is convex, $X \equiv \sum_{t=1}^q \alpha_t I_{\pi^t}$ is in $P_I^{(L,U)}$. So, $AX = A_\pi$; hence, condition (d) implies that $\underline{J}X = \underline{J}_\pi$, that is, $\underline{J}_\pi = \underline{J}X = \underline{J}\left(\sum_{t=1}^q \alpha_t I_{\pi^t}\right) = \sum_{t=1}^q \alpha_t \underline{J} I_{\pi^t} = \sum_{t=1}^q \alpha_t \underline{J}_{\pi^t}$. As $\underline{J}_\pi$ is assumed to be a vertex of $P_{\underline{J}}^{(L,U)}$, as all the α_t's are positive and all the $\underline{J}_{\pi^t}$'s are in $P_{\underline{J}}^{(L,U)}$, we conclude that for $t = 1, \ldots, q$, $\underline{J}_{\pi^t} = \underline{J}_\pi$ and therefore $A_{\pi^t} = \ddot{A}\underline{J}_{\pi^t} = \ddot{A}\underline{J}_\pi = A_\pi$. As each A_{π^t} is assumed to be a vertex of $P_{\underline{J}}^{(L,U)}$, whereas A_π is not, we reach a contradiction which proves the asserted implication. □

The next corollary provides a necessary condition for being a vertex by restricting condition (d) of Theorem 1.4.3 to vectors X associated with partitions.

Corollary 1.4.4. *Let $A \in \mathbb{R}^{d \times n}$, let $L_1, \ldots, L_p$, $U_1, \ldots, U_p$ be positive integers satisfying $L \le U$ and $\sum_{j=1}^p L_j \le n \le \sum_{j=1}^p U_j$, and let V be a vertex of $P_A^{(L,U)}$. Then $\underline{J}_\pi$ coincides for partitions $\pi \in \Pi^{(L,U)}$ with $A_\pi = V$, that is, all such partitions share $|\{i \in \pi_r : A^i = x\}|$ for each $x \in \mathbb{R}^d \setminus \{0\}$ and $r = 1, \ldots, p$. Further, if $L = U$, then $\underline{J}_\pi$ can be replaced by J_π and $x = 0$ need not be excluded.*

Proof. The implication $(a) \Rightarrow (b)$ in Theorem 1.4.3 implies that if a vertex $V = A_\pi = A_{\pi'}$ for $\pi, \pi' \in \Pi$, then $\underline{J}_\pi = \underline{J}_{\pi'}$; the interpretation of this condition follows directly from the comment following (8.4.4). Finally, when $L = U$, $\Gamma^{L,U}$ consists of a single shape and for each r, the specification of $|\{i \in \pi_r : A^i = x\}|$ for $x \ne 0$, determines $|\{i \in \pi_r : A^i = 0\}|$ (as the total number of elements in π_r is prescribed). □

Corollary 1.4.4 shows that in selecting a partition corresponding to a particular vertex of $P_A^{(L,U)}$, it is unique in the number of indices in a part that are associated with a common nonzero vector of A, while the zero vectors can be shifted.

Condition (b) of Theorem 1.4.3 is next used to describe a test for vectors associated with partitions to be vertices of given bounded-shape partition polytopes. The efficiency of the test slightly improves the one derived from Corollary 1.1.6 in Section 1.1.

Testing if a Vector A_π is a Vertex of the Bounded-Shape Partition Polytope

Let A, L, U and π be as in Theorem 1.4.3. Our test for determining whether A_π is a vertex of $P_A^{(L,U)}$ has two parts.

The first part of the test determines whether or not $\underline{J}_\pi$ is a vertex of $P_{\underline{J}}^{(L,U)}$. Observe that $\underline{J}_\pi$ is a vertex of $P_{\underline{J}}^{(L,U)}$ if and only if for every representation of $\underline{J}_\pi$ as $\frac{1}{2}(Y' + Y'')$ with Y' and Y'' in $P_{\underline{J}}^{(L,U)}$ we have that $Y' = Y''$. This condition holds if and only if for each $s = 0, 1, \ldots, \bar{n}$ and $j = 1, \ldots, p$ the maximum of $(Y' - Y'')_s^j$ over Y' and Y'' satisfying $Y' \in P_J^{(L,U)}$, $Y'' \in P_J^{(L,U)}$ and $Y = \frac{1}{2}(Y' + Y'')$ is zero, which is verifiable by solving $\bar{n}p$ linear programs where each has $\bar{n}p$ variables and $2(\bar{n}+p)+\bar{n}p$ constraints. When A has no zero vector, $\underline{J} = J$ and a test for being a vertex of $P_{\underline{J}}^{(L,U)} = P_J^{(L,U)}$ can be developed from the explicit representation of $P_J^{(L,U)}$ available from Theorem 1.4.2.

The second part of the test determines whether or not $\underline{Y} = \underline{J}_\pi$ for each matrix Y satisfying $Y \in P_J^{(L,U)}$ and $\dot{A}Y = A_\pi$. In view of Theorem 1.4.2 (with the expansion discussed in the proof of the lemma), the assertion that $Y \in P_J^{(L,U)}$ is characterized by a linear system having $(\bar{n}+1)(p+1)$ variables, $(\bar{n}+1)+p$ constraints, nonnegativity constraints and lower and upper bounds on p variables; also, the requirement $\dot{A}Y = A_\pi$ reduces to another dp constraints. Testing whether or not each solution Y to the joint system satisfies $\underline{Y} = \underline{J}_\pi$ can be achieved by maximizing and minimizing Y_s^j for $j = 1, \ldots, p$ and $s = 1, \ldots, \bar{n}$ over the joint system, that is, by solving $2\bar{n}p$ corresponding LP's. □

The above method for testing whether or not a vector A_π is a vertex of a bounded-shape partition polytope depends on condition (c) of Theorem 1.4.3 which reduces to condition (b) of Theorem 1.4.1 when A's columns are nonzero and distinct. Observe that conditions (b) and (c) of Theorem 1.4.1 yield alternative computational methods under the restricted assumption of that theorem. Indeed, condition (c) concerns solvability of a (sparse) linear system with np $\{0,1\}$-variables and $(n+1)(d+1)$ equality and weak inequality constraints, and condition (d) concerns solvability of a linear system having $(d+1)p$ variables and $(p-1)n$ strict inequality constraints. Each of these tests is obviously polynomial in d, n and p.

The results about characterizing partitions that are associated with vertices of bounded-shape partition polytopes are less satisfactory when A is allowed to have multiple columns than when it does not (see Theorems 1.3.3 and 1.3.10). Here the use of the Strong Complementary Theorem of linear programming yields a characterization that allows one to have equalities $(C^r - C^s)A^v = \alpha_s - \alpha_r$ on circuits, rather than to restrict them to specific inequalities. The resulting condition lacks a simple geometric/algebraic motivation. Further, the polynomial test for vertices described in Section 1.2 is simpler.

Can the vertex-enumeration method introduced in Section 1.3 extend to the case where the partitioned vectors need not be distinct? It is first noted that results of Hwang, Onn and Rothblum [52] demonstrate that the uniqueness of the representations of vertices of bounded-shape partition polytopes extends to such situations, except for switching identical vectors and moving zero vectors among parts (as long as permitted by the shape-bounds). So, uniqueness is preserved when partitions are viewed as vector-partitions rather than index-partitions. But, the next example demonstrates that Theorem 1.3.6 does not extend to situation of duplicate vectors.

Example 1.4.1. Let $d = 1, n = 4, A = [1, 1, 2, 2]$ and $p = 2$. Then

$$P_A^2 = \text{conv}\{(0,6), (1,5), (2,4), (3,3), (4,2), (5,1), (6,0)\},$$

$$P_{\widetilde{A}}^2 = \text{conv}\left\{ \begin{array}{l} \binom{0,4}{0,6}, \binom{1,3}{1,5}, \binom{1,3}{2,4}, \binom{2,2}{2,4}, \binom{2,2}{3,3}, \\ \binom{2,2}{4,2}, \binom{3,1}{5,1}, \binom{3,1}{4,2}, \binom{4,0}{6,0} \end{array} \right\},$$

$$V[P_{\widetilde{A}}^2] = \left\{ \begin{pmatrix} 0,4 \\ 0,6 \end{pmatrix}, \begin{pmatrix} 2,2 \\ 2,4 \end{pmatrix}, \begin{pmatrix} 2,2 \\ 4,2 \end{pmatrix}, \begin{pmatrix} 4,0 \\ 6,0 \end{pmatrix} \right\},$$

and

$$V[P_{\widetilde{A}}^{(1,3)}] = \left\{ \begin{pmatrix} 1,3 \\ 1,5 \end{pmatrix}, \begin{pmatrix} 1,3 \\ 2,4 \end{pmatrix} \right\}.$$

Then, $A_{(\{1\},\{2,3,4\})} = (1,5)$ is a vertex of $P_A^{(1,3)}$ but $P_{\widetilde{A}}^2$ contains no vertex of the form $\widetilde{A}_\pi$ with $\pi \in \Pi^{(1,3)}$, demonstrating that the conclusions of Theorem 1.3.6 need not hold. The sums of the columns of all matrices of partition polytopes corresponding to a given matrix A are the same (equaling $\sum_{i=1}^n A^i$); hence, when $p = 2$ it suffices to consider the first column of the matrices. Figure 1.4.2 illustrates the first column of the $\widetilde{A}_\pi$'s for $\pi \in \Pi^2$ and the first column of the corresponding single-size partition polytope. □

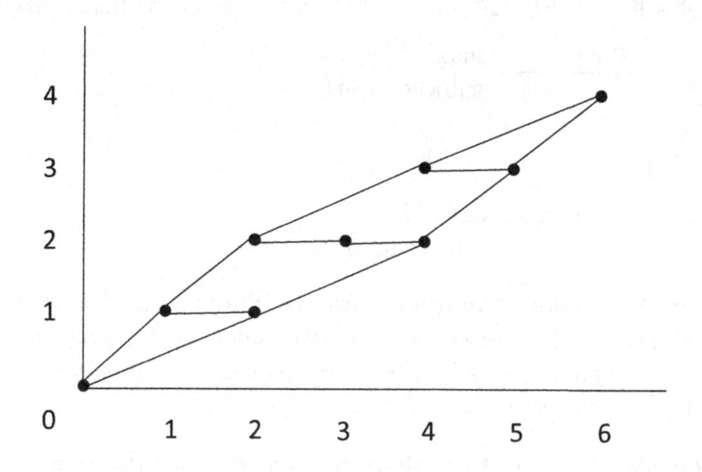

Figure 1.4.2 The partition polytope and the first column of $\widetilde{A}_\pi$ in Example 1.4.1.

The next result extends the bounds on the number of vertices of partition polytopes obtained in Corollary 1.3.7.

Theorem 1.4.5. *Let $A \in \mathbb{R}^{d\times n}$ and let Γ be a set of integer p-vectors $(n_1, \cdots, n_p)$ satisfying $\sum_{j=1}^{p} n_j = n$. Then $|V[P_A^\Gamma]| \le O(n^{(d+1)(p-1)-1})$.*

Proof. Consider a perturbation of the columns of A where column A^i is replaced by $A^i + \epsilon(i, i^2, \cdots, i^d)^T$ and denote the resulting matrix by $A(\epsilon)$. For all $\epsilon > 0$ sufficiently small, the columns of $A(\epsilon)$ are nonzero and distinct. Further, for $\epsilon > 0$ sufficiently small, a linear functional that attains a unique maximum at A_π over P_A^Γ attains a unique maximum at $A(\epsilon)_\pi$ over $\{A(\epsilon)_\sigma : \sigma \in \Pi^\Gamma\}$ and therefore also over $P_{A(\epsilon)}^\Gamma = \text{conv}\{A(\epsilon)_\sigma : \sigma \in \Pi^\Gamma\}$. So, if A_π is a vertex of P_A^Γ, then $A(\epsilon)_\pi$ is a vertex of $P_{A(\epsilon)}^\Gamma$ and $|V[P_A^\Gamma]| \le |V[P_{A(\epsilon)}^\Gamma]| \le O(n^{(d+1)(p-1)-1})$. □

Appendix A

In this appendix we summarize two results about linear programs which are known as the Weak and the Strong Complementarity Theorems. Their proof can be found in Schrijver [85, pp. 95–96].

Let $A \in \mathbb{R}^{m \times n}$, $b \in \mathbb{R}^m$ and $c \in \mathbb{R}^n$ and consider the linear program

$$\underline{\text{Program P}}: \max c^T x \quad \text{subject to: } Ax = b, \quad x \geq 0$$

and its dual

$$\underline{\text{Program D}}: \min y^T b \quad \text{subject to: } y^T A \geq c^T.$$

Theorem A.1 (Weak Complementarity Theorem). *Assume that for a set $J \subseteq \{1, \dots, n\}$ there exists an optimal solution x of Program P that satisfies $x_j > 0$ for all $j \in J$. Then for every optimal solution y of Program D, $(y^T A)_j = c_j$ for all $j \in J$.* □

Theorem A.2 (Strong Complementarity Theorem). *Assume that Program P has an optimal solution and that for a set $J \subseteq \{1, \dots, n\}$ every optimal solution x of Program P satisfies $x_j = 0$ for all $j \in J$. Then there exists an optimal solution y of Program D for which $(y^T A)_j > c_j$ for all $j \in J$.* □

Chapter 2

Constrained-Shape and Single-Size Sum-Partition Problems: Polynomial Approach

In this chapter, we consider two other sum-partition problems: the constrained-shape problems, a generalization, and the single-size problems, a special case of the bounded-shape problems. Although results applicable to bounded-shape partition problems are also applicable to single-size partition problems, often, stronger and more specific results can be obtained for the latter, thus worthy of a separate treatment. Again, empty parts are allowed, but the allowance is truly needed only in the last two sections in dealing with single-size problems.

In the first two sections, the constrained-shape problems are studied and the critical result that every vertex of the constrained-shape partition polytopes corresponds uniquely to an (almost-) separable partition is established (separable and almost-separable partitions are defined and characterized in Section 2.1). This result supports the searching of optimal partitions in Section 2.2 when the objective function is assumed to be (edge-)quasi convex (hence the set of partitions corresponding to the vertex set contains an optimal partition). However, the edge-direction method used in Section 1.2 to enumerate vertices applies only to the bounded-shape partition polytopes but not to the constrained-shape polytopes. The above-mentioned correspondence between vertices and almost-separable partitions is needed to replace the vertex set with the set of (almost-) separable partitions for searching optimal partitions. While the set of separable partitions can be directly bounded (by a polynomial function) and efficiently enumerated, that is not the case for the set of almost-separable partitions. A new class of partitions which contains the almost-separable class (in the sense that at least one representative of each equivalent classes of almost-separable partitions appears in the new class) is defined and shown to be bounded polynomially and enumerated efficiently.

Sections 2.3 and 2.4 deal with the single-size problems and mimic Sections 2.1 and 2.2 in structure. The critical result showing each vertex corresponds uniquely with a partition holding some geometric property is again obtained but the property is cone-separable, a more restrictive property than either separable or almost-separable, yet closely related to them. Again, when the objective function is (edge-)quasi-convex, the set of cone-separable partitions that contains an optimal partition is polynomially bounded and can be efficiently enumerated.

2.1 Constrained-Shape Partition Polytopes and (Almost-) Separable Partitions

In this section we introduce two geometric properties of partitions—separability and almost-separability. In particular, we determine conditions for these properties to be present in partitions whose associated vectors are vertices of a constrained-shape partition polytope; results of Chapter 3 of Vol. I can then be used to determine conditions which assure the existence of optimal partitions which are (almost-) separable.

We say that two subsets Ω^1 and Ω^2 of $\mathbb{R}^d$ are *separable* if there exists a nonzero d-vector C such that

$$C^T u^1 > C^T u^2 \quad \text{for all } u^1 \in \Omega^1 \text{ and } u^2 \in \Omega^2. \tag{2.1.1}$$

We say that two subsets Ω^1 and Ω^2 of $\mathbb{R}^d$ are *almost-separable* if there exists a nonzero d-vector C such that

$$C^T u^1 > C^T u^2 \quad \text{for all } u^1 \in \Omega^1 \text{ and } u^2 \in \Omega^2 \text{ with } u^1 \neq u^2. \tag{2.1.2}$$

In either case we refer to the vector C as a *separating vector*. Of course, separable sets are necessarily almost-separable and disjoint, and for disjoint sets almost-separability and separability coincide. When two sets are not disjoint, almost-separability (as defined by (2.1.2)) is the strongest possible form of separation by a linear functional, as the condition asserts strict separation of the values of the functional for all pairs of points at which the sets do not overlap. In particular, (2.1.2) implies that $\Omega^1 \cap \Omega^2$ contains at most a single point, for if u and w were distinct points in the intersection, we would have simultaneously $C^T u > C^T w$ and $C^T w > C^T u$.

Testing for (Almost) Separability of Finite Sets

Suppose Ω^1 and Ω^2 are two finite disjoint sets in $\mathbb{R}^d$. By possible scaling, we have that Ω^1 and Ω^2 are separable if and only if there exist a nonzero d-vector C and a scalar α such that

$$\begin{aligned} &C^T u^1 \geq \alpha + 1 \text{ for all } u^1 \in \Omega^1, \text{ and} \\ &C^T u^2 \leq \alpha \qquad \text{ for all } u^2 \in \Omega^2. \end{aligned} \tag{2.1.3}$$

Similarly (again, by the possible use of scaling), if Ω^1 and Ω^2 are as above except that they are not disjoint, then they are almost-separable if and only if they overlap at a single point, say u, and there exists a nonzero d-vector C such that

$$\begin{aligned} &C^T(u^1 - u) \geq 1 \quad \text{for all } u^1 \in \Omega^1 \setminus \{u\}, \text{ and} \\ &C^T(u^2 - u) \leq -1 \text{ for all } u^2 \in \Omega^2 \setminus \{u\}. \end{aligned} \tag{2.1.4}$$

As (2.1.3) and (2.1.4) are systems of linear inequalities in C and α, the question about (almost-) separability of Ω^1 and Ω^2 is reduced to determine feasibility of a system of linear inequalities – a task that can be resolved efficiently (e.g., Schrijver [85]). □

The next lemma characterizes separability and almost-separability of polytopes.

Lemma 2.1.1. *Let Ω^1 and Ω^2 be two polytopes in $\mathbb{R}^d$. Then:*

(a) Ω^1 and Ω^2 are separable if and only if $\Omega^1 \cap \Omega^2 = \emptyset$, and
(b) Ω^1 and Ω^2 are almost-separable if and only if $\Omega^1 \cap \Omega^2$ is either empty or contains a single point which is a common vertex of both Ω^1 and Ω^2.

Proof. Part (a) is standard; see Rockafellar [82] or Schrijver [85].

To prove (b) assume first that Ω^1 and Ω^2 are almost-separable, with vector $C \in \mathbb{R}^d \setminus \{0\}$ as a separating vector. In this case, $\Omega^1 \cap \Omega^2$ contains at most a single point (see the comments following (2.1.2)). Further, if w is a single point in $\Omega^1 \cap \Omega^2$, then $C^T u^1 > C^T w > C^T u^2$ for all $u^1 \in \Omega^1 \setminus \{w\}$ and $u^2 \in \Omega^2 \setminus \{w\}$, assuring that w is the unique minimizer over Ω^1 of the linear functional defined by the vector C and, in addition, is the unique maximizer of that linear functional over Ω^2. These conclusions imply that w is a vertex of both Ω^1 and Ω^2.

We next establish the reverse implication of (b). Assume that $\Omega^1 \cap \Omega^2$ is either empty or contains a single point which is a common vertex of both Ω^1 and Ω^2. In the former case, part (a) assures that Ω^1 and Ω^2 are separable; hence, they are almost-separable. Next, assume that $\Omega^1 \cap \Omega^2$

contains a single point which is a common vertex of both Ω^1 and Ω^2. A standard result shows that $\Omega^1 - \Omega^2 \equiv \{u^1 - u^2 : u^1 \in \Omega^1 \text{ and } u^2 \in \Omega^2\}$ is a polytope (see Rockafellar [82]), and we next show that 0 is a vertex of this polytope. Indeed, if $0 < \alpha < 1$, u^1 and w^1 in Ω^1, and u^2 and w^2 in Ω^2 satisfy $0 = \alpha(u^1 - u^2) + (1-\alpha)(w^1 - w^2)$, then

$$v \equiv \alpha u^1 + (1-\alpha)w^1 = \alpha u^2 + (1-\alpha)w^2;$$

so, $v \in \Omega^1 \cap \Omega^2$ and therefore (by our assumption) v is a vertex of both Ω^1 and Ω^2. It then follows that $u^1 = w^1 (= v)$ and $u^2 = w^2 (= v)$ and therefore $u^1 - w^1 = u^2 - w^2 = 0$. This proves that 0 is, indeed, a vertex of the polytope $\Omega^1 - \Omega^2$. It now follows that there exists a nonzero vector C with $0 = C^T 0 > C^T(u^1 - u^2)$ for every $u^1 \in \Omega^1$ and $u^2 \in \Omega^2$ with $u^1 - u^2 \neq 0$, that is, (2.1.2) is satisfied. □

The next figure demonstrates pairs of convex sets which are separable, almost-separable but not separable, and not almost-separable; see the equivalent conditions of Lemma 2.1.1 for verification of the statements.

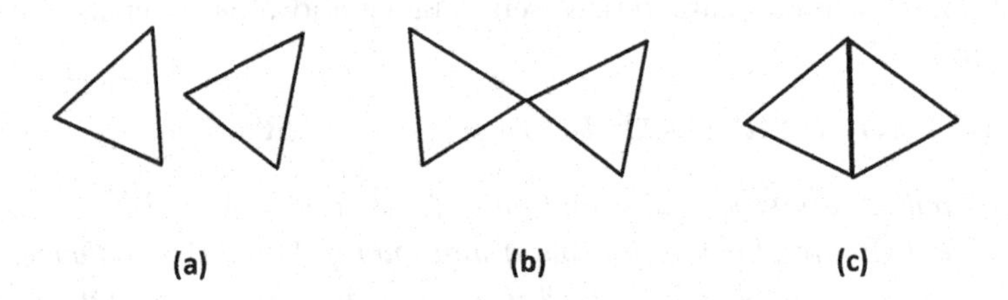

Figure 2.1.1 (a) separable (b) almost-separable but not separable (c) not almost-separable

The next lemma converts separability of finite sets to separability of their convex hulls, yielding as a corollary, a characterization of the former.

Lemma 2.1.2. *Let Ω^1 and Ω^2 be two finite sets in $\mathbb{R}^d$ and let C be a nonzero vector in $\mathbb{R}^d$. Then:*

(a) $C^T u^1 > C^T u^2$ for all $u^1 \in \Omega^1$ and $u^2 \in \Omega^2$ if and only if $C^T w^1 > C^T w^2$ for all $w^1 \in$ conv Ω^1 and $w^2 \in$ conv Ω^2, and

(b) $C^T u^1 > C^T u^2$ for all $u^1 \in \Omega^1$ and $u^2 \in \Omega^2$ with $u^1 \neq u^2$ if and only if $C^T w^1 > C^T w^2$ for all $w^1 \in$ conv Ω^1 and $w^2 \in$ conv Ω^2 with $w^1 \neq w^2$.

Proof. The "only if" (sufficiency) direction in both (a) and (b) is trivial because a set is always included in its convex hull. We next consider the inverse "if" (necessity) direction. Set $\alpha^* \equiv \min_{u \in \Omega^1} C^T u$ and $\alpha_* \equiv \max_{u \in \Omega^2} C^T u$. The finiteness of Ω^1 and Ω^2 assures that these values are finite; further: (i) if $\Omega^1 \cap \Omega^2 = \emptyset$, then $\alpha_* > \alpha^*$, and (ii) if $\Omega^1 \cap \Omega^2$ is not empty, then it contains a single element (see the comments following (2.1.2)), say v, and $C^T u^1 > C^T v > C^T u^2$ for all $u^1 \in \Omega^1 \setminus \{v\}$ and $u^2 \in \Omega^2 \setminus \{v\}$ assuring that $\alpha_* = C^T v = \alpha^*$.

(a) Assume that $C^T u^1 > C^T u^2$ for all $u^1 \in \Omega^1$ and $u^2 \in \Omega^2$. Then $\alpha^* = \min_{u^1 \in \Omega^1} C^T u^1 > \max_{u^2 \in \Omega^2} C^T u^2 = \alpha_*$, and $C^T w^1 \geq \alpha^* > \alpha_* \geq C^T w^2$ for all $w^1 \in \text{conv}\ \Omega^1$ and $w^2 \in \text{conv}\ \Omega^2$.

(b) Assume that $C^T u^1 > C^T u^2$ for all $u^1 \in \Omega^1$ and $u^2 \in \Omega^2$ with $u^1 \neq u^2$. Now, if $\Omega^1 \cap \Omega^2 = \emptyset$, then $C^T w^1 \geq \alpha^* > \alpha_* \geq C^T w^2$ for all $w^1 \in \text{conv}\ \Omega^1$ and $w^2 \in \text{conv}\ \Omega^2$. In the alternative case, $\Omega^1 \cap \Omega^2$ contains a single element, say v, and for $w^1 \in \Omega^1$ and $w^2 \in \Omega^2$ we have that $C^T w^1 > \alpha^* = C^T v = \alpha_* > C^T w^2$. Consider a vector $u^1 \in (\text{conv}\ \Omega^1) \setminus \{v\}$. As $u^1 \in \text{conv}\ \Omega^1$, u^1 is a convex combination of vertices of conv Ω^1 all of which must belong to Ω^1 (Proposition 3.1.1(a), (i) of Vol. I); further, as $u^1 \neq v$, not all of these vertices equal v (where it is not argued that v is a vertex of Ω^1). So, u^1 has a representation $\sum_{i=1}^q \beta_i w^i + (1 - \sum_{i=1}^q \beta_i) v$ where $w^1, \ldots, w^q$ are vertices of Ω^1 that are all distinct from v and $\beta_1, \ldots, \beta_q$ are nonnegative numbers with $0 < \sum_{i=1}^q \beta_i \leq 1$. With $\beta \equiv \sum_{i=1}^q \beta_i$ and $w \equiv \beta^{-1}(\sum_{i=1}^q \beta_i w^i)$, we have that $u^1 = \beta w + (1-\beta) v$, $w \in \text{conv}\ (\Omega^1 \setminus \{v\})$ and $0 < \beta \leq 1$. It then follows that $C^T w > \alpha^* = C^T v$ and therefore $C^T u^1 = \beta C^T w + (1-\beta) C^T v > C^T v$. A similar argument shows that if $u^2 \in (\text{conv}\ \Omega^2) \setminus \{v\}$, then $C^T u^2 < C^T v$. Now, if $u^1 \in \text{conv}\ \Omega^1$ and $u^2 \in \text{conv}\ \Omega^2$ with $u^1 \neq u^2$, then either $u^1 \neq v$ or $u^2 \neq v$; in the former case, we conclude that $C^T u^1 > C^T v \geq C^T u^2$, while in the latter case, we conclude that $C^T u^1 \geq C^T v > C^T u^2$. In either case we have that $C^T u^1 > C^T u^2$. □

Corollary 2.1.3. *Let Ω^1 and Ω^2 be two finite sets in $\mathbb{R}^d$. Then:*

(a) Ω^1 and Ω^2 are separable if and only if conv Ω^1 and conv Ω^2 are separable, and

(b) The following are equivalent:

(i) Ω^1 and Ω^2 are almost-separable,

(ii) conv Ω^1 and conv Ω^2 are almost-separable, and

(iii) conv Ω^1 and conv Ω^2 can have at most a single point in common

and such a point must be a vertex of both conv Ω^1 and conv Ω^2 (in particular, such a point must be in $\Omega^1 \cap \Omega^2$).

Proof. The equivalence of (almost-) separability of finite sets and (almost-) separability of their convex hulls is immediate from the definitions of these properties and the corresponding parts of Lemma 2.1.2. Next, the equivalence of conditions (ii) and (iii) of part (b) is immediate from the second part of Lemma 2.1.1. Finally, the parenthetic comment in part (iii) follows from the fact that a vertex of the convex hull of a finite set must be in the set (Proposition 3.1.1(i) of Vol. I). □

Corollary 2.1.4. *Let Ω be a finite set in $\mathbb{R}^d$ and let v be a vector in Ω. Then the following are equivalent:*

(a) v is a vertex of conv Ω,
(b) for some vector $C \in \mathbb{R}^d$, $C^T(v - u) > 0$ for every $u \in \Omega \setminus \{v\}$, and
(c) for some vector $C \in \mathbb{R}^d$, $C^T(v - u) \geq 1$ for every $u \in \Omega \setminus \{v\}$.

Proof. By definition, v is a vertex of conv Ω if v is a unique maximizer of a linear functional over conv Ω, that is, conv Ω and v are almost-separable. As (b) asserts almost-separability of Ω and $\{v\}$, the equivalence of (a) and (b) follows from Corollary 2.1.3(b) (with $\Omega^1 = \Omega$ and $\Omega^2 = \{v\}$). Also, trivially, a vector $C \in \mathbb{R}^d$ satisfies (b) if and only if some scalar multiple of C satisfies (c). □

Testing for a Point of a Finite Set to be a Vertex of the Convex Hull of that Set

Corollary 2.1.4 gives a characterization for an element of a finite set $\Omega \subseteq \mathbb{R}^d$ to be a vertex of conv Ω in terms of feasibility of a system of $|\Omega| - 1$ linear inequalities with d variables (the coefficients $C_1, \ldots, C_d$ of the vector C in part (c) of Corollary 2.1.4); it yields a test which is polynomial in d and $|\Omega|$. □

The above test is not generally useful for checking if a given vector is a vertex of a partition polytope as the size of the set Ω (consisting of vectors associated with the corresponding partitions) is large.

Attention is next turned to partition polytopes. Our main result demonstrates that sets of vectors corresponding to parts of a partition whose associated matrix is a vertex of a corresponding constrained-shape partition polytope are (almost-) separable. As in earlier sections we have A as the

$d \times n$ matrix whose columns are the vectors to be partitioned. For a subset S of $\mathcal{N} = \{1, \ldots, n\}$, denote by A^S the submatrix of A consisting of the columns of A indexed by S, ordered as in A. Recall that for a set Γ of positive integer p-vectors with coordinate-sum n, P_A^Γ is the constrained-shape partition polytope corresponding to the set of partitions Π^Γ – those with shape in Γ.

We say that a p-partition π is *separable*, if for each pair of distinct indices $r, s \in \{1, \ldots, p\}$, conv A^{π_r} and conv A^{π_s} are separable; we say that π is *almost-separable*, if for each pair of distinct indices $r, s \in \{1, \ldots, p\}$, conv A^{π_r} and conv A^{π_s} are almost-separable. Of course, almost-separable partitions are separable whenever the columns of A are distinct. Note that Corollary 2.1.3 implies that the requirement for (almost-) separability in the above definitions can be imposed on the sets A^{π_r} and A^{π_s} rather than on their convex hulls.

Testing for (Almost) Separability of Partitions

Suppose $\pi = (\pi_1, \ldots, \pi_p)$ is a partition. Using the method described earlier for testing (almost-) separability of finite sets, a test for (almost-) separability of π is available by testing the solvability of $\binom{p}{2}$ systems of linear inequalities corresponding to pairs of distinct parts of π. The system corresponding to π_r and π_s has $d+1$ variables and at most $|\pi_r|+|\pi_s|$ constraints. If these systems are combined, one obtains a block-diagonal system with $(d+1)\binom{p}{2}$ variables and at most $(p-1)n$ constraints where each block has $d+1$ variables. □

The next result and its corollary, given by Hwang, Onn and Rothblum [53], establish (almost-) separability of a partition π whose associated matrix A_π is a vertex of a corresponding constrained-shape partition polytope. The special case of bounded-shape polytopes with distinct columns of A was first given by Barnes, Hoffman and Rothblum [7]. Note that "separable" is called "disjoint" in both papers and "almost-separable" by "separable" in the former paper.

Theorem 2.1.5. *Let $A \in \mathbb{R}^{d \times n}$ and let Γ be a nonempty set of positive integer p-vectors with coordinate-sum n, and let π be a partition in Π^Γ where A_π is a vertex of P_A^Γ. Then for some matrix $C \in \mathbb{R}^{d \times p}$ (with columns $C^1, \ldots, C^p$)*

$$(C^r - C^s)^T u > (C^r - C^s)^T w \text{ for all } 1 \le r \ne s \le p, u \in \text{conv } A^{\pi_r} \text{ and } w \in \text{conv } A^{\pi_s} \text{ with } u \ne w. \tag{2.1.5}$$

Proof. As A_π is a vertex of P_A^Γ, it is the unique maximizer over P_A^Γ of some linear function, say, one that is determined by the matrix $C \in \mathbb{R}^{d\times p}$; so

$$\langle C, X\rangle < \langle C, A_\pi\rangle \text{ for each } X \in P_A^\Gamma \setminus \{A_\pi\}.$$

Let r and s be distinct indices in $\{1,\ldots,p\}$. Suppose $x \in \pi_r$ and $y \in \pi_s$ satisfy $A^x \neq A^y$ and let σ be the partition obtained from π by switching x from π_r to π_s and switching y from π_s to π_r; evidently, the shape of σ is in Γ which assures that $\sigma \in \Pi^\Gamma$. With e^r and e^s denoting the r- and s-unit vectors in $\mathbb{R}^p$, we see that $A_\sigma = A_\pi + (A^x - A^y)(e^s - e^r)^T \neq A_\pi$ (see (1.1.6)). As $A_\sigma \in P_A^\Gamma \setminus \{A_\pi\}$, we conclude that

$$\begin{aligned} 0 &> \langle C, A_\sigma\rangle - \langle C, A_\pi\rangle = \langle C, (A^x - A^y)(e^s - e^r)^T\rangle \\ &= \langle C(e^s - e^r), (A^x - A^y)\rangle = \langle (C^s - C^r), (A^x - A^y)\rangle \\ &= -(C^r - C^s)^T A^x + (C^r - C^s)^T A^y \end{aligned}$$

(with the second equality following from (1.1.1)). Thus, we proved a restricted version of (2.1.5) with u and w restricted to be in A^{π_r} and A^{π_s}, respectively. But, Lemma 2.1.2 shows that the conclusion extends to the convex hulls of these sets, establishing (2.1.5) in its full generality. □

Corollary 2.1.6. *Let Γ be a nonempty set of positive integer p-vectors with coordinate-sum n, and let π be a partition in Π^Γ where A_π is a vertex of P_A^Γ. Then π is almost-separable; further, if the columns of A are distinct, then π is separable.*

Proof. Let C be as in the conclusion of Theorem 2.1.5. Now, for a pair of distinct indices $r, s \in \{1,\ldots,p\}$ with $C^r \neq C^s$, Theorem 2.1.5 establishes the almost-separability of A^{π_r} and A^{π_s} with $C^r - C^s$ as a separating vector. Alternatively, if r and s are distinct indices with $C^r = C^s$, then either of the following conditions must hold: (i) conv A^{π_r} is either empty or consists of a single point which is also in conv A^{π_s} , or (ii) conv A^{π_s} is either empty or consists of a single point which is also in conv A^{π_r} ; in either case conv A^{π_r}, and conv A^{π_s} are almost-separable with any nonzero vector serving as a separating vector. Finally, when the columns of A are distinct, the almost-separability of π implies its separability. □

The next example demonstrates that when the columns of A are not distinct, almost-separability in the conclusion of Corollary 2.1.6 cannot be strengthened to separability.

Example 2.1.1. Let $d = 1, n = 4, p = 2, \Gamma = \{(1,3),(2,2),(3,1)\}$ and $A \in \mathbb{R}^{1\times 4} = (1,1,1,1)$. In this case every partition is almost-separable and no partition is separable; in particular, this is the case for partitions that correspond to the vertices of P_A^Γ (namely, $\binom{1}{3}$ and $\binom{3}{1}$). □

Theorem 2.1.5 establishes (almost-) separability of sets of vectors with separating vectors that are generated from the p columns of a matrix $C \in \mathbb{R}^{d \times p}$ through vector-subtraction, that is, the separating vector of A^{π_r} and A^{π_s} (and of conv A^{π_r} and conv A^{π_s}) is the difference of column r and column s of C; in particular, the separating vectors can be determined by selecting $d \times p$ coefficients. Corollary 2.1.6 provides one with much more freedom in the selection of the separating vectors – for every pair of indices r and s one can select a separating vector in $\mathbb{R}^d$, so the separating vectors can be determined by $d\binom{p}{2}$ coefficients, a larger number than dp. In particular, if $p = 3$ and C^1, C^2 and C^3 are three vectors as in the conclusion of Theorem 2.1.5, the separating vectors of conv A^{π_1} and conv A^{π_2}, of conv A^{π_2} and conv A^{π_3}, and of conv A^{π_3} and conv A^{π_1}, constructed in Corollary 2.1.6 are $C^{12} = C^1 - C^2$, $C^{23} = C^2 - C^3$ and $C^{31} = C^3 - C^1$ satisfying $C^{12} + C^{23} + C^{31} = 0$; in particular, C^{12}, C^{23} and C^{31} are linearly dependent.

The following example demonstrates that the separation derived in Theorem 2.1.5 is strictly stronger than regular (almost-) separability (stated in Corollary 2.1.4). The example further demonstrates that the matrix associated with a separable partition needs not be a vertex of any corresponding constrained-shape partition polytope. The example is a simplified version of one that appears in Barnes, Hoffman and Rothblum [7, p. 83]).

Example 2.1.2. Let $d = 3, n = 12, p = 3, n_1 = n_2 = n_3 = 4$ and $A \in \mathbb{R}^{3 \times 12}$ with its 12 columns given by

A^1	A^2	A^3	A^4	A^5	A^6	A^7	A^8	A^9	A^{10}	A^{11}	A^{12}
ϵ	ϵ	ϵ	1	$-\epsilon$	$-\epsilon$	-1	$-\epsilon$	1	-1	1	-1
1	-1	1	-1	ϵ	ϵ	ϵ	1	$-\epsilon$	$-\epsilon$	-1	$-\epsilon$
$-\epsilon$	$-\epsilon$	-1	$-\epsilon$	1	-1	1	-1	ϵ	ϵ	ϵ	1

where ϵ is a small positive number to be specified later. Consider the partition $\pi = (\pi_1, \pi_2, \pi_3)$ with $\pi_1 = \{1, \ldots, 4\}, \pi_2 = \{5, \ldots, 8\}, \pi_3 = \{9, \ldots, 12\}$. For $j = 1, 2, 3$ let e^j denote the standard j-unit vector in $\mathbb{R}^3$. With $\pi_4 = \pi_1$, we note that $(e^j)^T A^r > 0 > (e^j)^T A^s$ for every $j = 1, 2, 3$, $r \in \pi_j$ and $s \in \pi_{j+1}$. So, π is separable. We next show that π does not satisfy the conclusions of Theorem 2.1.5, implying that it does not satisfy its hypothesis; so, A_π is not a vertex of any constrained-shape partition polytope corresponding to a set of shapes that contains the shaper of π (namely, $(4,4,4)$).

By the paragraph following Example 2.1.1, it suffices to show that if $C^{12}, C^{23}, C^{31} \in \mathbb{R}^3$ satisfy

$$(C^{j,j+1})^T A^r > (C^{j,j+1})^T A^s \text{ for } j = 1,2,3,\ r \in \pi_j \text{ and } s \in \pi_{j+1}, \quad (2.1.6)$$

then $C^{12} + C^{23} + C^{31} \neq 0$. Indeed, assume C^{12}, C^{23} and C^{31} are vectors satisfying (2.1.6). With $C^{12} = (a,b,c)^T$, the application of (2.1.6) with $j = 1$ to A^1 and A^6 and to A^3 and A^8 implies the following inequalities:

values of r and s	derived inequality	equivalent form of derived inequality
$r = 1, s = 6$	$\epsilon a + b - \epsilon c > -\epsilon a + \epsilon b - c$	$(b+c)(1-\epsilon) > -2\epsilon a$
$r = 3, s = 8$	$\epsilon a + b - c > -\epsilon a + b - c$	$a > 0$

Let $e = (1,1,1)^T$. Taking $0 < \epsilon < \frac{1}{3}$, we then have that

$$e^T C^{12} = a + b + c > a - \frac{2a}{1-\epsilon} = \frac{a(1-3\epsilon)}{1-\epsilon} > 0.$$

Observing that for each $i = 1, \ldots, 12$, the coordinates of A^{i+4} and A^{i+8} (with the superscript to be taken modulus 12) are cyclic permutations of the coordinates of A^i, a similar analysis shows that $e^T C^{23} > 0$ and $e^T C^{31} > 0$. We conclude that $e^T(C^{12} + C^{23} + C^{31}) = e^T C^{12} + e^T C^{23} + e^T C^{31} > 0$, implying that $C^{12} + C^{23} + C^{31} \neq 0$. □

Example 2.1.2 demonstrates that the separating property of partitions that is described in Theorem 2.1.5 is more demanding than (almost-) separability. Still, the latter has the advantage that it addresses each pair of parts independently and is therefore easier to test and generate; in fact, in Section 2.2 we use separability as a tool for the efficient enumeration of vertices of bounded-shape partition polytopes.

The next two results establish (restricted) inverses of Corollary 2.1.6, showing that for single-shape partition problems with $p = 2$ or $d = 1$, almost-separability of a partition implies that its associated matrix is a vertex of the corresponding partition polytope.

Theorem 2.1.7. *Let $A \in \mathbb{R}^{d \times n}$, let n_1 and n_2 be positive integers with $n_1 + n_2 = n$ and let π be a 2-partition with shape (n_1, n_2). Then π is almost-separable if and only if A_π is a vertex of $P_A^{(n_1,n_2)} (= P_A^\Gamma$ for $\Gamma = \{(n_1, n_2)\})$.*

Proof. Corollary 2.1.6 implies that if A_π is a vertex of $P_A^{(n_1,n_2)}$, then π is almost-separable. To see the reverse implication assume that π is almost-separable and let $C \in \mathbb{R}^d$ be a separating vector between the vectors assigned to π_1 and π_2, that is,

$$C^T A^t > C^T A^s \text{ for each } t \in \pi_1 \text{ and } s \in \pi_2 \text{ with } A^t \neq A^s.$$

We will show that $\langle (C,0), A_\pi \rangle > \langle (C,0), A_\sigma \rangle$ for each 2-partition σ with shape (n_1, n_2) and $A_\sigma \neq A_\pi$; it will then follow (e.g., Proposition 3.1.1(i) of Vol. I) that A_π is a vertex of $P_A^{(n_1,n_2)} = \text{conv}\,\{A_\sigma : \sigma \text{ has shape } (n_1,n_2)\}$. So, let σ be a 2-partition with shape (n_1, n_2) and $A_\sigma \neq A_\pi$. As π and σ have the same shape, $\pi_1 \backslash \sigma_1$ and $\pi_2 \backslash \sigma_2$ have the same (nonzero) cardinality. Let $t_1, \ldots, t_q$ and $s_1, \ldots, s_q$ be enumerations of the elements in $\pi_1 \setminus \sigma_1$ and $\pi_2 \setminus \sigma_2$, respectively. As $A_\sigma \neq A_\pi$, $A^{t_u} \neq A^{s_u}$ for at least one u. It now follows from the selection of C that $C^T A^{t_u} \geq C^T A^{s_u}$ for $u = 1, \ldots, q$ with at least one inequality holding strictly, implying that

$$\begin{aligned} \langle (C,0), A_\pi \rangle - \langle (C,0), A_\sigma \rangle &= C^T A_{\pi_1} - C^T A_{\sigma_1} \\ &= \sum_{u=1}^{q} [C^T A^{t_u} - C^T A^{s_u}] > 0. \end{aligned}$$

□

Theorem 2.1.8. *Let $A \in \mathbb{R}^{1 \times n}$, let $n_1, \ldots, n_p$ be positive integers with $\sum_{j=1}^p n_j = n$ and let π be a partition with shape $(n_1, \ldots, n_p)$. Then π is almost-separable if and only if A_π is a vertex of $P_A^{(n_1,\ldots,n_p)} (= P_A^\Gamma$ for $\Gamma = \{(n_1, \ldots, n_p)\})$.*

Proof. Corollary 2.1.6 implies that if A_π is a vertex of $P_A^{(n_1,\ldots,n_p)}$, then π is almost-separable. To see the reverse implication assume that π is almost-separable. Without loss of generality permute the columns of A (which are numbers) so that $A^1 \leq \cdots \leq A^n$; in particular, for each $\alpha \in A^1, \ldots, A^n$, the integers $\{i = 1, \ldots, n : A^i = \alpha\}$ are then consecutive. As π is almost-separable, for each $1 \leq r < s \leq p$ there exists a number $C^{r,s} \neq 0$ such that

$$C^{r,s} A^u > C^{r,s} A^w \text{ for each } u \in \pi_r \text{ and } w \in \pi_s \text{ with } A^u \neq A^w.$$

Depending on whether $C^{r,s} > 0$ or $C^{r,s} < 0$ we conclude that either

$$u < w \text{ for each } u \in \pi_r, w \in \pi_s \text{ with } A^u \neq A^w, \text{ or}$$

$$u > w \text{ for each } u \in \pi_r, w \in \pi_s \text{ with } A^u \neq A^w.$$

It follows that it is possible to reassign indices i that are associated with the same values of A^i so that the parts of the resulting partition, say σ, consist of consecutive integers and $|\{i \in \sigma_j : A^i = \alpha\}| = |\{i \in \pi_j : A^i = \alpha\}|$ for each α. It then follows from part (a) of Theorem 5.1.9 of Vol. I that $A_\sigma = A_\pi$ is a vertex of $P_A^{(n_1,\ldots,n_p)}$. □

Theorems 2.1.7, 2.1.8 and 2.1.5 imply that for single-shape partition problems with $p = 2$ or $d = 1$, the stronger separability property of Theorem 2.1.5 (cast in (2.1.5)) is equivalent to almost-separability.

Example 2.1.2 demonstrates that the conclusions of Theorems 2.1.7 and 2.1.8 (that is, the restricted inverse of Corollary 2.1.6) need not hold for $p = d = 3$. Recently, Liu and Pan [66] constructed an example that shows that the inverse of Corollary 2.1.6 is not true for $d = 2$ and $p \geq 3$, thus removing the last uncertainty of the inversibility issue. The next example demonstrates that with $p = 2$ and $d = 1$ the "single-shape" assumption in Theorems 2.1.7 and 2.1.8 cannot be replaced with "bounded-shape".

Example 2.1.3. Let $d = 1, n = 3, p = 2, \Gamma = \{(1,2),(2,1)\}$ and $A = (1,2,3) \in \mathbb{R}^{1\times 3}$. Then $\{A_\pi : \pi \text{ has shape in } \Gamma\} = \{(i, 6-i) : i = 1,\ldots,5\}$ and for the separable partition $\pi = (\{1,2\},\{3\})$, $A_\pi = (3,3)$ is not a vertex of P_A^Γ. □

Theorem 2.1.5 implies that given a partition π with A_π as a vertex of a corresponding constrained-shape partition polytope, the vectors associated with distinct parts of π can be separated with the separating vectors having a particular structure, namely, there is a matrix $C \in \mathbb{R}^{d\times p}$ such that for each pair of distinct indices r and s in $\{1,\ldots,p\}$, $(C^r - C^s)^T A^x > (C^r - C^s)^T A^y$ for all $x \in \pi_r$ and $y \in \pi_s$ with $A^x \neq A^y$. Letting α_{rs} be a value between $\min\{(C^r - C^s)^T A^x : x \in \pi_r\}$ and $\max\{(C^r - C^s)^T A^y : y \in \pi_s\} = -\min\{(C^s - C^r)^T A^y : y \in \pi_s\}$, we have that $(C^r - C^s)^T A^x \geq \alpha_{rs}$ for every $x \in \pi_r$ and $(C^s - C^r)^T A^y \geq \alpha_{sr} \equiv -\alpha_{rs}$ for every $y \in \pi_s$; further, if A^{π_r} and A^{π_s} are disjoint, the inequalities are strict. In Theorem 1.3.3 we see that for bounded-shape partition problems, this conclusion can be sharpened by showing that the inequalities are satisfied with α_{rs}'s as differences of values associated with the indices of the parts. This will be accomplished by deriving characterizations, through feasibility of linear inequality systems, of partitions whose vectors are vertices of the corresponding bounded-shape partition polytopes.

Next we use the enumeration of vertices of single-size partition polytopes given in Section 1.3 to enumerate vertices of constrained-shape and bounded-shape partition polytopes.

It is well known that if P and P' are polytopes, then

$$P \subseteq P' \Rightarrow (V[P'] \cap P) \subseteq V[P]. \tag{2.1.7}$$

Recall that $\widetilde{A}$ is defined to be $\binom{1}{A}$. The next result demonstrates that, with underlying matrix $\widetilde{A}$, (2.1.7) holds with equality for constrained-shape subpolytopes of single-size partition polytopes.

Theorem 2.1.9. *Let $A \in \mathbb{R}^{d\times n}$ have nonzero and distinct columns and let Γ be a set of integer p-vectors $(n_1, \cdots, n_p)$ satisfying $\sum_{j=1}^{p} n_j = n$. Then $V[P^{\Gamma}_{\widetilde{A}}] = V[P^{p}_{\widetilde{A}}] \cap P^{\Gamma}_{\widetilde{A}}$, in particular, $|V[P^{\Gamma}_{\widetilde{A}}]| \leq |V[P^{p}_{\widetilde{A}}]|$.*

Proof. As $P^{\Gamma}_{\widetilde{A}} \subseteq P^{p}_{\widetilde{A}}$, the inclusion $V[P^{p}_{\widetilde{A}}] \cap P^{\Gamma}_{\widetilde{A}} \subseteq V[P^{\Gamma}_{\widetilde{A}}]$ follows from (2.1.7). Next assume that $v^* \in V[P^{\Gamma}_{\widetilde{A}}]$; in particular, v^* has a representation $\widetilde{A}_{\pi^*}$ with $\pi^* \in \Pi^{\Gamma}$. So, $v^* = \widetilde{A}_{\pi^*} \in P^{\Gamma}_{\widetilde{A}}$. Thus, it suffices to show that $v^* \in V[P^{p}_{\widetilde{A}}]$.

Let $(n_1, \cdots, n_p) \in \Gamma$ be the shape of π^*. Then

$$\begin{aligned} P^{(n_1,\cdots,n_p)}_{\widetilde{A}} &= \operatorname{conv}\{\widetilde{A}_\pi : \pi \in \Pi^{(n_1,\cdots,n_p)}\} \\ &= \operatorname{conv}\{\binom{(n_1,\cdots,n_p)}{A_\pi} : \pi \in \Pi^{(n_1,\cdots,n_p)}\} \\ &= \{\binom{(n_1,\cdots,n_p)}{X} : X \in P^{(n_1,\cdots,n_p)}_{A}\}, \end{aligned}$$

implying that

$$V[P^{(n_1,\cdots,n_p)}_{\widetilde{A}}] = \{\begin{pmatrix} (n_1, \cdots, n_p) \\ X \end{pmatrix} : X \in V[P^{(n_1,\cdots,n_p)}_{A}]\}.$$

As $\binom{(n_1,\cdots,n_p)}{A_{\pi^*}} = \widetilde{A}_{\pi^*} = v^* \in V[P^{\Gamma}_{\widetilde{A}}] \cap P^{(n_1,\cdots,n_p)}_{\widetilde{A}} \subseteq V[P^{(n_1,\cdots,n_p)}_{\widetilde{A}}]$ (the last inclusion following from (2.1.7)), it follows that $A_{\pi^*} \in V[P^{(n_1,\cdots,n_p)}_{A}]$. Thus, by Theorem 1.3.6, $v^* = \widetilde{A}_{\pi^*} \in V[P^{p}_{\widetilde{A}}]$. □

Note that when the vectors need not be distinct, then Example 1.4.1 gave a counterexample to the conclusions in Theorem 2.1.9.

Enumerating the Vertices of Constrained-Shape and Bounded-Shape Partition Polytopes with Underlying Matrix $\widetilde{A}$

Theorem 2.1.9 implies that the vertices of $P^{\Gamma}_{\widetilde{A}}$ can be enumerated using any algorithm that enumerates the vertices of single-size partition polytopes along with representing partitions (which are unique), apply the algorithm to $P^p_{\tilde{A}}$ and then test whether the shapes of the representing partitions of the vertices of $P^p_{\tilde{A}}$ are in Γ; see the discussion following Theorem 1.3.6.

When Γ is expressed as $\Gamma^{(L,U)}$, the test of each shape is simple (checking whether it satisfies the $2p$ lower and upper bound constraints), so, in this case the tests do not increase the computational complexity. In the more general case, the testing may increase the computational complexity, for example, if Γ is given by a list of shapes, the computational effort may be increased by a (multiplicative) factor $|\Gamma|$. Still, given n and p, the number of possible shapes equals the number of ways that n identical balls can be split into p marked urns, i.e., $\binom{n+p}{p} = O(n^p)$; hence, for any set of shapes Γ, $|\Gamma| \leq \binom{n+p}{p} = O(n^p)$. □

The proof of Theorem 2.1.9 builds on the property that for any integer p-vector with $\sum_{j=1}^p n_j = n$, $P^{(n_1,\cdots,n_p)}_{\tilde{A}}$ is the image of $P^{(n_1,\cdots,n_p)}_A$ obtained by adding the $(n_1, \cdots, n_p)$ as the first row to each matrix in $P^{(n_1,\cdots,n_p)}_A$. Consequently, $P^{(n_1,\cdots,n_p)}_A$ and $P^{(n_1,\cdots,n_p)}_{\tilde{A}}$ have the same geometry and their vertices are in one-to-one correspondence. This structure does not generalize to constrained-shape problems where the shape of representing partitions is not uniform and P^Γ_A and $P^\Gamma_{\tilde{A}}$ may have different geometry. The next lemma, given by Rothblum (unpublished), shows a weaker result for general Γ's.

Lemma 2.1.10. *Let $A \in \mathbb{R}^{d\times n}$ have nonzero and distinct columns and let Γ be a set of integer p-vectors $(n_1, \cdots, n_p)$ satisfying $\sum_{j=1}^p n_j = n$. Then the set of matrices obtained by removing the top row of the vertices of $P^\Gamma_{\tilde{A}}$ contains all vertices of P^Γ_A; in particular, $|V[P^\Gamma_A]| \leq |V[P^\Gamma_{\tilde{A}}]| \leq |V[P^p_{\tilde{A}}]|$.*

Proof. Let $T(\cdot)$ be the (linear) operator on $\mathbb{R}^{(d+1)\times p}$ that removes the first row of each matrix in $\mathbb{R}^{(d+1)\times p}$. If $\pi^1, \cdots, \pi^q$ are partitions in Π^Γ with shapes $\nu^1, \cdots, \nu^q$, respectively, and $\alpha_1, \cdots, \alpha_q$ are nonnegative numbers with $\sum_{j=1}^q \alpha_j = 1$, then

$$T[\sum_{j=1}^q \alpha_j \tilde{A}_{\pi^j}] = T[\begin{pmatrix} \sum_{j=1}^q \alpha_j \nu^j \\ \sum_{j=1}^q \alpha_j A_{\pi^j} \end{pmatrix}] = \sum_{j=1}^q \alpha_j A_{\pi^j}.$$

It follows that $T(\cdot)$ maps $P^\Gamma_{\tilde{A}}$ onto P^Γ_A and a standard fact implies that each vertex of the (image) polytope P^Γ_A is the image under $T(\cdot)$ of a vertex of the (domain) polytope $P^\Gamma_{\tilde{A}}$. In particular, $|V[P^\Gamma_A]| \leq |V[P^\Gamma_{\tilde{A}}]| \leq |V[P^p_{\tilde{A}}]|$, where the second inequality follows from Theorem 2.1.9. □

The following continuation of Example 2.1.4 illustrates that the asserted inclusion in Lemma 2.1.10 needs not hold with equality.

Example 2.1.4. Let $d = 1, n = 3, A = [1, 2, 3]$ and $p = 2$ with $L = (1, 1)$ and $U = (2, 2)$. In this case $\Gamma^{(L,U)} = \{(1, 2), (2, 1)\}$, $V[P_A^{(L,U)}] = \{(1, 5), (5, 1)\}$ and $V[P_{\tilde{A}}^{(L,U)}] = \{\binom{1,2}{1,5}, \binom{1,2}{3,3}, \binom{2,1}{3,3}, \binom{2,1}{5,1}\}$; eliminating the first row of vertices of $P_{\tilde{A}}^{(L,U)}$ yields the matrix $(3, 3)$ in addition to the vertices of $P_A^{(L,U)}$. □

Generating the Vertices of Bounded-Shape and Constrained-Shape Partition Polytopes

Assume that A and Γ are as in the statement of Lemma 2.1.10. With $\tilde{A}$ as the underlying matrix, it was already discussed how the vertices of bounded-shape and constrained-shape polytopes can be enumerated along with all representing partitions in Γ. Lemma 2.1.10 shows that eliminating the first row of each such vertex will yield a list of matrices in P_A^Γ that contains all its vertices. A particular vertex may have multiple occurrences in this list (when Γ is not expressible by lower and upper bounds), still, it will appear with all its representing partitions. □

The next result concerns only lists that contain all vertices (without any deletion), as these are enough for solving corresponding convex maximization problems.

Corollary 2.1.11. *Let $A \in \mathbb{R}^{d\times n}$ have nonzero and distinct columns and let Γ be a set of integer p-vectors $(n_1, \cdots, n_p)$ satisfying $\sum_{j=1}^p n_j = n$. Then $|V[P_A^\Gamma]| \le |V[P_{\tilde{A}}^p]| \le O(n^{(d+1)(p-1)-1})$ and a strongly polynomial algorithm using $O(n^{(d+1)(p-1)-1}|\Gamma|) \le O(n^{(d+2)(p-1)})$ arithmetic operations for producing a list containing all vertices of P_A^Γ is available. Further, if Γ is expressed by lower and upper bounds, then the above complexity bound can be reduced to $O(n^{(d+1)(p-1)-1})$.*

Proof. The proof is immediate from Lemma 2.1.10, Theorem 2.1.9, Corollary 1.3.7, the discussion preceding the corollary and the fact that the total number of all p-shapes equals $\binom{n+p}{p} = O(n^p)$. □

Corollary 2.1.11 provides a method for solving constrained-shape partition problems with convex objective.

Note that a method to enumerate all facets of a constrained-shape partition polytope was already given in Section 1.2.

2.2 Enumerating Separable and Limit-Separable Partitions of Constrained-Shape

In the preceding section, we studied the constrained-shape partition polytope and in particular, the properties of its vertex-set. The following result provides a motivation of this study.

Theorem 2.2.1. *Let Γ be a nonempty set of positive integer p-vectors with coordinate-sum n and let $f(\cdot)$ be an (edge-)quasi-convex function on the constrained-shape partition polytope P_A^Γ. Then there exists an optimal partition π which is almost-separable and has A_π as a vertex of P_A^Γ. Further, if $f(\cdot)$ is strictly (edge-)quasi-convex, then every optimal partition π is almost-separable and has A_π as a vertex of P_A^Γ. If A's columns are distinct, the quantifier "almost" can be dropped in the above statements.*

Proof. Corollary 3.3.7 of Vol. I assures that $f(\cdot)$ attains a maximum over P_A^Γ at one of its vertices, and if $f(\cdot)$ is strictly (edge-)quasi-convex, then each maximizer of $f(\cdot)$ over P_A^Γ is a vertex. As each vertex of P_A^Γ has a representation A_π with $\pi \in \Pi^\Gamma$ (Proposition 3.1.1(i) of Vol. I), it follows that

$$\phi^* \equiv \max_{x \in P_A^\Gamma} f(x) = \max_{\pi \in \Pi_A^\Gamma} f(A_\pi) = \max_{\pi \in \Pi^\Gamma} F(\pi),$$

and a partition is optimal if and only if $F(\pi) = f(A_\pi) = \phi^*$. One can further conclude that there exists an optimal partition π that has A_π as a vertex of P_A^Γ and when $f(\cdot)$ is strictly (edge-)quasi-convex, each optimal partition has this property; Corollary 2.1.6 assures that such partitions are (almost-) separable. □

Such a result is useful for solving partition problems only when combined with an efficient method for finding the best partition in the restricted class that is guaranteed to include an optimal partition, say by enumeration.

Theorem 2.2.1 shows that when considering constrained-shape partition problems with $f(\cdot)$ (edge-)quasi-convex, it suffices to consider partitions that are (almost-) separable (and whose associated matrix is a vertex of the corresponding partition polytope). The current section introduces a new class of partitions that includes the separable ones and has the property that the set of their associated matrices includes the vertices of all constrained-shape partition polytopes. We then develop a method for enumerating the partitions in that class. Further, as separable partitions constitute a subset of the new class, a by-product is a method of enumerating the separable partitions as well. In addition, we obtain a method for efficiently enumerating all vertices of constrained-shape partition polytopes and for solving corresponding partition problems.

The methods we develop in the current section enumerate (through the construction of separating hyperplanes) all partitions with corresponding properties, regardless of shape. We then solve constrained-shape partition problems, by testing the shape of the generated partitions and evaluating those whose shape is found feasible. Thus, our approach (and many of our results) addresses, in effect, the single-size problem and provides a solution to constrained-shape problems by simply restricting attention to subclasses of partitions. In particular, the approach provides one with a list of partitions that include optimal ones for each constrained-shape problem! This is particularly useful when problems with varying sets of feasible shapes are to be addressed (or when these sets are subject to changes).

Recall that a matrix $A \in \mathbb{R}^{d\times n}$ is *generic* if its columns are in general position, that is, for $k = 1, \ldots, d$ there are no $k+1$ of columns of A that lie in a common affine set of dimension less than k. Lemma 3.1.7 of Vol. I provides a characterization of such matrices; precisely, it shows that A is generic if and only if any set of distinct $k \leq d+1$ vectors from $\{\widetilde{A}^i : i = 1, \ldots, n\}$ are linearly independent. Of course, if A is generic, then the columns of A are distinct. For indices $i_1, \ldots, i_{d+1}$ from $\mathcal{N} = \{1, \ldots, n\}$ (not necessarily distinct or ordered), denote

$$\operatorname{sign}_A(i_1, \ldots, i_{d+1}) \equiv \operatorname{sign}(\det[\widetilde{A}^{i_1}, \ldots, \widetilde{A}^{i_{d+1}}]) \in \{-1, 0, 1\}. \qquad (2.2.1)$$

It follows that if $n \geq d+1$, A is generic if and only if all of these signs are nonzero; of course, it suffices to check the signs only for sets of indices $i_1, \ldots, i_{d+1}$ that are distinct and ordered, that is, $1 \leq i_1 < \cdots < i_{d+1} \leq n$.

We next provide a representation of the set of separable 2-partitions when A is generic. The case where $n \leq d+1$ is simple.

Lemma 2.2.2. *Let $A \in \mathbb{R}^{d\times n}$ be generic and $n \leq d+1$. Then every p-partition is separable.*

Proof. It suffices to consider the case $p = 2$. As A is generic and $n \leq d+1$, $\widetilde{A}^1, \ldots, \widetilde{A}^n$ are linearly independent vectors in $\mathbb{R}^{d+1}$ and rank $[\widetilde{A}^1, \ldots, \widetilde{A}^n]^T$ = rank $[\widetilde{A}^1, \ldots, \widetilde{A}^n] = n$. It follows that the range of the matrix $[\widetilde{A}^1, \ldots, \widetilde{A}^n]^T$ is $\mathbb{R}^n$. In particular, given a 2-partition $\pi = (\pi_1, \pi_2)$ of $\mathcal{N}$, there is a vector $\mu \in \mathbb{R}^{d+1}$ with $\mu^T \widetilde{A}^i = 1 > 0$ for each $i \in \pi_1$ and $\mu^T \widetilde{A}^u = -1 < 0$ for each $u \in \pi_2$; with C obtained from μ by truncating its first coordinate μ_1 we then have $C^T A^i > -\mu_1 > C^T A^u$ for all $i \in \pi_1$ and $u \in \pi_2$ proving that π is separable. □

Let $A \in \mathbb{R}^{d\times n}$ be generic with $n \geq d+1$ and let $I = \{i_1, \ldots, i_d\}$ be a d-subset of $\mathcal{N}$ with $i_1 < \cdots < i_d$. Define

$$\begin{aligned} I_A^- &\equiv \{i \in \mathcal{N} : \operatorname{sign}_A(i_1, \ldots, i_d, i) = -1\}, \text{ and} \\ I_A^+ &\equiv \{i \in \mathcal{N} : \operatorname{sign}_A(i_1, \ldots, i_d, i) = 1\}. \end{aligned} \tag{2.2.2}$$

As A is generic, Lemma 2.2.2 assures that $\{i \in \mathcal{N} \setminus I : \operatorname{sign}_A(i_1, \ldots, i_d, i) = 0\} = \emptyset$, implying that $\{I_A^-, I_A^+\}$ is a 2-partition of $\mathcal{N}\setminus I$. Further, if (J^-, J^+) is a 2-partition of I, we define the 2-partitions of $\mathcal{N}$ *associated with* A, I and (J^-, J^+) as the two 2-partitions $(I_A^- \cup J^-, I_A^+ \cup J^+)$ and $(I_A^+ \cup J^+, I_A^- \cup J^-)$.

Lemma 2.2.3. *Suppose $A \in \mathbb{R}^{d\times n}$ is generic and $n \geq d+1$. Then the set of separable 2-partitions is the set of 2-partitions associated with A, d-sets $I \subseteq \mathcal{N}$ and 2-partitions (J^-, J^+) of I; in particular, there are at most $2^{d+1}\binom{n}{d} = O(n^d)$ separable 2-partitions.*

Proof. We will show that for each d-set $I \subseteq \mathcal{N}$ and 2-partition (J^-, J^+) of I, the two 2-partitions associated with A, I and (J^-, J^+) are separable and that each separable 2-partition is generated in this way. The asserted bound on the number of separable 2-partitions follows from the fact that there are $\binom{n}{d}$ subsets I of $\mathcal{N}$ and 2^d partitions (J^-, J^+) of each such I.

First, let $I \subseteq \mathcal{N}$ have d elements, say $i_1 < \cdots < i_d$, and let (J^-, J^+) be a 2-partition of I. Evidently, the determinant $\det[\widetilde{A}^{i_1}, \ldots, \widetilde{A}^{i_d}, \widetilde{x}]$ is a real-valued affine function of $x \in \mathbb{R}^d$, that is, there exist $h \in \mathbb{R}^d$ and $\gamma \in \mathbb{R}$ such that $\det[\widetilde{A}^{i_1}, \ldots, \widetilde{A}^{i_d}, \widetilde{x}] = h^T x - \gamma$ for all $x \in \mathbb{R}^d$. Let $H \equiv \{x \in \mathbb{R}^d : h^T x = \gamma\}$. Evidently, $A^i \in H$ for $i \in I$. Also, for $i \in \mathcal{N} \setminus I$, as $\widetilde{A}^{i_1}, \ldots, \widetilde{A}^{i_d}, \widetilde{A}^i$ are linearly independent, $h^T A^i - \gamma = \det[\widetilde{A}^{i_1}, \ldots, \widetilde{A}^{i_d}, \widetilde{A}^i] \neq 0$ and therefore $A^i \notin H$. So, $h \neq 0$ (as $\mathcal{N} \setminus I \neq \emptyset$) and $I = \{i \in \mathcal{N} : A^i \in H\}$, while $I_A^- = \{i \in \mathcal{N} : h^T A^i < \gamma\}$ and $I_A^+ = \{i \in \mathcal{N} : h^T A^i > \gamma\}$; in particular, $h^T A^u < h^T A^v < h^T A^w$ for all $u \in I_A^-, v \in I$ and $w \in I_A^+$. Observe that $A^{\{i_1,\ldots,i_d\}} = [A^{i_1}, \ldots, A^{i_d}]$ is generic; hence Lemma 2.2.2 assures that the 2-partition $\{j : i_j \in J^-\}$, $\{j : i_j \in J^+\}$ of $\{1, \ldots, d\}$ is $A^{\{i_1,\ldots,i_d\}}$-separable (meaning separable with $A^{i_1}, \ldots, A^{i_d}$ as the partitioned vectors). Thus, there exists a vector $f \in \mathbb{R}^d$ with $f^T A^u < f^T A^w$ for all $u \in J^-$ and $w \in J^+$. For sufficiently small positive t, we then have that $(h+tf)A^u < (h+tf)^T A^w$ for all $u \in I_A^- \cup J^-$ and $w \in I_A^+ \cup J^+$, proving that $(I_A^- \cup J^-, I_A^+ \cup J^+)$ is separable. A symmetric argument shows that $(I_A^+ \cup J^+, I_A^- \cup J^-)$ is separable too, proving that the two 2-partitions of $\mathcal{N}$ associated with A, I and 2-partition (J^-, J^+) of I are separable.

Next assume that π is a separable 2-partition. Then there exists a vector $C \in \mathbb{R}^d$ that strictly separates A^{π_1} and A^{π_2}. In particular, it follows that for some $\gamma \in \mathbb{R}$, $\pi_1 = \{i \in \mathcal{N} : C^T A^i > \gamma\}$ and $\pi_2 = \{i \in \mathcal{N} : C^T A^i < \gamma\}$. Repeated use of Lemma 3.1.8 of Vol. I (with the v^j's at the columns of A) shows that C and γ can be modified so that $C^T A^i \geq \gamma$ for each $i \in \pi_1$, $C^T A^i \leq \gamma$ for each $i \in \pi_2$ and $H \equiv \{x \in \mathbb{R}^d : C^T x = \gamma\}$ containing exactly d columns of A; let these columns be $A^{i_1}, \ldots, A^{i_d}$ where $i_1 < \cdots < i_d$. With $I \equiv \{i_1, \ldots, i_d\}$, we then have that

$$\pi_1 = \{i \in \mathcal{N} : C^T A^i > \gamma\} \cup (\pi_1 \cap I) \tag{2.2.3}$$

and

$$\pi_2 = \{i \in \mathcal{N} : C^T A^i < \gamma\} \cup (\pi_2 \cap I). \tag{2.2.4}$$

There is only one hyperplane that contains $A^{i_1}, \ldots, A^{i_d}$ (for if there were two different such hyperplanes, their intersection would be an affine set of dimension less than $d-1$ that contains $A^{i_1}, \ldots, A^{i_d}$, a contradiction to the assertion that A's columns are in general position). Let $H_+ \equiv \{x \in \mathbb{R}^d : C^T x > \gamma\}$ and $H_- \equiv \{x \in \mathbb{R}^d : C^T x < \gamma\}$. Arguments used earlier in our proof showed that $\{x \in \mathbb{R}^d : \det[\widetilde{A}^{i_1}, \ldots, \widetilde{A}^{i_d}, \widetilde{x}] = 0\}$ is a hyperplane that contains $A^{i_1}, \ldots, A^{i_d}$. It follows that this hyperplane must coincide with H, and further,

$$\begin{aligned}\{H_+, H_-\} = \{\ &\{\ x \in \mathbb{R}^d : \det[\widetilde{A}^{i_1}, \ldots, \widetilde{A}^{i_d}, \widetilde{x}] > 0\},\\ &\{\ x \in \mathbb{R}^d : \det[\widetilde{A}^{i_1}, \ldots, \widetilde{A}^{i_d}, \widetilde{x}] < 0\}\},\end{aligned}$$

implying that

$$\{\{i \in \mathcal{N} : C^T A^i > \gamma\}, \{i \in \mathcal{N} : C^T A^i < \gamma\}\} = \{I_A^+, I_A^-\}.$$

Now, if $\{i \in \mathcal{N} : C^T A^i > \gamma\} = I_A^+$, then (2.2.3)–(2.2.4) show that $\pi = (I_A^+ \cup J^+, I_A^- \cup J^-)$, where $(J^+, J^-) = (\pi_1 \cap I, \pi_2 \cap I)$. Alternatively, if $\{i \in \mathcal{N} : C^T A^i > \gamma\} = I_A^-$, then (2.2.3)–(2.2.4) show that $\pi = (I_A^- \cup J^-, I_A^+ \cup J^+)$, where $(J^-, J^+) = (\pi_1 \cap I, \pi_2 \cap I)$. □

Lemma 2.2.3 gives a bound on the number of separable 2-partitions and a method for constructing all such partitions. Under the construction, different sets of points and/or partitions may produce the same 2-partition.

Enumerating all Separable 2-Partitions when A is Generic

Assume that d is fixed. Let $A \in \mathbb{R}^{d \times n}$ be generic. If $n \leq d+1$, the set of separable 2-partitions is the set of all partitions, of which there are $p^n \leq$

p^d; these can be generated using a fixed number of arithmetic operations. Henceforth we assume that $n > d+1$. For a d-set $I = \{i_1, \ldots, i_d\} \subseteq \mathcal{N}$ with $i_1 < \cdots < i_d$, I_A^- and I_A^+ are available from the signs of A at $(i_1, \ldots, i_d, i)$ for each $i \in \mathcal{N} \setminus I$. Each such sign is the determinant of a matrix of order $d+1$ and can be computed, say, by Gaussian Elimination using $O(d^3)$ arithmetic operations. So, all $\binom{n}{d+1}$ signs can be computed with $O\left(\binom{n}{d+1}d^4\right) = O(n^{d+1})$ arithmetic operations (permutations that put $\widetilde{A}^i$ into the right location may be applied). Next, from Lemma 2.2.3, the set of separable 2-partitions is the set of partitions of $\mathcal{N}$ of the form $(I_A^- \cup J^-, I_A^+ \cup J^+)$ or $(I_A^+ \cup J^+, I_A^- \cup J^-)$ where I is a d-subset of $\mathcal{N}$ and (J^-, J^+) is a 2-partition of I. For each d-set $I \subseteq \mathcal{N}$, the 2-partitions (I_A^-, I_A^+) have been determined; hence an enumeration of all the separable 2-partitions is available (the construction may contain duplicates). The number of such partitions is then $2^{d+1}\binom{n}{d}$ and the total effort to generate them is bounded by $O(n^{d+1})$ operations. Further, the algorithm can be executed in parallel by $\binom{n}{d}$ processors, each considering a set of d points of the n given points of $\mathbb{R}^d$ and requiring $O(n)$ arithmetic operations. □

The complexity bound of the above algorithm focused on arithmetic operations and did not count for operations needed to eliminate duplication.

Whenever there is more than one possible underlying matrix we index "separability" by a prefix that indicates the underlying matrix, e.g., we shall refer to *A-separable* partitions.

Let $p \geq 2$. We shall consider $\binom{p}{2}$-tuples of 2-partitions of $\mathcal{N}$, indexed by $\{(r,s) : 1 \leq r < s \leq p\}$; we refer to such a tuple as a *list of* $\binom{p}{2}$ *2-partitions* and its elements (which are 2-partitions) will be typically denoted by $\pi^{r,s} = (\pi_r^{r,s}, \pi_s^{r,s})$ where $1 \leq r < s \leq p$. For each such list we associate a p-tuple $\pi = (\pi_1, \ldots, \pi_p)$ of subsets of $\mathcal{N}$ as follows: for $j = 1, \ldots, p$ put

$$\pi_j \equiv \left(\cap_{t=j+1}^{p} \pi_j^{j,t}\right) \bigcap \left(\cap_{t=1}^{j-1} \pi_j^{t,j}\right). \tag{2.2.5}$$

This definition assures that for all $1 \leq r < s \leq p$, $\pi_r \subseteq \pi_r^{r,s}$ and $\pi_s \subseteq \pi_s^{r,s}$, implying that $\pi_1, \ldots, \pi_p$ are pairwise disjoint. So, if it happens that $\pi = (\pi_1, \ldots, \pi_p)$, defined by (2.2.5), satisfies $\cup_{j=1}^{p} \pi_j = \mathcal{N}$, then π is a p-partition which will be called *the partition associated with* $\{\pi^{r,s} = (\pi_r^{r,s}, \pi_s^{r,s}) : 1 \leq r < s \leq p\}$.

Hwang, Onn and Rothblum [53] proved the following result.

Theorem 2.2.4. *Suppose $A \in \mathbb{R}^{d \times n}$. Then the set of A-separable p-partitions equals the set of p-partitions associated with lists of $\binom{p}{2}$ A-separable 2-partitions. Further, if A is generic, then there are at most*

$[2^{d+1}\binom{n}{d}]^{\binom{p}{2}} = O[n^{d\binom{p}{2}}]$ *A-separable p-partitions.*

Proof. Consider a p-partition $\pi = (\pi_1, \ldots, \pi_p)$ associated with a list of $\binom{p}{2}$ A-separable 2-partitions whose elements are $\{\pi^{r,s} = (\pi_r^{r,s}, \pi_s^{r,s}) : 1 \leq r < s \leq p\}$. For each $1 \leq r < s \leq p$, $\pi_r \subseteq \pi_r^{r,s}$ and $\pi_s \subseteq \pi_s^{r,s}$ and therefore the A-separability of $\pi^{r,s} = (\pi_r^{r,s}, \pi_s^{r,s})$ implies that (π_r, π_s) is also A-separable.

Conversely, let $\pi = (\pi_1, \ldots, \pi_p)$ be an A-separable p-partition. Consider any pair $1 \leq r < s \leq p$. Then there exist a vector $C^{rs} \in \mathbb{R}^d$ and a constant γ^{rs} such that

$$(C^{rs})^T A^u > \gamma^{rs} > (C^{rs})^T A^v \text{ for each } u \in \pi_r \text{ and } v \in \pi_s.$$

Let $\pi^{r,s} \equiv (\pi_r^{r,s}, \pi_s^{r,s})$ be the A-separable 2-partition with $\pi_r^{r,s} \equiv \{i \in \mathcal{N} : (C^{rs})^T A^i > \gamma^{rs}\}$ and $\pi_s^{r,s} \equiv \{i \in \mathcal{N} : (C^{rs})^T A^i < \gamma^{rs}\}$. Now, let $\pi' = (\pi_1', \ldots, \pi_p')$ be the p-tuple associated with the list of $\binom{p}{2}$ separable 2-partitions whose elements are the $\pi^{r,s}$'s. Then the sets of π' are pairwise disjoint and we have

$$\pi_j \subseteq \left(\cap_{t=j+1}^{p} \pi_j^{j,t}\right) \bigcap \left(\cap_{t=1}^{j-1} \pi_j^{t,j}\right) = \pi_j' \text{ for } j = 1, \ldots, p.$$

Since $\mathcal{N} = \cup_{i=1}^{p} \pi_i \subseteq \cup_{i=1}^{p} \pi_i' \subseteq \mathcal{N}$, it follows that $\pi = \pi'$, that is, π is the p-partition associated with the constructed list of $\binom{p}{2}$ A-separable 2-partitions.

Finally, the asserted bound on the number of A-separable p-partitions when A is generic follows from the first part of the theorem and Lemma 2.2.3. □

The bound in Theorem 2.2.4 is not necessarily tight as different lists of separable 2-partitions may produce the same partition.

When $d = 1$, A is generic if and only if the elements in its only row are distinct. In this case, we get a tighter bound on the number of separable partitions than the one available from Theorem 2.2.4.

Lemma 2.2.5. *Suppose $A \in \mathbb{R}^{1 \times n}$ and its columns are distinct. Then the number of separable unlabeled p-partitions with empty parts allowed is $\sum_{j=1}^{p} \binom{n-1}{j-1} = O(n^{p-1})$, and if empty parts are prohibited the bound can be tightened to $\binom{n-1}{p-1} = O(n^{p-1})$. Corresponding bounds on the number of labeled partitions are $p! \sum_{j=1}^{p} \binom{n-1}{j-1}/(p-j)!$ and $p!\binom{n-1}{p-1}$, respectively*

Proof. The number of unlabeled separable p-partitions, with empty parts prohibited, is given in Theorem 6.2.1 of Vol. I. Specifically, the number is

$\binom{n-1}{p-1} = O(n^{p-1})$. When considering labeled partitions the above number is multiplied by $p!$, that is, the number of separable partitions equals $p!\binom{n-1}{p-1} = O(n^{p-1})$.

When allowing empty parts, we count the number of unlabeled separable p-partitions by conditioning on the number of empty parts; the j-th term of the sum represents the partitions of n distinct indices to j nonempty, unlabeled parts. For the labeled case, we multiply by $p!$, except that in the j-th term, corresponding to having $p-j$ empty parts, the term should be divided by $(p-j)!$, since interchanging empty parts among labeled parts does not lead to a different partition. □

We next consider the case where $d = 2$. We derive a bound on the number of separable p-partitions that is tighter than the one in Theorem 2.2.4.

Suppose $A \in \mathbb{R}^{d \times n}$ and π is a separable p-partition of A. A set S of hyperplanes is called a critical set if every two parts of π can be separated by at least one hyperplane in S. Alon and Onn [3] proved

Lemma 2.2.6. *Suppose $A \in \mathbb{R}^{d \times n}$ is generic and for every separable p-partition of A, there exists a critical set $\mathcal{H}$ with cardinality bounded by k. Then the number of separable p-partitions of A is bounded by $O(n^{dk})$.*

Proof. Given $A \in \mathbb{R}^{d \times n}$, consider any p-partition π of $A \in \mathbb{R}^{d \times n}$. By assumption, there exists a critical set $\mathcal{H}$ of hyperplanes (which we may assume to contain no A^i by a perturbation argument) that separate each pair of parts of π. Given $\mathcal{H}$ with the information of the location of each π_j (not knowing which elements are in π_j) with respect to each $H \in \mathcal{H}$, i.e., on one side of H, on the other side, or on both sides, we can easily reconstruct the whole partition $\pi = (\pi_1, \pi_2, \cdots, \pi_p)$ since each π_i is simply the intersection of A and all the corresponding half spaces supported by hyperplanes $H \in \mathcal{H}$ which contain it.

By Lemma 2.2.3, there are $O(n^d)$ ways of choosing a separable 2-partition, or equivalently, choosing a separating hyperplane (which we may assume to contain no A^i by a perturbation argument). Hence there are $O(n^{dk})$ ways of choosing the k hyperplanes in $\mathcal{H}$. For each $\mathcal{H}$, each $H \in \mathcal{H}$ and each part π_j, there are 3 cases of the location of π_j with respect to H, i.e., on either side of H or on both. Thus there are 3^{pk} such choices, a constant when p and k are fixed. The bound $O(n^{dk})$ follows immediately from the above discussion. □

Alon and Onn also used Lemma 2.2.6 to prove the following result.

Theorem 2.2.7. *Suppose $A \in \mathbb{R}^{2\times n}$. Then the number of A-separable p-partitions is $O(n^{6p-12})$ for $p > 2$ and $O(n^2)$ for $p = 2$.*

Proof. Assume that $\pi = (\pi_1, \cdots, \pi_p)$ is a separable p-partition of A. The convex hulls of its parts are then disjoint convex polygons that lie in some bounded set, say a big square. We grow the polygons in the square to maximize their area such that they have overlapping boundaries but disjoint interiors. This can be conducted by moving the sides of the polygons one by one. The expansion process is done by moving, arbitrarily, a side of a polygon in parallel and away from the polygon's interior. The side stops moving further only when it touches another polygon's vertex or it reaches the boundary of the square or when it shrinks to a point and vanishes. Notice that when a vertex of the moving side touches a side of another polygon, the moving side moves continuously with its touching vertex moving along the touched side, creating a new side overlapping the touched side. Eventually, the moving side shrinks to become a vertex of its polygon, either located on the touched line or its extension. This is the case that a moving side vanishes sometimes. When all moves end, each of the enlarged polygons contains only elements of one part π_i. Let $\mathcal{L}$ be the set of lines that are generated by the sides (excluding those in the boundaries of the square) of the enlarged polygons. Note that if two sides overlap in some segment then they generate the same line. Clearly, $\mathcal{L}$ is a critical set for the separable p-partition π of A.

Define a graph G by taking the set of p enlarged polygons as vertices and two such polygons P_x and P_y have an edge (P_x, P_y) if their boundaries touch (except when they touch only at a point which is a vertex of both polygons). We will say that the edge (P_x, P_y) crosses the side x if it either touches a vertex of P_y, or x and y overlap (then (P_x, P_y) also crosses y). Edelsbrunner, Robison and Shen [32] proved that G is planar and hence has at most $3p-6$ edges. According to the expansion process, each side touches a vertex of another polygon or overlaps with another side in some segment, or shrinks to a point and vanishes. Hence every side is crossed by an edge. Note that two convex polygons, which do not touch on two vertices and none strictly penetrates the other, cannot touch on two disjoint pairs of sides. Hence each side is crossed by a unique edge. On the other hand, the only time an edge (P_x, P_y) can cross two sides (x and y) is when x and y overlap, but then the lines generated by x and y is the same. Thus each such line crosses a unique side. Consequently, there is a onto mapping from the lines generated by the sides to the edges of G. Therefore, $|\mathcal{L}| \leq 3p-6$. Theorem 2.2.7 follows directly from Lemma 2.2.6. □

Alon and Onn [3] examined lower bounds on the maximal number of separable partitions. They showed a tight bound $\Theta(n^{6p-12})$ for $d=2$ and $p \geq 3$. They also proved a lower bound $\Omega(n^{d\binom{p}{2}})$ for $d \geq 3$ or $p \geq 3$, which, when combined with the upper bound (of the same complexity) in Theorem 2.2.4, yields the tight bound $\Theta(n^{d\binom{p}{2}})$.

Enumerating all Separable p-Partitions when A is Generic

Assume that d is fixed and $A \in \mathbb{R}^{d\times n}$ is generic. If $p=1$ there is only one partition that is separable. Also, an algorithm for enumerating the separable 2-partitions was provided following Lemma 2.2.3; in particular, there are $O(n^d)$ such partitions and the required effort of the algorithm is $O(n^{d+1})$ arithmetic operations. It next follows from Theorem 2.2.4 that the set of separable p-partitions is precisely the set of partitions associated with lists of $\binom{p}{2}$ separable 2-partitions of $\mathcal{N}$ (indexed by $\{(r,s) : 1 \leq r < s \leq p\}$). Specifically, to construct this set, produce all such lists of which there are $(O(n^d))^{\binom{p}{2}} = O\left(n^{d\binom{p}{2}}\right)$. For each list, one has to form the associated p-tuple (using (2.2.5)) and then test if the constructed tuple is a partition (if the tuple $\pi = (\pi_1, \ldots, \pi_p)$ is generated, its sets are assured to be disjoint and all one has to test is if $\cup_{i=1}^{p} \pi_i = \mathcal{N}$). For a tuple π that is found to be a partition, computation of A_π requires at most nd arithmetic operations. As there are $O(n^{d\binom{p}{2}})$ lists, all this work can be easily done using $O(n^{dp^2})$ arithmetic operations which subsumes the work for constructing the set of separable 2-partitions. Recall that the construction of the set of separable 2-partitions can be executed in parallel. We further observe that one can process in parallel lists of $\binom{p}{2}$ separable 2-partitions. The case $d=1$ refers to Chapter 6 of Vol. I. □

The complexity analysis of the above algorithm does not count operations needed to merge lists by intersections and unions.

The proof of the improved bound of Theorem 2.2.7 for $d=2$ suggests an enumeration method with complexity bound of $O(n^{6p-12})$. Although the method can be implemented in parallel, it requires the listing of all planar graphs with p vertices.

Next we discuss the enumeration of almost-separable partitions where A is not generic. Let $A \in \mathbb{R}^{d\times n}$. For each $\epsilon > 0$ define the *ϵ-perturbation* $A(\epsilon) \in \mathbb{R}^{d\times n}$ *of* A as follows: for $i = 1, \ldots, n$ let the i-th column of $A(\epsilon)$

be $A(\epsilon)^i \equiv A^i + \epsilon M_d^i$ where $M_d^i \equiv (i, i^2, \ldots, i^d)^T$ (M_d^i is called *the image of i on the moment curve in* $\mathbb{R}^d$).

For distinct indices $i_1, \ldots, i_{d+1} \in \mathcal{N}$, the determinant

$$D_{(i_1,\ldots,i_{d+1})}(\epsilon) \equiv \det[\widetilde{A}(\epsilon)^{i_1}, \ldots, \widetilde{A}(\epsilon)^{i_{d+1}}]$$

is a polynomial of degree d in ϵ, having a representation $\sum_{j=0}^{d} D_j \epsilon^j$ with D_d being the Van der Monde determinant $\det[\widetilde{M_d^{i_1}}, \ldots, \widetilde{M_d^{i_{d+1}}}]$ which is known to be nonzero. We conclude that for all sufficiently small $\epsilon > 0$, $\text{sign}_{A(\epsilon)}(i_0, \ldots, i_d) = \text{sign}[D_{(i_0,i_1,\ldots,i_d)}(\epsilon)]$ equals the sign of the first nonzero coefficient among $D_0, \ldots, D_d$; so, it is either -1 or 1 and is independent of ϵ. Define the *generic sign* of A at $(i_1, \ldots, i_{d+1})$, denoted by $\chi_A(i_1, \ldots, i_{d+1})$, as the common value of $\text{sign}_{A(\epsilon)}(i_1, \ldots, i_{d+1})$ for all sufficiently small positive ϵ.

Computing Generic Signs

Consider $A \in \mathbb{R}^{d\times n}$ and distinct indices $1 \leq i_1 < \cdots < i_{d+1} \leq n$ (which assures that $n > d+1$). We shall compute the generic sign $\chi_A(i_1, \ldots, i_{d+1})$ as follows. Evaluate the polynomial

$$D(\epsilon) \equiv \det[\widetilde{A}(\epsilon)^{i_1}, \ldots, \widetilde{A}(\epsilon)^{i_{d+1}}] = \sum_{j=0}^{d} D_j \epsilon^j$$

at $\epsilon = 0, 1, \ldots, d$. Each evaluation involves the computation of the determinant of a matrix of order $d+1$ and can be done, say by Gaussian elimination, using $O(d^3)$ arithmetic operations. Then, $D_0, D_1, \ldots, D_d$ can be determined by solving the following linear system of equations

$$\sum_{j=0}^{d} \epsilon^j D_j = D(\epsilon), \quad \epsilon = 0, \ldots, d.$$

This can be done by inverting the nonsingular Van der Monde matrix of coefficients of this system, again by Gaussian elimination. The total effort for executing this conversion is then $O(d^4)$. The generic sign $\chi_A(i_1, \ldots, i_{d+1})$ is then the sign of the first nonzero D_i. Finally, the extension of the results to $\epsilon = 0$ when A is generic is immediate from the observation that if a polynomial is nonzero at 0, its sign for all sufficiently small positive ϵ is the same as its sign at 0. □

Lemma 2.2.8. *Let $A \in \mathbb{R}^{d\times n}$. For all sufficiently small $\epsilon > 0$: (i) $A(\epsilon)$ is generic, (ii) for every d-set $I \subseteq \mathcal{N}$, the sets $I^-_{A(\epsilon)}$ and $I^+_{A(\epsilon)}$ are independent of ϵ, and (iii) the set of $A(\epsilon)$-separable p-partitions is invariant to ϵ. Further, if A is generic, "$\epsilon > 0$" can be replaced by "$\epsilon \geq 0$".*

Proof. Assume first that $n > d+1$. From the arguments preceding the current lemma, for all sufficiently small $\epsilon > 0$, $\text{sign}_{A(\epsilon)}(i_1, \ldots, i_{d+1})$ equals the nonzero generic sign $\chi_A(i_1, \ldots, i_{d+1})$ for all $1 \leq i_1 < \cdots < i_{d+1} \leq n$. It follows that for all sufficiently small ϵ, the matrix $A(\epsilon)$ is generic and for every d-set I, the sets $I^-_{A(\epsilon)}$ and $I^+_{A(\epsilon)}$ are independent of ϵ. By Lemma 2.2.3, the set of $A(\epsilon)$-separable 2-partitions is the set of all pairs of 2-partitions of $\mathcal{N}$ associated with $A(\epsilon)$, d-sets $I \subseteq \mathcal{N}$ and 2-partitions (J^-, J^+) of I. But each such pair depends only on $I^-_{A(\epsilon)}, I^+_{A(\epsilon)}$, J^- and J^+, hence is the same for all sufficiently small $\epsilon > 0$. We conclude that for sufficiently small positive ϵ, the set of $A(\epsilon)$-separable 2-partitions is independent of ϵ. By Theorem 2.2.4, the set of $A(\epsilon)$-separable p-partitions is entirely determined by the set of $A(\epsilon)$-separable 2-partitions; thus, it follows that the set of $A(\epsilon)$-separable partitions is independent of such ϵ.

Next assume that $n \leq d+1$. In this case we can augment A with $d+1-n$ zero vectors to obtain a matrix $A' \in \mathbb{R}^{d \times (d+1)}$. The arguments of the above paragraph show that for sufficiently small positive ϵ, $A'(\epsilon)$ is generic; it follows that for such ϵ, the submatrix $A(\epsilon)$ of $A'(\epsilon)$ is also generic and, by Lemma 2.2.2, the set of $A(\epsilon)$-separable 2-partitions is the set of all 2-partitions. In particular, we have that the set of $A(\epsilon)$-separable 2-partitions is independent of ϵ for all sufficiently small positive ϵ. As in the above paragraph, by Theorem 2.2.4, for such ϵ the set of $A(\epsilon)$-separable partitions is independent of such ϵ. □

Let $A \in \mathbb{R}^{d \times n}$. A p-partition of $\mathcal{N}$ is *A-limit-separable* if it is $A(\epsilon)$-separable for all sufficiently small $\epsilon > 0$. Denote by $\Pi^p_{A-lim-sep}$ the set of A-limit-separable p-partitions. Lemma 2.2.8 shows that for all sufficiently small $\epsilon > 0$, the set of $A(\epsilon)$-separable p-partitions is invariant of ϵ and equals $\Pi^p_{A-lim-sep}$. The relation between the new property of partitions and separability is clarified in the next lemma.

Lemma 2.2.9. *Let $A \in \mathbb{R}^{d \times n}$. Every A-separable partition is A-limit-separable; further, if A is generic, then the set of A-separable and A-limit-separable partitions coincide.*

Proof. The first conclusion of the lemma is immediate from the fact that the strict inequalities are preserved under small perturbations. The second conclusion is immediate from the extended version of Lemma 2.2.8 which applies to the case where A is generic. □

The next example demonstrates that limit-separable partitions need not be (almost-) separable.

Example 2.2.1. Let $d = 2, n = 3, n_1 = 2, n_2 = 1$ and $A = \binom{1\ 2\ 3}{0\ 0\ 0} \in \mathbb{R}^{2\times 3}$. The partition $\pi = (\{1,3\},\{2\})$ is not (almost-) separable for a vector $C = (C_1, C_2)$ satisfies $C^T A^1 > C^T A^2$ if and only if $C_1 < 0$ and it satisfies $C^T A^3 > C^T A^2$ if and only if $C_1 > 0$. To see that π is A-limit-separable, consider for each $\epsilon > 0$ the vector $C(\epsilon) \equiv \binom{4\epsilon}{-1-\epsilon}$; as

$$C(\epsilon)^T A(\epsilon) = (4\epsilon, -1-\epsilon) \begin{bmatrix} 1+\epsilon & 2+2\epsilon & 3+3\epsilon \\ \epsilon & 4\epsilon & 9\epsilon \end{bmatrix} = [3\epsilon+3\epsilon^2, 4\epsilon+4\epsilon^2, 3\epsilon+3\epsilon^2]$$

we have that $C(\epsilon)^T A(\epsilon)^1 = C(\epsilon)^T A(\epsilon)^3 = 3\epsilon+3\epsilon^2 < 4\epsilon+4\epsilon^2 = C(\epsilon)^T A(\epsilon)^2$, assuring that π is $A(\epsilon)$-separable. So, π is A-limit-separable. □

The next example demonstrates that almost-separable partitions need not be limit-separable.

Example 2.2.2. Let $d = 2, n = 4, n_1 = n_2 = 2$ and $A = \binom{0\ 0\ 0\ 0}{0\ 0\ 0\ 0} \in \mathbb{R}^{2\times 4}$. Every partition for this problem is almost-separable, in particular, so is $\pi = (\{1,3\},\{2,4\})$. Next, for $\epsilon > 0$ and $i = 1,2,3,4$, $A(\epsilon)^i = \epsilon\binom{i}{i^2}$ and a simple calculation shows that $\epsilon\binom{2.5}{7} \in [\text{conv}\ \{A(\epsilon)^1, A(\epsilon)^3\}] \cap [\text{conv}\ \{A(\epsilon)^2, A(\epsilon)^4\}]$, assuring that π is not $A(\epsilon)$-separable. So, π is not limit-separable. □

Two sets $\Omega^1, \Omega^2 \subseteq \mathbb{R}^d$ are *weakly separable* if they satisfy the separability condition (2.1.1) with weak inequality replacing the strict inequality. A p-partition π is *weakly separable*, if for each pair of distinct indices $r, s \in \{1,\ldots,p\}$, conv A^{π_r} and conv A^{π_s} are weakly separable. A simple continuity argument shows that a limit-separable partition must be weakly separable. But, the next example shows that a weakly separable partition having disjoint parts need not be limit-separable.

Example 2.2.3. Let $d = 2, n = 4, n_1 = n_2 = 2$ and $A = \binom{1\ 2\ 3\ 4}{0\ 0\ 0\ 0} \in \mathbb{R}^{2\times 4}$. The partition $\pi = (\{1,3\},\{2,4\})$ is clearly weakly separable (with $C = \binom{0}{1}$ as a weakly separating vector between A^{π_1} and A^{π_2}). Next, for $\epsilon > 0$ and $i = 1,2,3,4$, $A(\epsilon)^i = \binom{i+i\epsilon}{i^2\epsilon}$ and a simple calculation shows that any half-space that contains $A(\epsilon)^1$ and $A(\epsilon)^3$ excludes $A(\epsilon)^2$ if and only if it includes $A(\epsilon)^4$, assuring that π is not $A(\epsilon)$-separable. So, π is not A-limit-separable. □

The forthcoming Theorem 2.2.15 demonstrates a further relation between the sets of almost-separable and limit-separable 2-partitions.

The next result provides a polynomial bound on the number of limit-separable partitions. It is followed by a description of a method for their enumeration.

Theorem 2.2.10. *Let $A \in \mathbb{R}^{d \times n}$. Then*

$$\left|\Pi^p_{A-lim-sep}\right| \leq \left[2^{d+1}\binom{n}{d}\right]^{\binom{p}{2}} = O(n^{d\binom{p}{2}});$$

when $d = 1$, the bound can be tightened to $p!\sum_{j=1}^{p}\binom{n-1}{j-1}/(p-j)! = O(n^{p-1})$ and when $d = 2$ and $p > 2$, the bound can be tightened to $O(n^{6p-12})$.

Proof. For sufficiently small positive ϵ, the set $\Pi^p_{A-lim-sep}$ of A-limit-separable p-partitions equals the set of $A(\epsilon)$-separable p-partitions; further, by Lemma 2.2.8, ϵ can be selected so that $A(\epsilon)$ is generic. So, Theorem 2.2.4 shows $|\Pi^p_{A-lim-sep}| \leq [2^{d+1}\binom{n}{d}]^{\binom{p}{2}} = O(n^{d\binom{p}{2}})$. Next, Lemma 2.2.5 and Theorem 2.2.7 provide the tighter bounds when $d = 1$ and $d = 2$. □

Corollary 2.2.11. *Let $A \in \mathbb{R}^{d \times n}$. Then the number of separable p-partitions is bounded by $[2^{d+1}\binom{n}{d}]^{\binom{p}{2}} = O(n^{d\binom{p}{2}})$; when $d = 1$ the bound can be tightened to $p!\sum_{j=1}^{p}\binom{n-1}{j-1}/(p-j)! = O(n^{p-1})$ and when $d = 2$ and $p > 2$ the bound can be tightened to $O(n^{6p-12})$.*

Proof. The bounds follow from Theorem 2.2.10 and Lemma 2.2.9. □

Enumerating all A-Limit-Separable Partitions

Consider $A \in \mathbb{R}^{d \times n}$. If $n \leq d+1$, the set of A-limit-separable p-partitions is the set of all partitions, of which there are $p^n \leq p^d$; these can be generated using a fixed number of arithmetic operations. Also, if $n > d + 1$ and $p = 1$, then $\Pi^p_{A-lim-sep} \equiv \{(\mathcal{N})\}$ consists of the single p-partition $(\mathcal{N})$. Henceforth assume that d is fixed, $n > d + 1$ and $p \geq 2$. By Lemma 2.2.8, for sufficiently small positive ϵ, for each d-set $I \subseteq \mathcal{N}$, $I^-_{A(\epsilon)}$ and $I^+_{A(\epsilon)}$ are independent of ϵ. For a d-set $I = \{i_1, \ldots, i_d\} \subseteq \mathcal{N}$ with $i_1 < \cdots < i_d$, $I^-_{A(\epsilon)}$ and $I^+_{A(\epsilon)}$ for sufficiently small positive ϵ are available from the generic signs of A at $(i_1, \ldots, i_d, i)$ for each $i \in \mathcal{N} \setminus I$. A method described earlier in this section shows how each generic sign can be computed with $O(d^4)$ arithmetic operations; so, all $\binom{n}{d+1}$ generic signs can be computed with $O\left(\binom{n}{d+1}d^4\right) = O(n^{d+1})$ arithmetic operations (permutations that put $\widetilde{A}^i$ into the right location may be applied).

Next, from Lemmas 2.2.8 and 2.2.3, $\Pi^p_{A-lim-sep}$ is the common set of $A(\epsilon)$-separable p-partitions for sufficiently small positive ϵ; for such ϵ the set of $A(\epsilon)$-separable 2- and p-partitions, $p > 2$, can be determined from the generic signs by the earlier method for enumerating the separable partitions corresponding to generic matrices. As the method will only use the generic signs, the complexity bounds will be $O(n^{dp^2})$, the same as for a fixed generic matrix; further, this bound subsumes the work for computing the generic signs. □

Enumerating all A-Separable Partitions

Consider $A \in \mathbb{R}^{d \times n}$. Use the above algorithm to compute all A-limit-separable partitions and their associated matrices. By Lemma 2.2.9, the list contains all separable partitions. Now, an algorithm for testing separability of a partition was described in Section 2.1 (requiring the solution of a block diagonal system with $(d+1)\binom{p}{2}$ variables and $(p-1)n$ linear inequalities). The algorithm can be used to test which of the (limit-separable) partitions is separable. An alternative approach for enumerating all separable p-partitions is to generate them from $\binom{p}{2}$ lists of separable 2-partitions (relying on Lemma 2.2.4); the set of separable 2-partitions can be generated by testing the limit-separable 2-partitions for separability. □

The next result shows that every vertex of a constrained-shape partition polytope has a representation as a vector associated with a limit-separable partition.

Theorem 2.2.12. *Let $A \in \mathbb{R}^{d \times n}$, Γ be a nonempty set of positive integer p-vectors with coordinate-sum n and v a vertex of P^Γ_A. Then $v = A_\pi$ for some A-limit-separable p-partition π with shape in Γ.*

Proof. Let $v \in \mathbb{R}^{d \times p}$ be a vertex of P^Γ_A. It follows that v is the unique maximizer over P^Γ_A of some linear function, say, one that is determined by the matrix $C \in \mathbb{R}^{d \times p}$; Let $\Pi^* \equiv \{\pi \in \Pi^\Gamma : A_\pi = v\}$; in particular, (by Proposition 3.1.1(i) of Vol. I) Π^* is nonempty. As $\langle C, A_\pi \rangle > \langle C, A_\sigma \rangle$ for all $\pi \in \Pi^*$ and $\sigma \in \Pi^\Gamma \setminus \Pi^*$, we have that for such π and σ and all sufficiently small $\epsilon > 0$, $\langle C, A(\epsilon)_\pi \rangle > \langle C, A(\epsilon)_\sigma \rangle$ for all $\pi \in \Pi^*$ and $\sigma \in \Pi \setminus \Pi^*$. Further, Lemma 2.2.8 guarantees that, in addition, for sufficiently small positive ϵ a p-partition is $A(\epsilon)$-separable if and only if it is A-limit-separable. Consider such a (sufficiently small positive) ϵ. The linear function $\langle C, \cdot \rangle$ is then maximized over the perturbed polytope $P^\Gamma_{A(\epsilon)}$ at a vertex of the form $A(\epsilon)_{\pi^*}$ for some $\pi^* \in \Pi^*$. By Theorem 2.2.1, π^* is $A(\epsilon)$-separable (since $A(\epsilon)$

is generic, it has distinct columns and therefore $A(\epsilon)$-almost-separability and $A(\epsilon)$-separability coincide). As π^* is $A(\epsilon)$-separable, it is also A-limit-separable. This proves that Π^* contains an A-limit-separable partition. □

With A and Γ as in Theorem 2.2.12, Corollary 2.1.6 implies that every partition π with shape in Γ and A_π a vertex of P_A^Γ is almost-separable (separable when the columns of A are distinct—but, when the columns of A are not distinct, Example 2.1.1 demonstrates that "separable" cannot replace "almost-separable"). Thus, the almost-separable partitions generate the vertices of P_A^Γ in a stronger way than was just established for the limit-separable partitions. The next example demonstrates that the stronger representation needs not hold under the set of limit-separable partitions.

Example 2.2.4. Let $d = 1, n = 3, \Gamma = \{(1,2)\}$ and $A = (0,0,1) \in \mathbb{R}^{1\times 3}$. Then $\pi = (\{2\},\{1,3\})$ is not A-limit-separable while $A_\pi = (0,1)$ is a vertex of $P_A^\Gamma = \{(x, 1-x) : 0 \leq x \leq 1\}$. Note that $\sigma = (\{1\},\{2,3\})$ is A-limit-separable and $A_\sigma = (0,1) = A_\pi$; so, the vertex $(0,1)$ has a representation of the form predicted by Theorem 2.2.12 (cf. Theorem 2.2.15). □

The inverse of the implication established in Theorem 2.2.12 needs not hold. Indeed, Example 2.1.2 demonstrates a separable partition π for which A_π is not a vertex of the corresponding single-shape partition polytope. Observing that the matrix A for that example is generic, the A-separable partition π of that example is A-limit-separable (Lemma 2.2.8).

As is the case for bounded-shape partition polytopes, the number of matrices in the set $\{A_\pi : |\pi| \in \Gamma\}$ is typically exponential in n, even for fixed d, p. Therefore, although the dimension of P_A^Γ is bounded by dp, this polytope can potentially have exponentially many vertices and facets as well. But, Theorems 2.2.10 and 2.2.12 yield the following polynomial bound on the number of vertices of constrained-shape partition polytopes.

Corollary 2.2.13. *For any $A \in \mathbb{R}^{d\times n}$ and $\Gamma \neq \emptyset$ of positive integer p-vectors with coordinate-sum n, the number of vertices of P_A^Γ is bounded by $[2^{d+1}\binom{n}{d}]^{\binom{p}{2}} = O(n^{d\binom{p}{2}})$; further, when $d = 1$ the bound can be tightened to $p!\sum_{j=1}^{p}\binom{n-1}{j-1}/(p-j)!$ and when $d = 2$ the bound can be tightened to $O(n^{6p-12})$.*

Proof. The bounds on the number of vertices of P_A^Γ are immediate from Theorems 2.2.10 and 2.2.12. □

Theorem 2.2.12 is next used to establish the optimality of limit-separable partitions.

Theorem 2.2.14. *Let Γ be a nonempty set of positive integer p-vectors with coordinate-sum n and let $f(\cdot)$ be an (edge-)quasi-convex function on the constrained-shape partition polytope P_A^Γ. Then there exists an optimal partition π which is A-limit-separable and has A_π as a vertex of P_A^Γ.*

Proof. As in the proof of Theorem 2.2.1, there is a vertex v of P_A^Γ that maximizes $f(\cdot)$ and each partition π with $A_\pi = v$ is optimal for the underlying constrained-shape partition problem; Theorem 2.2.12 assures that there exists such an A-limit-separable partition. □

Solving Constrained-Shape Partition Problems with $f(\cdot)$ (Edge-)Quasi-Convex by Enumerating Limit-Separable Partitions

Let $A \in \mathbb{R}^{d\times n}$, let Γ be a nonempty set of positive integer p-vectors with coordinate-sum n and let $f(\cdot)$ be an (edge-)quasi-convex function on the constrained-shape partition polytope P_A^Γ. Also assume that a membership oracle is given for Γ which, given a positive integer p-vector λ with coordinate-sum n, determines whether or not $\lambda \in \Gamma$, and that an evaluation oracle is given for the function $f(\cdot)$ which, given a partition π with shape in Γ, returns $F(A_\pi)$. To solve the partition problem, use the algorithm described earlier in this section to construct the set $\Pi^p_{A-lim-sep}$ of A-limit-separable p-partitions along with their associated matrices, using $O(n^{dp^2})$ arithmetic operations. Next, test shapes of the partitions in the list to obtain the subset $\Pi^\Gamma_{A-lim-sep} \equiv \Pi^\Gamma \cap \Pi^p_{A-lim-sep}$ of A-limit-separable partitions with shape in Γ by querying the Γ-oracle on each of the $|\Pi^p_{A-lim-sep}| = O(n^{d\binom{p}{2}})$ partitions in $\Pi^\Gamma_{A-lim-sep}$. Next query the f-oracle for the value $f(A_\pi)$ for each $\pi \in \Pi^\Gamma_{A-lim-sep}$ and pick the best. The number of operations involved and queries to the f-oracle is $O(n^{dp^2})$. Since $f(\cdot)$ is (edge-)quasi-convex on P_A^Γ, Theorem 2.2.14 assures that any partition π in $\Pi^\Gamma_{A-lim-sep}$ achieving $\max\{f(A_\pi) : \pi \in \Pi^\Gamma_{A-lim-sep}\}$ is an optimal solution to the underlying constrained-shape partition problem. □

We next use the enumeration of limit-separable partitions to produce all the vertices of given constrained-shape partition polytopes. We recall that the vertex-enumeration methods described in Section 1.2 are based on the availability of edge-directions, as such they are relevant only to bounded-shape partition polytopes. To describe an efficient version of the new method we need some further definitions. Recall the notation $\widetilde{x}$ for an m-vector x, denoting the $(m+1)$-vector obtained from x by augmenting it with a single coordinate that equals 1. For a set $W \subseteq R^m$, let $\widetilde{W} \equiv \{\widetilde{x} : x \in W\}$. The affine dimension of W is defined as $\dim [\text{lin } \widetilde{W}] - 1$, and an affine basis of W is a set of vectors $\Omega \subseteq W$ where $\widetilde{\Omega}$ is a basis of $\text{lin } \widetilde{W}$. The Carathéodory Theorem (e.g., Rockafellar [82]) assures that a point w is in the convex hull of W if and only if it is in the convex hull of vectors that form an affine basis of W.

Enumerating the Vertices of Constrained-Shape Partition Polytopes Using Limit-Separable Partitions

Let $A \in \mathbb{R}^{d\times n}$ and let Γ be a nonempty set of positive integer p-vectors with coordinate-sum n. Also assume that a membership oracle is given for Γ which, given a positive integer p-vector λ with coordinate-sum n, determines whether or not $\lambda \in \Gamma$. Using the method described within the above description of the algorithm for solving constrained-shape partition problems, enumerate all partitions in $\Pi^{\Gamma}_{A-lim-sep} \equiv \Pi^{p}_{A-lim-sep} \cap \Pi^{\Gamma}$ along with the associated matrices A_π, using $O(n^{dp^2})$ arithmetic operations and calls to the Γ-oracle. Let $U = \{A_\pi : \pi \in \Pi^{\Gamma}_{A-lim-sep}\}$. The set U is contained in P^{Γ}_{A} and Theorem 2.2.12 assures that it contains the set of vertices of P^{Γ}_{A}. So $u \in U$ is a vertex of P^{Γ}_{A} precisely when it is not a convex combination of the elements of $U \setminus \{u\}$. This could be tested using any linear programming algorithm to solve a system with $|U|$ nonnegative variables and $dp+1$ constraints.

The test of whether or not a point u in $U = \{A_\pi : \pi \in \Pi^{\Gamma}_{A-lim-sep}\}$ is a vertex of P^{Γ}_{A} can be executed more efficiently by breaking the system up to smaller systems of linear equations without the nonnegative constraints, systems that can be solved by Gauss elimination. Let a be the affine dimension of U. As the sum of the columns of matrices associated with partitions is constant (equaling $\sum_{i=1}^{n} A^i$), the affine dimension of P, hence that of U (namely a), is bounded from above by $d(p-1)$. Now, consider $u \in U$. If the affine dimension of $U \setminus \{u\}$ is less than a, then u is not in the convex hull of $U \setminus \{u\}$ and u is a vertex of U. So assume that the affine dimension of $U \setminus \{u\}$ equals a. For each $(a+1)$-subset $\{u_0, \ldots, u_a\}$ of $U \setminus \{u\}$, test if $\{\widetilde{u}_0, \ldots, \widetilde{u}_a\}$ are linearly independent and if they are found the unique solution of the linear system $u = \sum_{i=0}^{a} \mu_i u_i$ and $\sum_{i=0}^{a} \mu_i = 1$ (with variables $\mu_0, \ldots, \mu_a$); u is in the convex hull of $\{u_0, \ldots, u_a\}$ if and only if the corresponding unique solution $\mu_0, \ldots, \mu_a$ of the system is nonnegative. By the aforementioned conclusion of the Carathéodory Theorem, u is a vertex of P_A^Γ if and only if the solution of each such system has a negative coordinate. Computing the affine dimension a of U, testing if an $(a+1)$-subsets of U is an affine basis, and computing the unique μ_i's for such subsets can all be done by Gaussian elimination. Since we have to perform the entire procedure for each of the $|U| \leq |\Pi^\Gamma| = O(n^{d\binom{p}{2}})$ elements $u \in U$, and for each such u the number of affine bases of $U \setminus \{u\}$ is at most $\binom{|U|-1}{d(p-1)+1}$, the number of arithmetic operations involved is $O(|U|\binom{|U|-1}{d(p-1)+1}) = O(n^{d^2p^3})$ which absorbs the work for constructing U. $\square$

The second paragraph of the above algorithm has broader applicability. Given a polytope $P \subset \mathbb{R}^{d'}$ of affine dimension a with V as its set of vertices and a set $V \subseteq U \subseteq P$, it shows how one can test if a vector $u \in U$ is in V by solving at most $\binom{|U|-1}{a+1}$ linear equality systems of dimension $(d'+1) \times (a+1)$ (where each is guaranteed to have a unique solution). So, an efficient method for producing a set U that contains the vertices of a polytope P can be enhanced to produce the set of vertices of that polytope. We shall refer to this method in the forthcoming sections without repeating its details.

Recall from Section 1.2 that a method for enumerating vertices of a polytope can be used to enumerate faces. Also, characterization of partitions whose associated matrix is a vertex of a corresponding bounded shape partition polytope is provided in Theorems 1.3.1, 1.3.10 and 1.4.3. These can replace the test that is used in the above algorithm.

Lemma 2.2.9 shows that separable partitions are limit-separable and Example 2.2.2 demonstrates that this property does not extend to almost-separable partitions (namely, an almost-separable partition needs not be limit-separable). We shall next demonstrate a weaker relation between being almost-separable and being limit-separable.

Theorem 2.2.15. *Let $A \in \mathbb{R}^{d\times n}$ and suppose π is an almost-separable p-partition. If either $p = 2$ or $d = 1$, then there exists an A-limit-separable partition σ which is equivalent to π; in particular, $A_\pi = A_\sigma$.*

Proof. We first consider the case where $p = 2$. If π is separable, Lemma 2.2.9 assures that it is limit-separable. So assume π is not separable. As π is almost-separable, this assumption assures that $A^{\pi_1} \cap A^{\pi_2}$ consists of a single vector, say $v \in \mathbb{R}^d$. Let $J \equiv \{i \in \mathcal{N} : A^i = v\}$. The almost-separability of π further assures that for some vector $C \in \mathbb{R}^d$,

$$C^T A^t > C^T v > C^T A^s \text{ for each } t \in \pi_1 \setminus J \text{ and } s \in \pi_2 \setminus J. \qquad (2.2.6)$$

Recall the M_d^i's were defined earlier in this section as the images on the moment curve in $\mathbb{R}^d$. As differences of the M_d^i's are nonzero, Lemma 3.1.9 of Vol. I with $\mathcal{L} = \{0\}$ implies that there is a vector $H \in \mathbb{R}^d$ satisfying

$$H^T[M_d^t - M_d^s] \neq 0 \text{ for each pair of distinct indices } t, s \in J. \qquad (2.2.7)$$

For sufficiently small positive δ, (2.2.6) is satisfied with $C+\delta H$ replacing C and (2.2.7) is satisfied with $C+\delta H$ replacing H; by replacing C with $C+\delta H$ for such a δ, we may and will assume that (2.2.6) is satisfied and that the values $\{C^T M_d^i : i \in J\}$ are distinct. It then follows that the indices in J can be enumerated as $i_1, i_2, \ldots, i_m$ so that $C^T M_d^{i_1} > C^T M_d^{i_2} > \cdots > C^T M_d^{i_m}$. For each $\epsilon > 0$ and $i \in J$, $C^T A(\epsilon)^i = C^T(A^i + \epsilon M_d^i) = C^T v + \epsilon C^T M_d^i$, implying that $C^T A(\epsilon)^{i_1} > C^T A(\epsilon)^{i_2} > \cdots > C^T A(\epsilon)^{i_m}$; using (2.2.6), it follows that for sufficiently small positive ϵ,

$$C^T A(\epsilon)^t > C^T A(\epsilon)^{i_1} > C^T A(\epsilon)^{i_2} > \cdots > C^T A(\epsilon)^m > C^T A(\epsilon)^s$$
$$\text{for each } t \in \pi_1 \setminus J \text{ and } s \in \pi_2 \setminus J.$$

Let $k \equiv |\pi_1 \cap J|$ and let $\sigma \equiv (\sigma_1, \sigma_2)$ be the 2-partition with $\sigma_1 \equiv (\pi_1 \setminus J) \cup \{i_1, \ldots, i_k\}$ and $\sigma_2 \equiv (\pi_2 \setminus J) \cup \{i_{k+1}, \ldots, i_m\}$. We then have that for $r = 1, 2$ and $x \in \mathbb{R}^d$, $|\{i \in \pi_r : A^i = x\}| = |\{i \in \sigma_r : A^i = x\}|$ (in fact, the corresponding sets are equal whenever $x \neq v$); so, π and σ are equivalent. Further, for sufficiently small positive ϵ,

$$C^T A(\epsilon)^t > C^T A(\epsilon)^s \text{ for each } t \in \sigma_1 \text{ and } s \in \sigma_2,$$

implying that σ is $A(\epsilon)$-separable; so, it is limit-separable.

The above arguments apply for any pair of parts, say π_r and π_s with A^{π_r} and A^{π_s} either separable or containing a single vector that is not present in any other A^{π_t}. Next observe that when $d = 1$, a split of the indices corresponding to a vector v into $q > 2$ parts can occur only if all indices of $q - 2$ of the parts correspond to that vector. It follows that the argument used for the case $p = 2$ apply, except that $i_1, i_2, \ldots, i_m$ have to be split into q parts. Finally, for equivalent partitions π and σ, $A_\pi = A_\sigma$ follows immediately. □

We next use Corollary 1.4.4 to obtain a weaker version of Theorem 2.2.15 for cases where $p > 2$ and $d > 1$.

Theorem 2.2.16. *Let $A \in \mathbb{R}^{d \times n}$, Γ be a nonempty set of positive integer p-vectors with coordinate-sum n and π an almost-separable p-partition with shape in Γ and A_π as a vertex of P_A^Γ. Then there exists an A-limit-separable partition σ which is equivalent to π; in particular, $A_\pi = A_\sigma$.*

Proof. Let $v = A_\pi$ and let $(n_1, \ldots, n_p) \in \Gamma$ be the shape of π. Evidently, A_π is a vertex of $P_A^{n_1,\ldots,n_p}$ (the linear function that is maximized by A_π over P_A^Γ is also maximized by A_π over the smaller polytope $P_A^{n_1,\ldots,n_p}$). By Theorem 2.2.12, there is a limit-separable partition σ with $A_\sigma = v = A_\pi$. It then follows from the concluding part of Corollary 1.4.4 that σ and π are equivalent. Finally, the equality $A_\pi = A_\sigma$ is immediate from the equivalence of π and σ. □

Under the assumptions of Theorem 2.2.16, the existence of a limit-separable partition σ with $A_\pi = A_\sigma$ is immediate from Theorem 2.2.12.

The "in particular" conclusion of Theorem 2.2.15 with $p = 2$ can be proved directly. Specifically, consider the single-shape partition polytope P corresponding to the set of partitions having the same shape as π. Now, Theorem 2.1.7 assures that A_π is a vertex of P and Theorem 2.2.12 assures that there is an A-limit-separable partition σ with $A_\sigma = A_\pi$.

To demonstrate the construction of Theorem 2.2.15 with $p = 2$, we reexamine Example 2.2.2.

Example 2.2.2 (continued) Reconsider the data of Example 2.2.2, namely, let $d = 2, n = 4, n_1 = n_2 = 2$ and $A = \begin{pmatrix} 0&0&0&0 \\ 0&0&0&0 \end{pmatrix} \in \mathbb{R}^{2\times 4}$. We have already seen that the partition $\pi = (\{1,3\},\{2,4\})$ is almost-separable (like every other partition) but it is not limit-separable. Next, consider the partition $\sigma = (\{1,2\},\{3,4\})$. We note that σ is obtained from π by switching the indices 2 and 3 and these indices correspond to (the single) vector in $\{A^i : i \in \pi_1\} \cap \{A^i : i \in \pi_2\}$; so, σ is related to π via the requirements of Theorem 2.2.15. Next, the vector $C = \begin{pmatrix} 3 \\ -1 \end{pmatrix}$ satisfies for each $\epsilon > 0$: $C^T A(\epsilon)^1 = C^T A(\epsilon)^2 = 2\epsilon$, $C^T A(\epsilon)^3 = 0 < 2\epsilon$ and $C^T A(\epsilon)^4 = -4\epsilon < 2\epsilon$, assuring that σ is $A(\epsilon)$-separable; so σ is A-limit-separable. Also, it is obvious that $A_\sigma = A_\pi = \begin{pmatrix} 0&0 \\ 0&0 \end{pmatrix}$. □

Corollary 1.4.4 shows that all partitions representing a vertex of a single-shape partition polytope are equivalent; we then use this result to extend Theorem 2.2.15 to $p > 2$ and $d > 1$ for almost-separable partitions π for which A_π is a vertex of a constrained-shape partition polytope. The generalization of Theorem 2.2.15 to $p > 2$ and $d > 1$ beyond Corollary 1.4.4 is posed as an open question, that is,

> "determine other conditions under which an almost-separable partition π must be equivalent to a limit-separable partition σ, i.e., $A_\pi = A_\sigma$."

To understand the difficulty in such extensions, observe that Theorem 2.2.15 implies that if an almost-separable p-partition π splits a set of indices

i that share the same vector A^i between two parts, then there is a split of the perturbed vectors which preserves cardinality. But, if π splits such a set of indices among $q > 2$ parts, the pairwise splits of the indices need not be consistent (and $i \in \pi_1$ may be assigned to π_2 when splitting $\pi_1 \cup \pi_2$ and to π_3 when splitting $\pi_1 \cup \pi_3$). The simple proof of the implied conclusion of Theorem 2.2.15 when $p = 2$ (the existence of an A-limit-separable partition with $A_\pi = A_\sigma$) does not generalize to $p > 2$ (as Theorem 2.1.7 does not generalize to $p > 2$).

Enumerating all Almost-Separable 2-Partitions

The enumeration consists of two steps: enumerating the set of separable partitions and enumerating the set of almost-separable but not separable partitions. A method for enumerating the first set is already given in this section under the title "Enumerating all A-separable Partitions". We now give a method for enumerating the second set.

Call $v \in A$ a multi-vector if the vector appears $\nu^A(v) > 1$ times in A. Then a 2-partition in the second set must have its separating hyperplane going through a multi-vector v with both parts containing some copies of v. Then we can enumerate all members in the second set by first enumerating all multi-vectors and for each multi-vector v, enumerate the set of separable partitions of $A \setminus \{v\}$. Finally, we enumerate all 2-partitions of the $\nu^A(v)$ copies of v into two nonempty parts. The combination of the above three enumerations yields all members of the second set.

We have shown that the time complexity of enumerating the first set before. To compute the time complexity of enumerating the second set, observe that there are at most n multi-vectors, and at most $O(n^{d\binom{p}{2}})$ separable partitions for each fixed multi-vector v lying in the separating hyperplane. Finally, there are $\nu^A(v)-1$ ways in dividing $\nu^A(v)$ copies into two nonempty parts.

Enumerating all Almost-Separable p-Partitions

This construction of almost-separable 2-partitions is next modified to generate all almost-separable p-partitions. Similar to the construction of separable p-partitions, we use lists of $\binom{p}{2}$ almost-separable 2-partitions of a given multiset A to generate its almost-separable p-partitions. But, given a list of $\binom{p}{2}$ almost-separable 2-partitions of a multiset A, the modified construction has to overcome a tricky problem of how to assign all copies of a vector x that appear in A to the p parts of the constructed partition (each of the 2-partitions in the list assigns all copies of x to its two parts). This problem does not appear in the constructions of separable partitions since all copies of x must go to the same part for separable partitions.

We next introduce a formal framework for multisets and Boolean operations on them (described in an informal language in the first paragraph of the Introduction). Given a multiset U in $\mathbb{R}^d$, the *multiplicity* of a vector $v \in \mathbb{R}^d$, denoted $\nu^U(v)$, is a nonnegative integer indicating the number of occurrences of v in U. We say that v *occurs in* U, written $v \in U$, if $\nu^U(v) \geq 1$; *finiteness* of U means that there are finitely many such vectors. If U and V are two finite multisets of $\mathbb{R}^d$ and $v \in \mathbb{R}^d$, then $U \cap V$ and $U \uplus V$ are the multisets with $\nu^{U \cap V}(v) = \min\{\nu^U(v), \nu^V(v)\}$ and $\nu^{U \uplus V}(v) = \nu^U(v) + \nu^V(v)$. Also, we write $U \subseteq V$ if $U \cap V = U$ (which is not equivalent to $U \uplus V = V$).

Consider a multiset A consisting of n vectors in $\mathbb{R}^d$ and integer $p \geq 2$. We shall consider $\binom{p}{2}$-tuples of almost-separable 2-partitions of A, indexed by $\{(r,s) : 1 \leq r < s \leq p\}$; we refer to such an indexed tuple as a *list of* $\binom{p}{2}$ *almost-separable 2-partitions*, denote it by $\pounds$ and denote its elements (which are almost-separable 2-partitions) by $\pi^{r,s} = (\pi_r^{r,s}, \pi_s^{r,s})$ where $1 \leq r < s \leq p$. With such a list, say $\pounds = \{\pi^{r,s} : 1 \leq r < s \leq p\}$, we associate a p-tuple $\pi = (\pi_1, \ldots, \pi_p)$ of multi-subsets of $\mathcal{N}$ by a two-step construction (which is motivated in the next paragraph). First, (using multiset-intersection) set

$$\pi_j' \equiv \left(\cap_{t=j+1}^{p} \pi_j^{j,t}\right) \bigcap \left(\cap_{t=1}^{j-1} \pi_j^{t,j}\right) \quad \text{for } j = 1, \ldots, p. \tag{2.2.8}$$

We next produce each multiset π_j from the multiset π_j' by (possibly) reducing some of the $\nu^{\pi_j'}(v)$'s. Formally, for each $v \in \{A^1, \ldots, A^n\}$, let $j^{\pi'}(v) \equiv \max\{j = 1, \ldots, p : x \in \pi_j'\}$. If $j \neq j^{\pi'}(v)$, we let $\nu^{\pi_j}(v) = \nu^{\pi_j'}(v)$. Alternatively, if $j = j^{\pi'}(v)$ we set

$$\nu^{\pi_j}(v) = \min\{\nu^{\pi_j'}(v), [\nu^A(v) - \sum_{t=1}^{j-1} \nu^{\pi_t'}(v)]_+\}$$

(where ζ_+ for $\zeta \in \mathbb{R}$ stands for $\max\{\zeta, 0\}$). For a vector $v \in \{A^1, \ldots, A^n\}$, the number $\sum_{j=1}^{p} \nu^{\pi_j}(v)$ may be larger or smaller than $\nu^A(v)$ (in fact, the sum may be 0, indicating that v does not appear in either of the multisets $\pi_1, \ldots, \pi_p$). If it happens that $\sum_{j=1}^{p} \nu^{\pi_j}(v) = \nu^A(v)$ for each vector $v \in A$, then $\pi = (\pi_1, \ldots, \pi_p)$ is a p-partition of A which will be called *the p-partition associated with* $\pounds$.

We next motivate the construction of the above paragraph. The goal is to facilitate the generation of each almost-separable p-partition (formally accomplished in the proof of Theorem 2.2.18 below). Given an almost-separable partition σ, each pair of parts constitutes an almost-separable 2-partition of a sub-multiset of A in which vectors may have fewer occurrences than in A. The idea is then to extend the two parts of each pair to a 2-partition of A so that for each j the intersection of the $p-1$ extended parts containing σ_j coincides with σ_j. A key problem is the handling of situations where multiple copies of a vector v are shared by more than 2 parts of σ. For example, assume that 12 copies of v are present and σ_1, σ_2 and σ_3 contain, respectively, 2, 3 and 7 copies. The above construction of the p-partition of A sends the access copies of v to the part with the larger index. So, we get 2-partitions $\pi^{1,2}, \pi^{1,3}, \pi^{2,3}$ of A such that $\pi_1^{1,2}$ has 2 copies of v, $\pi_2^{1,2}$ has 10 copies, $\pi_1^{1,3}$ has 2 copies, $\pi_3^{1,3}$ has 10 copies, $\pi_2^{2,3}$ has 3 copies and $\pi_3^{2,3}$ has 9 copies. Next, the intersection formula (2.2.8) produces π_1', π_2', π_3' having, respectively, 2, 3, 9 copies of v. In general, we get $\nu^{\pi_j'}(v) = \nu^{\sigma_j}(v)$ for $j \neq j^* \equiv j^{\pi'}(v)$ and $\nu^{\pi_{j^*}'}(v) \geq \nu^{\sigma_{j^*}}(v)$. In our example, $j^* = 3$ and the number of occurrences of v in the third part is reduced in the second step from 9 to $\nu^A(v) - [\nu^{\pi_1'}(v) + \nu^{\pi_2'}(v)] = 12 - (2+3) = 7 = \nu^{\sigma_3}(v)$. It is shown below in Theorem 2.2.18 that the above construction produces all almost-separable p-partitions, using almost-separable 2-partitions as building blocks.

We shall need the following Lemma for our development.

Lemma 2.2.17. *Let $v^1, \ldots, v^q$ be vectors in $\mathbb{R}^d$ and $\emptyset \neq L \subset \mathbb{R}^d$ a linear set with $L \cap \{v^1, \ldots, v^q\} = \emptyset$. Then there exists a vector $H \in \mathbb{R}^d$ with $H^T x = 0$ for each $x \in L$ and $H^T v^i \neq 0$ for each $i = 1, \ldots, q$.*

Proof. For each $i = 1, \ldots, q$, let $v^i = u^i + w^i$ be the (unique orthogonal) decomposition of v^i with $u^i \in L$ and $w^i \in L^\perp = \{z : z^T x = 0$ for each $x \in L\}$; in particular, $w^j \neq 0$ and $(w^j)^T u^i = w^j w^i$ for each $j, i = 1, \ldots, q$. Next, for each $\epsilon > 0$, let $H(\epsilon) = \sum_{j=1}^q \epsilon^j w^j$; of course, $H(\epsilon)^T x = 0$ for each $x \in L$. Also, for $i = 1, \ldots q$, $H(\epsilon)^T v^i = \sum_{j=1}^q \epsilon^j (w^j)^T w^i$ is a polynomial in ϵ of degree q or less which is not identically zero (as its i-th coefficient is $\|w^i\|^2$). Each of these polynomials can have at most q roots; hence, for some $\epsilon > 0$, none of these polynomials has a root in the interval $(0, \epsilon)$, assuring that each preserves the sign on $(0, \epsilon)$. Selecting ϵ^* positive and sufficiently small, $H \equiv H(\epsilon^*)$ will satisfy the conclusions of the lemma. □

Theorem 2.2.18. *Suppose $A \in \mathbb{R}^{d \times n}$ and $p \geq 2$. The set of almost-separable p-partitions of A equals the set of p-partitions of A that are associated with lists of $\binom{p}{2}$ almost-separable 2-partitions of A.*

Proof. Consider a p-partition $\pi = (\pi_1, \ldots, \pi_p)$ associated with a list of

$\binom{p}{2}$ almost-separable 2-partitions whose elements are $\{\pi^{r,s} = (\pi_r^{r,s}, \pi_s^{r,s}) : 1 \leq r < s \leq p\}$. We will show that π is almost-separable. Indeed, let $1 \leq r < s \leq p$. As each $\pi^{r,s}$ is almost-separable and (2.2.8) and the construction of π from π' imply that $\pi_r \subseteq \pi_r^{r,s}$ and $\pi_s \subseteq \pi_s^{r,s}$, it immediately follows from the characterization of almost-separability that π_r and π_s are almost-separable.

Next, let $\sigma = (\sigma_1, \ldots, \sigma_p)$ be an almost-separable p-partition and we will construct a list of $\binom{p}{2}$ almost-separable 2-partitions which has an associated partition and this partition is σ. We start by considering any specific pair $1 \leq r < s \leq p$ and constructing 2-partition $\pi^{r,s} = (\pi_r^{r,s}, \pi_s^{r,s})$ of A that has the following properties:

(i) $\sigma_r \subseteq \pi_r^{r,s}$ and $\sigma_s \subseteq \pi_s^{r,s}$.

(ii-a) If σ_r and σ_s are separable, then so are $\pi_r^{r,s}$ and $\pi_s^{r,s}$.

(ii-b) If σ_r and σ_s are not separable, then $\pi_r^{r,s}$ and $\pi_s^{r,s}$ are almost-separable but not separable; further, in this case the unique vector x occurring in σ_r and σ_s is the unique vector occurring in both $\pi_r^{r,s}$ and $\pi_s^{r,s}$,

$$\nu^{\pi_r^{r,s}}(x) = \nu^{\sigma_r}(x) \quad \text{and} \quad \nu^{\pi_s^{r,s}}(x) \geq \nu^{\sigma_s}(x). \tag{2.2.9}$$

First consider the case where σ_r and σ_s are separable. In this case there exist a vector $C^{rs} \in \mathbb{R}^d$ and a constant γ^{rs} such that

$$(C^{rs})^T y > \gamma^{rs} > (C^{rs})^T z$$

whenever $y \in \sigma_r$ and $z \in \sigma_s$. By possibly perturbing γ^{rs} we may and will assume that $(C^{rs})^T A^i \neq \gamma^{rs}$ for each $i \in \mathcal{N}$. Let $\pi^{r,s} \equiv (\pi_r^{r,s}, \pi_s^{r,s})$ be the 2-partition with $\pi_r^{r,s}$ and $\pi_s^{r,s}$ as, respectively, the multisets $\{y \in A : (C^{rs})^T y > \gamma^{rs}\}$ and $\{z \in A : (C^{rs})^T z < \gamma^{rs}\}$ (counting for multiplicities). Then $\pi^{r,s}$ is separable, $\sigma_r \subseteq \pi_r^{r,s}$ and $\sigma_s \subseteq \pi_s^{r,s}$.

Next assume that σ_r and σ_s are almost-separable, but not separable. Then there exist a vector $C^{rs} \in \mathbb{R}^d$ and a unique point x in A that occurs in both σ_r and σ_s such that

$$(C^{rs})^T y > (C^{rs})^T x > (C^{rs})^T z \quad \text{if } y \in \sigma_r, z \in \sigma_s \text{ and } y, z \neq x. \tag{2.2.10}$$

By Lemma 2.2.17, there exists a vector $H \in \mathbb{R}^d$ with $H^T(A^i - x) \neq 0$ for each $i = 1, \ldots, n$ with $A^i \neq x$. It then follows that for $\epsilon > 0$ sufficiently small, $(C^{rs} + \epsilon H)^T(A^i - x) \neq 0$ for all $i = 1, \ldots, n$ with $A^i \neq x$; in particular, ϵ can be selected so that (2.2.10) is preserved when C^{rs} is replaced by $C^{rs} + \epsilon H$. It follows that we may and will assume that C^{rs} and x satisfy (2.2.10) and $\gamma^{r,s} \equiv (C^{rs})^T x \neq (C^{rs})^T A^i$ for each $i = 1, \ldots, n$ with $A^i \neq x$. Next, let $\pi^{r,s} = (\pi_r^{r,s}, \pi_s^{r,s})$ be the partition with $\pi_r^{r,s}$ as the multiset consisting of the union $\{y \in A : (C^{rs})^T y > \gamma^{r,s}\}$ (counting for multiplicities) and $\nu^{\sigma_r}(x)$ copies of x, while $\pi_s^{r,s}$ as the multiset consisting of the union $\{z \in A : (C^{rs})^T z < \gamma^{r,s}\}$ (counting for multiplicities) and $\nu^A(x) - \nu^{\sigma_r}(x) \geq \nu^{\sigma_s}(x)$ copies of x. It is easily verified that $\pi^{r,s}$ satisfies (i) and (ii-b).

Our construction produces a list $\mathcal{L} = \{\pi^{r,s} : 1 \leq r < s \leq p\}$ of $\binom{p}{2}$ almost-separable 2-partitions of A. Let $\pi = (\pi_1, \ldots, \pi_p)$ be the p-tuple of multisets associated with $\mathcal{L}$ and let $\pi' = (\pi'_1, \ldots, \pi'_p)$ be the intermediary construction. We next show that π is a partition which coincides with σ. As the elements of the σ_j's and of the π_j's come from A, it suffices to show that $\nu^{\pi_j}(x) = \nu^{\sigma_j}(x)$ for each x occurring in A. Consider any such vector x. Property (i) of our construction assures (the multiset-inclusion)

$$\sigma_j \subseteq \left(\cap_{t=j+1}^{p} \pi_j^{j,t}\right) \bigcap \left(\cap_{t=1}^{j-1} \pi_j^{t,j}\right) = \pi'_j \quad \text{for } j = 1, \ldots, p, \tag{2.2.11}$$

implying that

$$\nu^{\sigma_j}(x) \leq \nu^{\pi'_j}(x) \quad \text{for } j = 1, \ldots, p. \tag{2.2.12}$$

Let

$$J \equiv \{j = 1, \ldots, p : x \in \sigma_j\} \subseteq \{j = 1, \ldots, p : x \in \pi'_j\} \tag{2.2.13}$$

and let s be the maximal element in J.

We next argue that the last inclusion in (2.2.13) holds as equality. Indeed, suppose $j \notin J$. Then $x \notin \sigma_j$. As σ is a partition of the columns of A, there is an index $t \in \{1, \ldots, p\}$ such that $x \in \sigma_t$. Depending on whether $j < t$ or $j > t$, our construction implies that $x \in \pi_t^{j,t}$ and $x \notin \pi_j^{j,t}$, or $x \in \pi_t^{t,j}$ and $x \notin \pi_j^{t,j}$; in either case, (2.2.8) implies that $x \notin \pi'_j$. So, indeed, equality holds in (2.2.13); in particular,

$$s = \max\{j = 1, \ldots, p : x \in \pi'_j\} = j^{\pi'}(x).$$

Now, if $j \in J \setminus \{s\}$ then $j < s$ and x occurs in both σ_j and σ_s, implying that these parts are not separable. It then follows from (2.2.9), (2.2.8) and the construction of π from π' that

$$\nu^{\sigma_j}(x) = \nu^{\pi_j^{j,s}}(x) \geq \nu^{\pi'_j}(x) = \nu^{\pi_j}(x).$$

This inequality combines with (2.2.12) to show that

$$\nu^{\sigma_j}(x) = \nu^{\pi_j}(x) \quad \text{for} \quad j \in J \setminus \{s\}. \tag{2.2.14}$$

We next extend (2.2.14) for $j = s$. From (2.2.12), $\sum_{j \in J} \nu^{\pi'_j}(x) \geq \sum_{j \in J} \nu^{\sigma_j}(x) = \nu^A(x)$ and therefore

$$\nu^{\pi'_s}(x) \geq \nu^A(x) - \sum_{j \in J \setminus \{s\}} \nu^{\pi'_j}(x) = \nu^A(x) - \sum_{j \in J \setminus \{s\}} \nu^{\sigma_j}(x) = \nu^{\sigma_s}(x) \geq 0.$$

Next, the rules of constructing π from π' imply that

$$\nu^{\pi_s}(x) = \min\{\nu^{\pi'_s}(x), [\nu^A(x) - \sum_{j=1}^{s-1} \nu^{\pi'_j}(x)]_+\} = \nu^{\sigma_s}(x).$$

As (2.2.14) was extended to $j = s$, the π_j's coincide (as multisets) with the parts of σ; thus, π is a partition and coincides with σ. □

2.3 Single-Size Partition Polytopes and Cone-Separable Partitions

In this section we introduce and study the single-size partition polytope and show that each vertex corresponds to a cone-separable partition, which is closely related to the (almost-) separable partition. It is emphasized that the results of the current section require the assumption that empty parts are allowed. Of course, single-size problems with empty parts prohibited are instances of constrained-shape problems corresponding to the set of all positive shapes to which the results of Sections 2.1 and 2.2 apply. In particular, single-size problems with parts required to be nonempty have optimal solutions that are (almost-) separable and enumerating (almost-) separable partitions can be used to solve these problems. But, the stronger conclusions of the current section do not apply to single-size problem with empty parts prohibited (see the forthcoming Example 2.4.1).

We say that two subsets Ω^1 and Ω^2 of $\mathbb{R}^d$ are *cone-separable* if there exists a nonzero d-vector C such that

$$C^T u^1 > 0 > C^T u^2 \text{ for all } u^1 \in \Omega^1 \setminus \{0\} \text{ and } u^2 \in \Omega^2 \setminus \{0\} \qquad (2.3.1)$$

(if either $\Omega^1 \setminus \{0\}$ or $\Omega^2 \setminus \{0\}$ is empty and the other set is not, then (2.3.1) is to be interpreted as a requirement just on the elements of the nonempty set).

Testing for Cone-Separability of Finite Sets

By possible scaling, we have that two nonempty finite sets in $\mathbb{R}^d$, say Ω^1 and Ω^2 are cone-separable if and only if there exists a nonzero d-vector C such that

$$\begin{aligned} &C^T u^1 \geq 1 \quad \text{for all } u^1 \in \Omega^1 \setminus \{0\}, \text{ and} \\ &C^T u^2 \leq -1 \text{ for all } u^2 \in \Omega^2 \setminus \{0\}. \end{aligned} \qquad (2.3.2)$$

As (2.3.2) is a system of linear inequalities in C, the question about cone-separability of Ω^1 and Ω^2 is reduced to determine feasibility of a system of linear inequalities—a task that can be resolved efficiently (e.g., Schrijver [85]). □

The relation between cone-separability and (almost-) separability is determined in the next lemma.

Lemma 2.3.1. *Suppose Ω^1 and Ω^2 are subsets of $\mathbb{R}^d$ that are cone-separable. Then Ω^1 and Ω^2 are almost-separable with the vector C satisfying (2.3.1) as a separating vector; in this case, 0 is the only possible vector in $\Omega^1 \cap \Omega^2$. Further, if $0 \notin \Omega^1$ or $0 \notin \Omega^2$, then Ω^1 and Ω^2 are separable.*

Proof. The asserted cone-separability of Ω^1 and Ω^2 means that (2.3.1) is satisfied, immediately implying that $\Omega^1 \cap \Omega^2$ cannot contain any nonzero vector. Now, if $u^1 \in \Omega^1$ and $u^2 \in \Omega^2$ are distinct, then one of these vectors is different from 0. It now follows from (2.3.1) that:
(i) if both u^1 and u^2 are nonzero, then $C^T u^1 > 0 > C^T u^2$,
(ii) if $u^1 = 0 \neq u^2$, then $C^T u^1 = 0 > C^T u^2$, and
(iii) if $u^1 \neq 0 = u^2$, then $C^T u^1 > 0 = C^T u^2$. So, the almost-separability requirement has been established in all possible cases. Finally, if $0 \notin \Omega^1$ or $0 \notin \Omega^2$, then the above arguments show that necessarily $\Omega^1 \cap \Omega^2 = \emptyset$, in either case almost-separability of Ω^1 and Ω^2 assures their separability. □

The following example demonstrates that (almost-) separability does not imply cone-separability.

Example 2.3.1. For $k = 1, 2$, let $\Omega^k = \{\binom{k}{0}, \binom{k}{2}\} \subseteq \mathbb{R}^2$. Then Ω^1 and Ω^2 are separable (with $C = \binom{1}{0}$ as a separating vector), but they are not cone-separable as a vector $C \in \mathbb{R}^2$ satisfies $C^T \binom{1}{0} > 0$ if and only if $C^T \binom{2}{0} > 0$. □

Recall that the *conic hull* of a set $\Omega \subseteq \mathbb{R}^d$, denoted cone Ω, is defined as the set of linear combinations $\sum_{t=1}^q \gamma_t x^t$ with $\gamma_t \geq 0$ and $x^t \in \Omega$ for $t = 1, \ldots, q$ (with cone $\emptyset = \{0\}$). A set is a *cone* if it equals its conic hull. A cone is *pointed* if for any nonzero vector x in the cone, $-x$ is not in the cone. A fundamental result about cones (see Rockafellar [82] or Schrijver [85] or Ziegler [92]) shows that a set is the conic hull of a finite set in $\mathbb{R}^m$ if and only if it has a representation $\{x \in \mathbb{R}^m : Ax \leq 0\}$ with A as a real matrix having m columns; such a cone is called a *polyhedral cone*. The following (standard) result provides a characterization of pointed polyhedral cones in terms of cone-separability.

Lemma 2.3.2. *A polyhedral cone Ω is pointed if and only if Ω and $\{0\}$ are cone-separable.*

Proof. See Rockafellar [82] or Schrijver [85] or Ziegler [92]. □

Lemma 2.3.3. *Two cones Ω^1 and Ω^2 in $\mathbb{R}^d$ are cone-separable if and only if they are almost-separable; further, in this case, Ω^1 and Ω^2 are pointed and $\Omega^1 \cap \Omega^2 = \{0\}$.*

Proof. Lemma 2.3.1 proves the "only if" part. Next assume that Ω^1 and Ω^2 are two cones in $\mathbb{R}^d$ which are almost-separable with C as a separating vector. As 0 is in every cone, we have that $0 \in \Omega^1 \cap \Omega^2$; further, as almost-separable sets can contain at most a single vector, no other vector is in $\Omega^1 \cap \Omega^2$, assuring that $\Omega^1 \cap \Omega^2 = \{0\}$. Now, consider $u^1 \in \Omega^1 \setminus \{0\}$ and $u^2 \in \Omega^2 \setminus \{0\}$. Then $u^1 \neq u^2$ and the almost-separability of Ω^1 and Ω^2 and the fact that $0 \in \Omega^1 \cap \Omega^2$ imply that $C^T u^1 > C^T 0 = 0$ and $0 = C^T 0 > C^T u^2$, establishing (2.3.1). Finally, (2.3.1) implies that if $x \neq 0$ is in either Ω^1 or Ω^2 then $-x$ cannot be in that set, proving that Ω^1 or Ω^2 are pointed. □

The next example demonstrates that the necessary condition for cone-separability given in Lemma 2.3.3 (Ω^1 and Ω^2 pointed and $\Omega^1 \cap \Omega^2 = \{0\}$) is not sufficient (but, the forthcoming Lemma 2.3.5 will establish the sufficiency for polyhedral cones).

Example 2.3.2. Let $\Omega^1 = \{\binom{x}{0} : x \geq 0\} \subseteq \mathbb{R}^2$ and $\Omega^2 = \{\binom{x}{y} : x \in \mathbb{R}, y > 0\} \cup \{0\} \subseteq \mathbb{R}^2$. These sets are pointed cones that satisfy $\Omega^1 \cap \Omega^2 = \{0\}$ but are not cone-separable. □

Lemma 2.3.4. *Let Ω^1 and Ω^2 be two finite sets in $\mathbb{R}^d$ and let C be a nonzero vector in $\mathbb{R}^d$. Then $C^T u^1 > 0 > C^T u^2$ for all $u^1 \in \Omega^1 \setminus \{0\}$ and $u^2 \in \Omega^2 \setminus \{0\}$ if and only if $C^T w^1 > 0 > C^T w^2$ for all $w^1 \in (\text{cone } \Omega^1) \setminus \{0\}$ and $w^2 \in (\text{cone } \Omega^2) \setminus \{0\}$.*

Proof. The "only if" (sufficiency) direction is trivial because a set is always included in its conic hull. We next consider the inverse "if" (necessity) direction. If either $\Omega^1 \setminus \{0\} = \emptyset$ or $\Omega^2 \setminus \{0\} = \emptyset$, the necessity direction is obvious. So, assume that Ω^1 and Ω^2 are cone-separable with C as the separating vector, that $\Omega^1 \setminus \{0\} \neq \emptyset$ and that $\Omega^2 \setminus \{0\} \neq \emptyset$. Set $\alpha^* \equiv \min_{u \in \Omega^1 \setminus \{0\}} C^T u$ and $\alpha_* \equiv \max_{u \in \Omega^2 \setminus \{0\}} C^T u$; the finiteness of Ω^1 and Ω^2 assures that these values are finite and the cone-separability of Ω^1 and Ω^2 assures that $\alpha^* > 0 > \alpha_*$. A vector $w^1 \in (\text{cone } \Omega^1) \setminus \{0\}$ has a representation $\sum_{t=1}^q \gamma_t x^t$ with $q > 0$, $\gamma_1, \ldots, \gamma_q$ as positive numbers and $\{x^1, \ldots, x^q\} \subseteq \Omega^1$, assuring that $\langle C, w^1 \rangle = \sum_{t=1}^q \gamma_t \langle C, x^t \rangle \geq$

$(\sum_{t=1}^{q} \gamma_t)\alpha_* > 0$. Similar arguments show that $\langle C, w^2 \rangle < 0$ for every $w^2 \in (\text{cone } \Omega^2) \setminus \{0\}$. □

The next result provides characterizations of cone-separability for polyhedral cones.

Lemma 2.3.5. *Let Ω^1 and Ω^2 be two finite sets in $\mathbb{R}^d$. Then the following are equivalent:*

(a) Ω^1 and Ω^2 are cone-separable,
(b) cone Ω^1 and cone Ω^2 are cone-separable,
(c) cone Ω^1 and cone Ω^2 are almost-separable, and
(d) cone Ω^1 and cone Ω^2 are pointed cones and $(\text{cone } \Omega^1) \cap (\text{cone } \Omega^2) = \{0\}$.

Proof. The equivalences (a)⇔(b) and (b)⇔(c)⇒(d) follow, respectively, from Lemmas 2.3.4 and 2.3.3. Finally, to see that (d)⇒(c) assume that (d) holds. Let $\Omega \equiv \{x^1 - x^2 : x^1 \in \Omega^1 \text{ and } x^2 \in \Omega^2\}$. Standard results (e.g., Rockafellar [82]) show that Ω is a polyhedral cone. Further, we observe that Ω is pointed. Indeed, if $u^1 - u^2 = -(v^1 - v^2)$, then $u^1 + v^1 = u^2 + v^2$; as this vector is in $(\text{cone } \Omega^1) \cap (\text{cone } \Omega^2)$ and it is the zero vector, implying that $u^1 = -v^1$ and $u^2 = -v^2$. By the pointedness of Ω^1 and Ω^2, it then follows that $u^1 = v^1 = u^2 = v^2 = 0$ and $u^1 - u^2 = v^1 - v^2 = 0$. It now follows from Lemma 2.3.2 that there exists a vector $C \in \mathbb{R}^d$ with $C^T(x^1 - x^2) > 0$ for every $x^1 \in \text{cone } \Omega^1$ and $x^2 \in \text{cone } \Omega^2$ with $x^1 \neq x^2$, implying (c). □

With one set consisting of $\{0\}$, Lemma 2.3.5 specializes to the following (standard) testable characterization that the conic hull of a finite set has 0 as its vertex.

Corollary 2.3.6. *Let Ω be a finite subset of $\mathbb{R}^d$. Then the following are equivalent:*

(a) Ω and $\{0\}$ are cone-separable,
(b) cone Ω and $\{0\}$ are cone-separable,
(c) cone Ω and $\{0\}$ are almost-separable,
(d) cone Ω is a pointed cone, and
(e) for some vector $C \in \mathbb{R}^d$, $C^T u \geq 1$ for every $u \in \Omega \setminus \{0\}$.

Proof. The equivalences of (a), (b), (c) and (d) are immediate from Lemma 2.3.5 with $\Omega^1 = \Omega$ and $\Omega^2 = \{0\}$. Further, (b) asserts that for some vector $C \in \mathbb{R}^d$, $C^T u > 0$ for every $u \in \Omega \setminus \{0\}$; as Ω is finite, this condition holds if and only if some scalar multiple of C satisfies (e). □

We next turn our attention to partition problems. As is explained in the second paragraph of this section, we consider fixed-size problems with a relaxation of the assumption that partitions' parts are nonempty. Formally, let $\widehat{\Pi^p}$ denote the set of p-partitions which allow empty parts. For $\pi \in \widehat{\Pi^p}$, A_π has the natural definition with the empty sum taken as zero. The partition polytope corresponding to $\widehat{\Pi^p}$ is defined by $P_A^{\widehat{\Pi^p}} \equiv \text{conv}\,\{A_\pi : \pi \in \widehat{\Pi^p}\}$. As we do in all sum-partition problems, we assume that the objective function F over $\widehat{\Pi^p}$ is given by $F(\pi) = f(A_\pi)$ with f as a real valued function on $P_A^{\widehat{\Pi^p}}$.

We say that π is *cone-separable*, if for each pair of distinct indices $r, s \in \{1, \ldots, p\}$, A^{π_r} and A^{π_s} are cone-separable, or equivalently (by Lemma 2.3.5), cone A^{π_r} and cone A^{π_s} are cone-separable. Also, Lemma 2.3.1 shows that cone-separability implies almost-separability, with 0 as the only potential overlapping vector for any pair of parts; moreover, a cone-separable partition with at most one part containing indices that correspond to the 0 vector is separable.

Testing for Cone-Separability of Partitions

Suppose $\pi = (\pi_1, \ldots, \pi_p)$ is a partition. Using the method described earlier for testing cone-separability of finite sets, a test for (almost-) separability of π is available by testing the solvability of $\binom{p}{2}$ systems of linear inequalities corresponding to pairs of distinct parts of π. The system corresponding to π_r and π_s has d variables and at most $|\pi_r|+|\pi_s|$ constraints. If these systems are combined, one obtains a block-diagonal system with $d\binom{p}{2}$ variables and at most $(p-1)n$ constraints where each block has d variables. □

We next establish useful characterizations of cone-separable partitions when $d = 1$ and $d = 2$ under the assumption that the underlying matrices have no zero vector (when zero vectors are present, they can be distributed arbitrarily among the parts).

Theorem 2.3.7. *Suppose $A \in \mathbb{R}^{1\times n}$ has no zero vector. If $p \geq 2$, then a p-partition π is cone-separable if and only if two of its parts are $\{i \in \mathcal{N} : A^i > 0\}$ and $\{i \in \mathcal{N} : A^i < 0\}$ (where either may be empty) and all other parts are empty. If $p = 1$, then the (only) 1-partition $(\mathcal{N})$ is cone-separable if and only if A contains either just positive elements or just negative elements.*

Proof. First consider the case where $p \geq 2$. Trivially, a partition with two of its parts $\{i \in \mathcal{N} : A^i > 0\}$ and $\{i \in \mathcal{N} : A^i < 0\}$ and all other parts empty is cone-separable. Next consider a cone-separable partition π.

Observing that any two points in $\mathbb{R}$ with the same strict sign (positive or negative) have the same set of conic combinations, we have that if a part of π contains one index i with $A^i > 0$ or $A^i < 0$, then it must contain all such indices, respectively. Further, the pointedness of each coneA^{π_j} implies that no part can contain indices of both types. This proves that π must have the asserted structure. The above arguments also prove that the only 1-partition is cone-separable if and only if the elements of A do no include a positive element and a negative element. $\square$

To study cone-separable partitions with $d = 2$, we associate each $x \in \mathbb{R}^2 \setminus \{0\}$ with the *angular coordinate* of its polar-coordinate representation, which we denote $\phi(x)$ (measured in degrees). Observe that for $\emptyset \neq C \subseteq \mathbb{R}^2 \setminus \{0\}$, $C \cup \{0\}$ is a pointed polyhedral cone if and only if for some $0 \leq \underline{\phi} \leq \overline{\phi}$ with $\underline{\phi} < 360°$ and $\overline{\phi} < \underline{\phi} + 180°$,

$$C = \begin{cases} \{x \in \mathbb{R}^2 \setminus \{0\} : \underline{\phi} \leq \phi(x) \leq \overline{\phi}\} & \text{if } \overline{\phi} < 360° \\ \{x \in \mathbb{R}^2 \setminus \{0\} : \underline{\phi} \leq \phi(x) < 360° \text{ or } 0 \leq \phi(x) \leq \overline{\phi} - 360°\} & \text{if } \overline{\phi} \geq 360°. \end{cases}$$

Theorem 2.3.8. *Suppose $A \in \mathbb{R}^{2\times n}$ has no zero vector. A p-partition π is cone-separable if and only if for some $q \leq p$ there exist $0 \leq \underline{\phi}_1 \leq \overline{\phi}_1 < \underline{\phi}_2 \leq \overline{\phi}_2 < \cdots < \underline{\phi}_q \leq \overline{\phi}_q < \underline{\phi}_1 + 360°$ such that $\underline{\phi}_t < 360°$, $\overline{\phi}_t - \underline{\phi}_t < 180°$ for each $t = 1, \ldots, q$, and the nonempty parts of π are*

$$\{i \in \mathcal{N} : \underline{\phi}_t \leq \phi(A^i) \leq \overline{\phi}_t\} \textit{ for } t = 1, \ldots, q-1$$

and

$$\begin{cases} \{i \in \mathcal{N} : \underline{\phi}_q \leq \phi(A^i) \leq \overline{\phi}_q\} & \textit{if } \overline{\phi}_q < 360° \\ \{i \in \mathcal{N} : \underline{\phi}_q \leq \phi(A^i) < 360° \textit{ or } 0 \leq \phi(A^i) \leq \overline{\phi}_q - 360°\} & \textit{if } \overline{\phi}_q \geq 360°; \end{cases}$$

further, the $\underline{\phi}_t$'s and $\overline{\phi}_t$'s can be selected from $\{\phi(A^i) : i \in \mathcal{N}\}$.

Proof. Assume that π is a p-partition that possesses the structure described in the theorem with corresponding values $\underline{\phi}_1, \overline{\phi}_1, \ldots, \underline{\phi}_q, \overline{\phi}_q$. Let

$$X_t \equiv \{x \in \mathbb{R}^2 \setminus \{0\} : \underline{\phi}_t \leq \phi(x) \leq \overline{\phi}_t\}, \quad t = 1, \ldots, q-1 \tag{2.3.3}$$

and

$$X_q \equiv \begin{cases} \{x \in \mathbb{R}^2 \setminus \{0\} : \underline{\phi}_q \leq \phi(x) \leq \overline{\phi}_q\} & \text{if } \overline{\phi}_q < 360° \\ \{x \in \mathbb{R}^2 \setminus \{0\} : \underline{\phi}_q \leq \phi(x) < 360° \text{ or} & \\ \qquad 0 \leq \phi(x) \leq \overline{\phi}_q - 360°\} & \text{if } \overline{\phi}_q \geq 360°. \end{cases} \tag{2.3.4}$$

Let $J \equiv \{j = 1, \ldots, p : \pi_j \neq \emptyset\}$; the assumption about π asserts that its nonempty parts are $\sigma_t \equiv \{i \in \mathcal{N} : A^i \in X_t\}$ for $t = 1, \ldots, q$. Thus, for each

$j \in J$, there is a (unique) index $t_j \in \{1, \ldots, q\}$ with $\pi_j = \sigma_{t_j}$; in particular, $A^{\pi_j} = A^{\sigma_{t_j}} \subseteq X_{t_j}$, implying that cone$A^{\pi_j} =$ cone$A^{\sigma_{t_j}} \subseteq X_{t_j}$. As the sets $\{X_t \cup \{0\} : t = 1, \ldots, q\}$ are pointed (polyhedral) cones and all of whose pairwise intersections equal $\{0\}$, the cones $\{\text{cone}A^{\pi_j} : j \in J\}$ have the same properties. So, Lemma 2.3.5 implies that π is cone-separable (adding the empty parts clearly preserves the asserted properties).

Next assume that π is a cone-separable partition, that is, the cones spanned by the vectors associated with its nonempty parts are pointed and all of their pairwise intersections equal $\{0\}$. For $j \in J \equiv \{j = 1, \ldots, p : \pi_j \neq \emptyset\}$, let

$$\underline{\phi}_j \equiv \begin{cases} \min\{\phi(A^i) : i \in \pi_j\} & \text{if } \max_{i,i' \in \pi_j} \phi(A^i) - \phi(A^{i'}) < 180^\circ \\ \max\{\phi(A^i) : i \in \pi_j\} & \text{otherwise .} \end{cases}$$

Let $j_1, \ldots, j_q$ be an enumeration of the indices in J such that $0 \leq \underline{\phi}_{j_1} < \cdots < \underline{\phi}_{j_q} < 360^\circ$. Evidently for each $j \in J$, with the possible exception of $j = j_q$, the first option of the definition of $\underline{\phi}_j$ applies. Also, for $j \in J$, let

$$\overline{\phi}_j \equiv \begin{cases} \max\{\phi(A^i) : i \in \pi_j\} & \text{if } \max_{i,i' \in \pi_j} \phi(A^i) - \phi(A^{i'}) < 180^\circ \\ \min\{\phi(A^i) : i \in \pi_j\} + 360^\circ & \text{otherwise ;} \end{cases}$$

here again, the first option of the definition of $\overline{\phi}_j$ applies to all indices with the possible exception of $j = j_q$. As all pairwise intersections of cone$(A^{\pi_{j_t}})$'s equal to $\{0\}$, we have that $\underline{\phi}_{j_1} \leq \overline{\phi}_{j_1} < \underline{\phi}_{j_2} \leq \overline{\phi}_{j_2} < \cdots < \underline{\phi}_{j_q} \leq \overline{\phi}_{j_q} < \underline{\phi}_{j_1} + 360^\circ$; also, as these are pointed cones, $\overline{\phi}_{j_t} - \underline{\phi}_{j_t} < 180^\circ$ for $t = 1, \ldots, q-1$. We finally observe that for $t = 1, \ldots, q$, the set X_t defined by (2.3.3)–(2.3.4) with the determined values $\underline{\phi}_{j_t}$ and $\overline{\phi}_{j_t}$ coincides with cone$(A^{\pi_{j_t}})$, implying that $\pi_{j_t} = \{i \in \mathcal{N} : A^i \in X_t\}$ for $t = 1, \ldots, q$. So, the parts of π have the asserted structure; in particular, the constructed values $\underline{\phi}_{j_1}, \overline{\phi}_{j_1}, \ldots, \underline{\phi}_{j_q}, \overline{\phi}_{j_q}$ are all from $\{\phi(A^i) : i \in \mathcal{N}\}$. □

The following Corollary of Theorem 2.3.8 shows that a cone-separable partition with q nonempty parts is characterized by q angular coordinates of columns of the underlying matrix A. The result will be useful for the forthcoming enumeration of cone-separable partitions (in Section 2.4).

Corollary 2.3.9. *Suppose $A \in \mathbb{R}^{2\times n}$ has no zero vector. A p-partition π is cone-separable if and only if there exist $0 \leq \phi_1 < \cdots < \phi_q < 360^\circ$ such that the nonempty parts of π are*

$$\{i \in \mathcal{N} : \phi_t \leq \phi(A^i) < \min\{\phi_t + 180^\circ, \phi_{t+1}\}\} \quad \textit{for} \quad t = 1, \ldots, q-1$$

and

$$\{i \in \mathcal{N} : \phi_q \leq \phi(A^i) < \min\{\phi_q + 180°, 360°\} \text{ or } 0 \leq \phi(A^i) < \min\{\phi_q - 180°, \phi_1\}\};$$

further, the ϕ_t's can be selected from $\{\phi(A^i) : i \in \mathcal{N}\}$.

Proof. First assume that π has the asserted structure. Then for each nonempty part π_j of π, $\max_{i,i' \in \pi_j} |\phi(A^i) - \phi(A^{i'})| < 180°$, implying that cone$(A^{\pi_j})$ is a pointed (polyhedral) cone; further, all pairwise intersections of these cones equal $\{0\}$. So, by Lemma 2.3.5, π is cone-separable.

Next assume that π is cone-separable. Let $0 \leq \underline{\phi}_1 \leq \overline{\phi}_1 < \underline{\phi}_2 \leq \overline{\phi}_2 < \cdots < \underline{\phi}_q \leq \overline{\phi}_q < \underline{\phi}_1 + 360°$ be as in the conclusion of Theorem 2.3.8. We will show that the asserted structure of the nonempty parts of π holds with $\phi_t = \underline{\phi}_t$ for $t = 1, \ldots, q$; in particular, these values can be selected from $\{\phi(A^i) : i \in \mathcal{N}\}$. For $t = 1, \ldots, q-1$, $\{i \in \mathcal{N} : \phi(A^i) \in (\overline{\phi}_t, \underline{\phi}_{t+1})\} = \emptyset$; as $\overline{\phi}_t < \underline{\phi}_t + 180°$, we have that

$$\{i \in \mathcal{N} : \phi(A^i) \in [\underline{\phi}_t, \overline{\phi}_t]\} = \{i \in \mathcal{N} : \phi(A^i) \in [\underline{\phi}_t, \min\{\underline{\phi}_t + 180°, \underline{\phi}_{t+1}\})\}.$$

By Theorem 2.3.8, these sets represent all but one of the nonempty parts of π. Let σ be the remaining set appearing in the statement of the theorem and we will show that it coincides with the remaining nonempty part of π. We consider two cases.

Case (i): $\overline{\phi}_q < 360°$. In this case, $\{i \in \mathcal{N} : \phi(A^i) \in (\overline{\phi}_q, 360°)\}$ and $\{i \in \mathcal{N} : \phi(A^i) \in [0, \underline{\phi}_1)\}$ are empty; as $\overline{\phi}_q < \underline{\phi}_q + 180°$, we conclude that the remaining nonempty part of π (using the representation of Theorem 2.3.8) is

$$\begin{aligned}
&\{i \in \mathcal{N} : \phi(A^i) \in [\underline{\phi}_q, \overline{\phi}_q]\} \\
&= \{i \in \mathcal{N} : \phi(A^i) \in [\underline{\phi}_q, \min\{\underline{\phi}_q + 180°, 360°\}) \cup [0, \min\{\underline{\phi}_q - 180°, \underline{\phi}_1\})\} \\
&= \sigma.
\end{aligned}$$

Case (ii): $\overline{\phi}_q \geq 360°$. In this case, $\underline{\phi}_q + 180° > \overline{\phi}_q \geq 360°$ and $\underline{\phi}_q - 180° > \overline{\phi}_q - 360° \geq 0$. In particular,

$$\{i \in \mathcal{N} : \phi(A^i) \in [\underline{\phi}_q, 360°)\} = \{i \in \mathcal{N} : \phi(A^i) \in [\underline{\phi}_q, \min\{\underline{\phi}_q + 180°, 360°\})\}.$$

Also, $\{i \in \mathcal{N} : \phi(A^i) \in (\overline{\phi}_q - 360°, \underline{\phi}_1)\}$ is empty; as $\underline{\phi}_q - 180° > 0$, we have that

$$\{i \in \mathcal{N} : \phi(A^i) \in [0, \overline{\phi}_q - 360°)\} = \{i \in \mathcal{N} : \phi(A^i) \in [0, \min\{\underline{\phi}_q - 180°, \underline{\phi}_1\})\}.$$

We conclude that the remaining nonempty part of π (using the representation of Theorem 2.3.8) is

$$\begin{aligned}&\{i \in \mathcal{N} : \phi(A^i) \in [\underline{\phi}_q, 360°) \cup [0, \overline{\phi}_q - 360°)\}\\ &= \{i \in \mathcal{N} : \phi(A^i) \in [\underline{\phi}_q, \min\{\underline{\phi}_q + 180°, 360°\}) \cup [0, \min\{\underline{\phi}_q - 180°, \underline{\phi}_1\})\}\\ &= \sigma.\end{aligned}$$

□

The next theorem and its corollary extend results of Barnes, Hoffman and Rothblum [7] (which considered the case where the A^i's are distinct).

Theorem 2.3.10. *If $\pi \in \widehat{\Pi^p}$ and A_π is a vertex of $P_A^{\widehat{\Pi^p}}$, then for some matrix $C \in \mathbb{R}^{d \times p}$*

$$(C^r)^T u > (C^s)^T u \quad \text{for all } 1 \le r \ne s \le p \text{ and } u \in (\text{cone } A^{\pi_r}) \setminus \{0\}. \tag{2.3.5}$$

Proof. Let $\widehat{P} \equiv P_A^{\widehat{\Pi^p}}$. As A_π is a vertex of $\widehat{P}$, it is the unique maximizer over $\widehat{P}$ of some linear function, say, one that is determined by the matrix $C \in \mathbb{R}^{d \times p}$; so

$$\langle C, X \rangle < \langle C, A_\pi \rangle \text{ for each } X \in \widehat{P} \setminus \{A_\pi\}.$$

Let $r, s \in \{1, \ldots, p\}$, $x \in \pi_r$ with $A^x \ne 0$ and σ be the partition obtained from π by switching x from π_r to π_s (as empty parts are allowed, such a reallocation is always feasible). With e^j denoting the j-unit vector in $\mathbb{R}^p$, we have that $A_\sigma = A_\pi + A^x(e^s - e^r)^T \ne A_\pi$, the inequality following from the assumption that $A^x \ne 0$. As $A_\sigma \in \widehat{P} \setminus \{A_\pi\}$, one can conclude that

$$\begin{aligned}0 &> \langle C, A_\sigma \rangle - \langle C, A_\pi \rangle = \langle C, A^x(e^s - e^r)^T \rangle\\ &= \langle C(e^s - e^r), A^x \rangle = \langle (C^s - C^r), A^x \rangle\\ &= (C^s - C^r)^T A^x\end{aligned}$$

(with the second equality following from (1.1.1)). Thus, we proved the restricted version of (2.3.5) with u restricted to vectors in $A^{\pi_r} \setminus \{0\}$. But Lemma 2.3.4 assures that the inequality $(C^r - C^s)^T u > 0$ extends to any vector $u \in (\text{cone } A^{\pi_r}) \setminus \{0\}$, establishing (2.3.5) in its full generality. □

Corollary 2.3.11. *If $p \ge 2$, $\pi \in \widehat{\Pi^p}$ and A_π is a vertex of $P_A^{\widehat{\Pi^p}}$, then π is cone-separable. Further, for every vertex v of $P_A^{\widehat{\Pi^p}}$, there exists a partition π which is both cone-separable and separable and has $A_\pi = v$.*

Proof. Let C be as in the conclusion of Theorem 2.3.10 and let r and s be two distinct indices in $\{1, \ldots, p\}$. If $C^r \ne C^s$, the conclusions of Theorem 2.3.10 (applied twice) show that for $u \in (\text{cone } A^{\pi_r}) \setminus \{0\}$ and

$w \in (\text{cone } A^{\pi_s})\setminus\{0\}$, $(C^r - C^s)^T u > 0$ and $(C^s - C^r)^T w > 0$, implying that cone A^{π_r} and cone A^{π_s} are cone-separable. Alternatively, if $C^r = C^s$, then the conclusions of Theorem 2.3.10 assure that cone $A^{\pi_r} =$ cone $A^{\pi_s} = \{0\}$ and the cone-separability of A^{π_r} and A^{π_s} is trite. Finally, let v be a vertex of $P_A^{\widehat{\Pi^p}}$. By Proposition 3.1.1(i) of Vol. I, v has a representation $v = A_{\pi'}$ with $\pi' \in \widehat{\Pi^p}$; the above arguments assure that π' must be cone-separable. By Lemma 2.3.1, π' is also almost-separable, with $A^{\pi_r} \cap A^{\pi_s} \subseteq \{0\}$. Now, shifting all indices i with $A^i = 0$ to any single part will create a partition π which is both cone-separable and separable and has $A_\pi = A_{\pi'} = v$. □

Observe that allowance of empty parts is crucial for Corollary 2.3.11. The conclusion of Corollary 2.3.11 holds for $p = 1$ if and only if the single 1-partition $(\mathcal{N})$ is cone-separable; this happens if and only if cone A is pointed (see Corollary 2.3.6).

Aviran, Lev-Tov, Onn and Rothblum [5] showed that the inverse of the main conclusion of Corollary 2.3.11 is true when $p = 2$ or $d \leq 2$, but not for $d = p = 3$. We first consider the case $p = 2$.

Theorem 2.3.12. *Let $A \in \mathbb{R}^{d\times n}$ and let π be a cone-separable 2-partition. Then A_π is a vertex of $P_A^{\widehat{\Pi^p}}$.*

Proof. As π is a cone-separable 2-partition, there exists a vector $C \in \mathbb{R}^d$ satisfying $C^T A^t > 0 > C^T A^s$ for each $t \in \pi_1$ and $s \in \pi_2$ with $A^t \neq 0$ and $A^s \neq 0$. Let σ be a 2-partition with $A_\sigma \neq A_\pi$. Let $\mathcal{N}_1 \equiv \pi_1 \setminus \sigma_1 = \sigma_2 \setminus \pi_2$ and $\mathcal{N}_2 \equiv \pi_2 \setminus \sigma_2 = \sigma_1 \setminus \pi_1$; we then have that $A^i \neq 0$ for some $i \in \mathcal{N}_1 \cup \mathcal{N}_2$, $C^T A^t > 0$ for each $t \in \mathcal{N}_1$ that satisfies $A^t \neq 0$, and $C^T A^s < 0$ for each $s \in \mathcal{N}_2$ that satisfies $A^s \neq 0$. We next observe that

$$A_{\pi_1} - A_{\sigma_1} = \sum_{t\in\mathcal{N}_1} A^t - \sum_{s\in\mathcal{N}_2} A^s,$$

implying that

$$\begin{aligned}\langle (C,0), A_\pi - A_\sigma\rangle &= \langle C, A_{\pi_1} - A_{\sigma_1}\rangle + \langle 0, A_{\pi_2} - A_{\sigma_2}\rangle \\ &= \textstyle\sum_{t\in\mathcal{N}_1} C^T A^t + \sum_{s\in\mathcal{N}_2} (-C^T A^s).\end{aligned}$$

Each of the terms in the sums of the right-hand side of the above expression is nonnegative with at least one of the terms positive, implying that

$$\langle (C,0), A_\pi - A_\sigma\rangle > 0.$$

So, $\langle (C,0), A_\pi\rangle > \langle (C,0), A_\sigma\rangle$ for each 2-partition σ with $A_\sigma \neq A_\pi$; and therefore (e.g., Proposition 3.1.1(i) of Vol. I), A_π is a vertex of $P_A^{\widehat{\Pi^p}} =$ conv $\{A_\sigma : \sigma \text{ is a 2-partition}\}$. □

We next derive the inverse of Corollary 2.3.11 when $d \leq 2$.

Theorem 2.3.13. *Let $A \in \mathbb{R}^{d \times n}$ where $d \leq 2$, let $p \geq 2$ and let π be a cone-separable p-partition. Then A_π is a vertex of $P_A^{\widehat{\Pi^p}}$.*

Proof. We shall temporarily restrict attention to the case where A has no zero vector, an assumption that will eventually be relaxed.

Consider first the case where $d = 1$. Theorem 2.3.7 shows that in this case there exist distinct indices $r, s \in \{1, \ldots, p\}$ with $\pi_r = \{i \in \mathcal{N} : A^i > 0\}$ and $\pi_s = \{i \in \mathcal{N} : A^i < 0\}$ (either part can be empty). Let e^r and e^s be the corresponding unit vectors in $\mathbb{R}^p$. It then follows that $(e^r - e^s)^T A_{\pi'} < (e^r - e^s)^T A_\pi$ for each partition $\pi' \neq \pi$ in $P_A^{\widehat{\Pi^p}}$, implying that A_π is a vertex of $P_A^{\widehat{\Pi^p}}$ (Proposition 3.1.1(i) of Vol. I).

Next let $d = 2$ (still, under the assumption that A has no zero vector). Assume that $\pi = (\pi_1, \ldots, \pi_p)$ has $q \geq 1$ nonempty parts. By Theorem 2.3.8, there exist $0 \leq \underline{\phi}_1 \leq \overline{\phi}_1 < \underline{\phi}_2 \leq \overline{\phi}_2 < \cdots < \underline{\phi}_q \leq \overline{\phi}_q < \underline{\phi}_1 + 360°$ such that $\underline{\phi}_q < 360°$, $\overline{\phi}_t - \underline{\phi}_t < 180°$ for $t = 1, \ldots, q$ and the nonempty parts of π are

$$\{\{i \in \mathcal{N} : \underline{\phi}_t \leq \phi(A^i) \leq \overline{\phi}_t\} : t = 1, \ldots, q\},$$

where for $i \in \mathcal{N}$ with $0 \leq \phi(A^i) < \underline{\phi}_1$, $\phi(A^i)$ is to be replaced by $\phi(A^i) + 360°$. Let the nonempty parts of π be $\pi_{j_1}, \ldots, \pi_{j_q}$ where $\pi_{j_t} = \{i \in \mathcal{N} : \underline{\phi}_t \leq \phi(A^i) \leq \overline{\phi}_t\}$. For each $t = 1, \ldots, q$, let

$$\underline{e}^t = \begin{pmatrix} \cos(\underline{\phi}_t) \\ sin(\underline{\phi}_t) \end{pmatrix}, \quad \overline{e}^t = \begin{pmatrix} \cos(\overline{\phi}_t) \\ sin(\overline{\phi}_t) \end{pmatrix}, \quad \gamma_t = (\underline{e}^t)^T \overline{e}^t \text{ and } C^t = \frac{\underline{e}^t + \overline{e}^t}{\sqrt{2 + 2\gamma}}, \tag{2.3.6}$$

that is, $\underline{e}^t$ and $\overline{e}^t$ are the unit vectors whose angular coordinate are $\underline{\phi}_t$ and $\overline{\phi}_t$, respectively, γ_t is the cosine of the angle between $\underline{e}^t$ and $\overline{e}^t$ and C^t is a unit vector whose angular coordinate is $\frac{\underline{\phi}_t + \overline{\phi}_t}{2}$. We further observe that, $(C^t)^T \underline{e}^t = (C^t)^T \overline{e}^t = 1$ and (using simple trigonometry) $x \in \mathbb{R}^2 \setminus \{0\}$ satisfies $\underline{\phi}_t \leq \phi(x) \leq \overline{\phi}_t$ if and only if the angle between x and C^t is less than or equal to $\frac{\overline{\phi}_t - \underline{\phi}_t}{2} < 90°$, or equivalently, $(C^t)^T \frac{x}{\|x\|} \geq 1$. So, for $i \in \mathcal{N}$,

$$[i \in \pi_{j_t}] \Leftrightarrow [\underline{\phi}_t \leq \phi(A^i) \leq \overline{\phi}_t] \Leftrightarrow [(C^t)^T A^i \geq \|A^i\|].$$

If $\pi_j \neq \emptyset$ and $j = j_t$, let $C^j \equiv C^{j_t}$. It then follows that

$$\pi_j = \{i \in \mathcal{N} : (C^j)^T A^i \geq \|A^i\|\}.$$

Further, with $C^j = 0$ for $j \in \{1, \ldots, p\} \setminus \{1, \ldots, q\}$ (that is $\pi_j = \emptyset$), we have (vacuously) that $i \in \pi_j$ if and only if $(C^j)^T A^i = 0 \geq \|A^i\|$ (recall

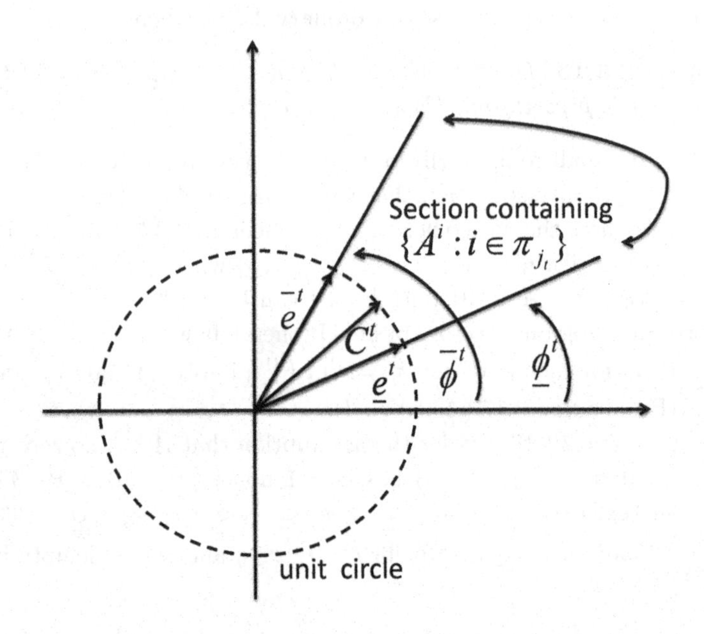

Figure 2.3.1 The Construction of C^t

that A^i has no zero vector); so, here too, $\pi_j = \{i \in \mathcal{N} : (C^j)^T A^i \geq \|A^i\|\}$. Let $C \equiv [C^1, \ldots, C^p] \in \mathbb{R}^{d \times p}$. We will show that A_π is a vertex of $P_A^{\widehat{\Pi^p}}$ by showing that for every p-partition $\pi' \neq \pi$, $\langle C, A_\pi \rangle > \langle C, A_{\pi'} \rangle$ (relying on Theorem 3.1.1(i) of Vol. I). Indeed, consider a p-partition $\pi' \neq \pi$. Let $\Delta' \neq \emptyset$ be the set of indices i which are not assigned to the same part under π and π' . For $i \in \Delta'$, let $j^\pi(i)$ and $j^{\pi'}(i)$ be, respectively, the indices of the parts of π and π' that contains i; it then follows that $(C^{j^\pi(i)})^T A^i \geq \|A^i\| > (C^{j^{\pi'}(i)})^T A^i$. Thus,

$$\langle C, A_\pi \rangle - \langle C, A_{\pi'} \rangle = \sum_{i \in \Delta'} (C^{j^\pi(i)} - C^{j^{\pi'}(i)})^T A^i > 0.$$

We finally consider the case where A has zero vectors. In this case, let $\mathcal{N}' \equiv \{i \in \mathcal{N} : A^i \neq 0\}$, $A' \equiv A^{N'}$ and consider the columns of A' indexed by the elements in $\mathcal{N}'$. Also, let $\widehat{(\Pi')^p}$ be the set of single-size partitions of $\mathcal{N}'$, allowing for empty parts; evidently, $\{A_\pi : \pi \in \widehat{\Pi^p}\} = \{A_{\pi'} : \pi' \in \widehat{(\Pi')^p}\}$ and therefore $P_A^{\widehat{\Pi^p}} = P_A^{\widehat{(\Pi')^p}}$. Now, if π is a cone-separable p-partition of $\mathcal{N}$, then the p-partition $\pi' \equiv (\pi_1 \cap \mathcal{N}', \ldots, \pi_p \cap \mathcal{N}')$ on $\mathcal{N}'$ is cone-separable (with A' as the underlying matrix) and has $A_{\pi'} = A_\pi$. So, the established conclusions of the theorem for matricesthat have no zero vector and the fact that A' is such a matrix imply that $A'_{\pi'} = A_\pi$ is a vertex of $P_{A'}^{\widehat{(\Pi')^p}} = P_A^{\widehat{\Pi^p}}$. □

The next example adapts Example 2.1.2 to show that the inverse of Corollary 2.3.11 needs not hold for $p = d = 3$.

Example 2.3.3. Consider Example 2.1.2 which exhibits a partition problem with $d = 3$, $n = 12$ and matrix $A \in \mathbb{R}^{3\times 12}$. With e^1, e^2 and e^3 as the unit vectors in $\mathbb{R}^3$, the example demonstrated that for the 3-partition $\pi = (\{1,2,3,4\},\{5,6,7,8\},\{9,10,11,12\})$, $(e^j)^T A^i > 0 > (e^j)^T A^u$ for each $i \in \pi_j$ and $u \in \pi_{j+1}$ (with $\pi_4 = \pi_1$), implying that π is cone-separable. It was further shown that A_π is not a vertex of any constrained-shape partition polytope where the corresponding set of shapes includes $(4,4,4)$; in particular, A_π is not a vertex of the single-size partition polytope $P_A^{\widehat{\Pi^p}}$ (the analysis was based on Theorem 2.1.5 which allows shapes in the corresponding set of shapes to have zero coordinates). □

We end this section with the observation that bounded- and constrained-shape partition problems can be solved by solving single-size partition problems for each of the feasible sizes. We emphasize that polyhedral methods are not fit for problems of non-uniform size as A_π's are then points (matrices) in Euclidean space of varied dimension and their convex hull cannot be defined. Still, it is sometimes (under further structural assumptions) useful to embed a bounded-size partition problem with p as the bound on the size of partitions into a single-size p problem by augmenting empty sets.

2.4 Enumerating (Limit-)Cone-Separable Partitions

Corollary 2.3.11 is next used to establish the optimality of cone-separable partitions (extending a result of Barnes, Hoffman and Rothblum [7] which considered the case where the A^i's are distinct).

Theorem 2.4.1. *Let $f(\cdot)$ be an (edge-)quasi-convex on $P_A^{\widehat{\Pi^p}}$. Then there exists an optimal partition π which is both cone-separable and (almost-) separable and has A_π as a vertex of $P_A^{\widehat{\Pi^p}}$. Further, if $f(\cdot)$ is strictly (edge-) quasi-convex, then every optimal partition π is cone-separable and has A_π as a vertex of $P_A^{\widehat{\Pi^p}}$.*

Proof. Again, let $\widehat{P} \equiv P_A^{\widehat{\Pi^p}}$. Corollary 3.3.7 of Vol. I assures that $f(\cdot)$ attains a maximum over $\widehat{P}$ at one of its vertices, and if $f(\cdot)$ is strictly (edge-)quasi-convex then each maximizer of $f(\cdot)$ over $\widehat{P}$ is a vertex. As each vertex of $\widehat{P}$ has a representation A_π with $\pi \in \widehat{\Pi^p}$ (Proposition 3.1.1(i) of Vol. I), it follows that

$$\phi^* \equiv \max_{x \in \widehat{P}} f(x) = \max_{\pi \in \widehat{\Pi^p}} f(A_\pi) = \max_{\pi \in \widehat{\Pi^p}} F(\pi),$$

and a partition π is optimal if and only if $F(\pi) = f(A_\pi) = \phi^*$. We can further conclude that there exists an optimal partition π that has A_π as a vertex of $\widehat{P}$ and when $f(\cdot)$ is strictly (edge-)quasi-convex, each optimal partition has this property; Corollary 2.3.11 assures that such partitions are cone-separable. Finally, if π is any optimal partition with A_π a vertex of $\widehat{P}$, Corollary 2.3.11 also implies that there is a partition π' which is both cone-separable and separable and has $A_{\pi'} = A_\pi$. As $F(\pi') = f(A_{\pi'}) = f(A_\pi) = \phi^*$, it then follows that π' is optimal. □

The next example demonstrates that the conclusions of Theorem 2.4.1 do not extend to single-size partition problems with parts required to be nonempty.

Example 2.4.1. Consider a single-size partition problem with $d = 1, p = 2, n \geq 2$, $A \in \mathbb{R}^{1 \times n}$ positive and $f(\cdot)$ strictly convex. This problem has precisely two cone-separable partitions – $\pi^1 \equiv (\mathcal{N}, \emptyset)$ and $\pi^2 \equiv (\emptyset, \mathcal{N})$. But, neither π^1 nor π^2 is feasible for the single-size partition problem with parts required to be nonempty, implying that no cone-separable partition is optimal for this problem. Of course, Theorem 2.4.1 assures that an optimal partition for the single-size partition problem which allows for empty parts must be either π^1 or π^2. □

In the current section we enhance the results of Section 2.3 and Theorem 2.4.1 by providing an algorithm for enumerating cone-separable partitions. As in Section 2.3, the partitions we consider allow for empty parts and the set of such p-partitions is denoted by $\widehat{\Pi^p}$. The approach we follow parallels the one used in Section 2.2 and relies on the introduction of a "limit-" version of cone-separability.

When studying cone-separability, it is useful to augment the matrix $A \in \mathbb{R}^{d \times n}$ of partitioned vectors with the 0 vector; the augmented matrix will be denoted $[0, A]$ and its columns will be indexed by $0, 1, \ldots, n$ with $[0, A]^0 = 0$. In particular, for $i_1 < \cdots < i_d$, we shall refer to $\text{sign}_{[0,A]}(0, i_1, \ldots, i_d)$. Also, we recall the notation "$\widetilde{v}$" which augments a vector v with 1 as the first coordinate.

Lemma 2.4.2. *Let $A \in \mathbb{R}^{d \times n}$ and $n \leq d$. If $[0, A]$ is generic, then every p-partition is cone-separable.*

Proof. It suffices to consider the case $p = 2$. As $[0, A]$ is generic and $n \leq d$, the $(d+1)$-vectors $\widetilde{0}, \widetilde{A}^1, \ldots, \widetilde{A}^n$ are linearly independent and rank $[\widetilde{0}, \widetilde{A}^1, \ldots, \widetilde{A}^n]^T = $ rank $[\widetilde{0}, \widetilde{A}^1, \ldots, \widetilde{A}^n] = n + 1$. It follows that the range of the matrix $[\widetilde{0}, \widetilde{A}^1, \ldots, \widetilde{A}^n]^T$ is $\mathbb{R}^{n+1}$. In particular, given a 2-partition $\pi = (\pi_1, \pi_2)$ of $\mathcal{N}$, there is a vector $\mu \in \mathbb{R}^{d+1}$ with $\mu^T \widetilde{0} = 0$, $\mu^T \widetilde{A}^i = 1 > 0$ for each $i \in \pi_1$ and $\mu^T \widetilde{A}^u = -1 < 0$ for each $u \in \pi_2$. As $\mu^T \widetilde{0} = 0$, the first coordinate of μ is 0; with C obtained from μ by truncating this first coordinate, we have that $C^T A^i = 1 > 0 > -1 = C^T A^u$ for all $i \in \pi_1$ and $u \in \pi_2$. So, π is cone-separable. □

Note that the assumption in Lemma 2.4.2 about $[0, A]$ being generic cannot be replaced by the (weaker) assumption that A is generic.

Example 2.4.2. Let $n = d = 2$, $A^1 = \binom{1}{1}$ and $A^1 = \binom{2}{2}$. Then the partition $\pi = (\{1\}, \{2\})$ is not cone-separable as a vector $C \in \mathbb{R}^2$ satisfies $C^T A^1 > 0$ if and only if it satisfies $C^T A^2 > 0$. □

Suppose $A \in \mathbb{R}^{d \times n}$ with $n \geq d - 1$ and $[0, A]$ generic, and let $I = \{i_1, \ldots, i_{d-1}\}$ be a $(d-1)$-subset of $\mathcal{N}$ with $i_1 < \cdots < i_{d-1}$. We adapt the notation used in (2.2.2) and set

$$\begin{aligned} I_A^- &\equiv \{i \in \mathcal{N} : \ \text{sign}_A(0, i_1, \ldots, i_{d-1}, i) = -1\}, \text{ and} \\ I_A^+ &\equiv \{i \in \mathcal{N} : \ \text{sign}_A(0, i_1, \ldots, i_{d-1}, i) = 1\}. \end{aligned} \tag{2.4.1}$$

(No confusion should occur with notation used in Section 2.2 as the set I herein is a $(d-1)$-set and not a d-set.) As $[0, A]$ is generic, Lemma 2.4.2 assures that $\{i \in \mathcal{N} \setminus I : \ \text{sign}_A(0, i_1, \ldots, i_{d-1}, i) = 0\} = \emptyset$, implying that $\{I_A^-, I_A^+\}$ is a 2-partition of $\mathcal{N} \setminus I$. Further, if (J^-, J^+) is a 2-partition of I, we define the 2-partitions of $\mathcal{N}$ *associated with* A, I and (J^-, J^+) as the two 2-partitions $(I_A^- \cup J^-, I_A^+ \cup J^+)$ and $(I_A^+ \cup J^+, I_A^- \cup J^-)$.

Lemma 2.4.3. *Suppose $A \in \mathbb{R}^{d \times n}$, $n \geq d - 1$ and $[0, A]$ is generic. Then the set of cone-separable 2-partitions is the set of all 2-partitions associated with A, $(d-1)$-sets $I \subseteq \mathcal{N}$ and 2-partitions (J^-, J^+) of I; in particular, there are at most $2^d \binom{n}{d-1} = O(n^{d-1})$ cone-separable 2-partitions.*

Proof. We will show that for each $(d-1)$-set $I \subseteq \mathcal{N}$ and 2-partition (J^-, J^+) of I, the two 2-partitions associated with A, I and (J^-, J^+) are cone-separable and that each cone-separable 2-partition is generated in this way. The asserted bound on the number of cone-separable 2-partitions follows from the fact that there are $\binom{n}{d-1}$ subsets I of $\mathcal{N}$ and 2^{d-1} partitions (J^-, J^+) of each such I.

First, let $I \subseteq \mathcal{N}$ have $(d-1)$ elements, say $i_1 < \cdots < i_{d-1}$, and let (J^-, J^+) be a 2-partition of I. Evidently, the determinant $\det[\widetilde{0}, \widetilde{A}^{i_1}, \ldots, \widetilde{A}^{i_{d-1}}, \widetilde{x}]$ is a linear function of $x \in \mathbb{R}^d$, that is, there exists $h \in \mathbb{R}^d$ such that

$$\det[\widetilde{0}, \widetilde{A}^{i_1}, \ldots, \widetilde{A}^{i_{d-1}}, \widetilde{x}] = h^T x \text{ for all } x \in \mathbb{R}^d$$

(the determinant is linear and not just affine as $x = 0$ is mapped to 0). It then follows that the hyperplane H in $\mathbb{R}^d$ that contains 0 and the columns of A indexed by I have the representation $H \equiv \{x \in \mathbb{R}^d : h^T x = 0\}$, while $I_A^- = \{i \in \mathcal{N} : h^T A^i < 0\}$ and $I_A^+ = \{i \in \mathcal{N} : h^T A^i > 0\}$. We next observe that $[0, A^{i_1}, \ldots, A^{i_{d-1}}]$ is generic, hence Lemma 2.4.2 assures that the 2-partition $(\{j : i_j \in J^-\}, \{j : i_j \in J^+\})$ of $\{1, \ldots, d-1\}$ is $A^{\{i_1, \ldots, i_{d-1}\}}$-cone-separable (meaning cone-separable with $A^{i_1}, \ldots, A^{i_{d-1}}$ as the partitioned vectors). Thus, there exists a vector $f \in \mathbb{R}^d$ with $f^T A^u < 0 < f^T A^w$ for all $u \in J^-$ and $w \in J^+$. For sufficiently small positive t, we then have that

$$(h + tf)A^u < 0 < (h + tf)^T A^w$$

for all $u \in I_A^- \cup J^-$ and $w \in I_A^+ \cup J^+$, proving that $(I_A^- \cup J^-, I_A^+ \cup J^+)$ is cone-separable. A symmetric argument shows that $(I_A^+ \cup J^+, I_A^- \cup J^-)$ is cone-separable too, proving that the two 2-partitions of $\mathcal{N}$ associated with A, I and 2-partition (J^-, J^+) of I are cone-separable.

Next assume that π is a cone-separable 2-partition. As in the proof of Lemma 2.2.3, repeated use of Lemma 3.1.8 of Vol. I shows that the hyperplane passing through the origin which separates A^{π_1} and A^{π_2} can be perturbed so that it "weakly separating" A^{π_1} and A^{π_2} while containing 0 and exactly $(d-1)$ columns of A. In particular, the hyperplane will then have the representation $H = \text{aff}\,\{0, A^{i_1}, \ldots, A^{i_{d-1}}\}$ with $1 \le i_1 < \cdots < i_{d-1} \le n$. It then follows that for $I \equiv \{i_1, \ldots, i_{d-1}\}$ either $I_A^- \subseteq \pi_1$ and $I_A^+ \subseteq \pi_2$ or $I_A^+ \subseteq \pi_1$ and $I_A^- \subseteq \pi_2$. In the former case we have $\pi = (I_A^- \cup J^-, I_A^+ \cup J^+)$ for $J^- = \pi_1 \cap I$ and $J^+ = \pi_2 \cap I$ and in the latter case $\pi = (I_A^+ \cup J^+, I_A^- \cup J^-)$ for $J^+ = \pi_1 \cap I$ and $J^- = \pi_2 \cap J$. $\square$

Lemma 2.4.3 gives a bound on the number of cone-separable 2-partitions and a method for constructing all such partitions; similar to the situation in Lemma 2.2.3, different sets of points and/or partitions may produce the same 2-partition. The following variant of Theorem 3.2.2 provides an exact count of the number of cone-separable 2-partitions when $[0, A]$ is generic (without improving on the order of the bound of Lemma 2.4.3); the result corrects an instance of Harding [45, Theorem 2].

As for separability, whenever there is more than one possible underlying matrix we index "cone-separability" by a prefix that indicates the underlying matrix, e.g., we shall refer to A-cone-separable partitions. When A has no zero vector, the definition of cone-separability in (2.3.1) implies that a partition π is A-cone-separable if and only if each pair of its parts is separable by a hyperplane that contains the origin.

Let $p \geq 2$. We recall from Section 2.2 that for each list of $\binom{p}{2}$ 2-partitions of $\mathcal{N}$ indexed as $\pi^{r,s} = (\pi_r^{r,s}, \pi_s^{r,s})$ where $1 \leq r < s \leq p$, we *associate* a p-tuple $\pi = (\pi_1, \ldots, \pi_p)$ of subsets of $\mathcal{N}$ such that for $j = 1, \ldots, p$

$$\pi_j \equiv \left(\cap_{t=j+1}^{p} \pi_j^{j,t}\right) \bigcap \left(\cap_{t=1}^{j-1} \pi_j^{t,j}\right). \tag{2.4.2}$$

In particular, $\pi_1, \ldots, \pi_p$ are pairwise disjoint and if it happens that $\cup_{j=1}^{p} \pi_j = \mathcal{N}$, then π is a p-partition which is called *the partition associated with* $\{\pi^{r,s} = (\pi_r^{r,s}, \pi_s^{r,s}) : 1 \leq r < s \leq p\}$.

The next result provides a bound on the number of cone-separable partitions, the bound is due to Hwang, Lee, Liu and Rothblum [51].

Theorem 2.4.4. *Suppose $A \in \mathbb{R}^{d \times n}$ and $p \geq 2$. Then the set of A-cone-separable p-partitions equals the set of p-partitions associated with lists of $\binom{p}{2}$ A-cone-separable 2-partitions; in particular, if $[0, A]$ is generic then there are at most $[2^d \binom{n}{d-1}]^{\binom{p}{2}} = O\left(n^{(d-1)\binom{p}{2}}\right)$ A-cone-separable p-partitions.*

Proof. Consider a p-partition $\pi = (\pi_1, \ldots, \pi_p)$ associated with a list $\{\pi^{r,s} = (\pi_r^{r,s}, \pi_s^{r,s}) : 1 \leq r < s \leq p\}$ of $\binom{p}{2}$ A-cone-separable 2-partitions. For each $1 \leq r < s \leq p$, $\pi_r \subseteq \pi_r^{r,s}$ and $\pi_s \subseteq \pi_s^{r,s}$ and therefore the cone-separability of $(\pi_r^{r,s}, \pi_s^{r,s})$ implies that (π_r, π_s) is cone-separable (see (2.3.1)). Conversely, let $\pi = (\pi_1, \ldots, \pi_p)$ be a cone-separable p-partition. Consider any pair r, s with $1 \leq r < s \leq p$. Then there exists a vector $C^{rs} \in \mathbb{R}^d$ such that

$$(C^{rs})^T A^u > 0 > (C^{rs})^T A^v \text{ for each } u \in \pi_r, v \in \pi_s \text{ with } A^u, A^v \neq 0.$$

By Lemma 3.1.9 of Vol. I with $\mathcal{L} = \{0\}$ there is a vector $F \in \mathbb{R}^d$ with $F^T A^i \neq 0$ for each $i \in \mathcal{N}$ with $A^i \neq 0$. It follows that by replacing C^{rs} with

$C^{rs} + \epsilon F$ for sufficiently small positive ϵ, we may and will assume that the aforementioned property of C^{rs} is preserved and, in addition, $(C^{rs})^T A^i \neq 0$ for each $i \in \mathcal{N}$ with $A^i \neq 0$. Let $\pi^{r,s} \equiv (\pi_r^{r,s}, \pi_s^{r,s})$ be any 2-partition satisfying $\pi_r \subseteq \pi_r^{r,s} \subseteq \{i \in \mathcal{N} : (C^{rs})^T A^i \geq 0\}$ and $\pi_s \subseteq \pi_s^{r,s} \subseteq \{i \in \mathcal{N} : (C^{rs})^T A^i < 0\}$ (the only ambiguity in the definition of the π^{rs}'s is in the assignment of indices $i \notin \pi_r \cup \pi_s$ for which $A^i = 0$). Evidently, π^{rs} is A-cone-separable 2-partition. As in the proof of Theorem 2.2.4, let π' be the p-tuple associated with the constructed $\pi^{r,s}$'s. Then the sets of π' are pairwise disjoint and for $j = 1, \ldots, p$ we have

$$\pi_j \subseteq \left(\cap_{t=j+1}^{p} \pi_j^{j,t}\right) \bigcap \left(\cap_{t=1}^{j-1} \pi_j^{t,j}\right) = \pi'_j.$$

Since $\mathcal{N} = \cup_{j=1}^{p} \pi_j \subseteq \cup_{j=1}^{p} \pi'_j \subseteq \mathcal{N}$, it follows that $\pi = \pi'$, that is, π is the p-partition associated with the constructed list of $\binom{p}{2}$ A-cone-separable 2-partitions.

Finally, if $[0, A]$ is generic, then the second conclusion of the theorem follows immediately from Lemmas 2.4.2, 2.4.3, and the first conclusion of the theorem. □

The proofs of Theorem 2.4.4 and Lemma 2.4.3 were first given by Hwang, Lee, Liu and Rothblum [51], but without the details provided here.

Enumerating All Cone-Separable Partitions when $[0, A]$ *is Generic*

Adapt the construction of Section 2.2 of all separable partitions when A is generic, except noting that there will only be $\binom{n}{d}$ signs to be computed and hence the bound on the required effort is then $O(n^d)$. □

We make a few comments (that parallel those following the algorithm of Section 2.2 for determining the set of separable partitions when A is generic). First, the above method can be executed in parallel. Second, the inductive proof of Lemma 2.2.3 yields an algorithm for enumerating the cone-separable 2-partitions, but it does not lend itself naturally to parallel computation. Finally, the complexity analysis of the above algorithm focuses on arithmetic operations and does not count operations needed to merge lists by intersections and unions.

Tighter bounds on the number of cone-separable partitions than the one given in Theorem 2.4.4 are available for $d \leq 2$ (and they apply to a broader class of matrices). We first consider the case where $d = 1$.

Theorem 2.4.5. *Suppose* $A \in \mathbb{R}^{1 \times n}$ *has no zero vector. If* A *has both* *positive and negative elements, then there are no cone-separable partitions if $p = 1$, and there are $p(p-1)$ cone-separable partitions if $p \geq 2$. If A has either only positive elements or only negative elements, then there are p cone-separable partitions.*

Proof. By Theorem 2.3.7, if A has both positive and negative elements, cone-separable partitions are specified by two (distinct) parts that equal $\{i \in \mathcal{N} : A^i > 0\}$ and $\{i \in \mathcal{N} : A^i < 0\}$, respectively; there are $2!\binom{p}{2}$ choices if $p \geq 2$ and none if $p = 1$. If A has only positive elements or only negative elements, Theorem 2.3.7 implies that cone-separable partitions are specified by a single nonempty part that equals $\mathcal{N}$; there are p choices. □

Recall the characterization of Theorem 2.3.8 and Corollary 2.3.9 for cone-separable partitions when $d = 2$ that used the angular coordinate $\phi(x)$ of vectors $x \in \mathbb{R}^2 \setminus \{0\}$. The next result, due to Aviran, Lev-Tov, Onn and Rothblum [5], tightens the bound of Theorem 2.4.4 on the number of cone-separable partitions when $d = 2$.

Theorem 2.4.6. *Suppose $A \in \mathbb{R}^{2\times n}$ has no zero vector. Then the number of cone-separable p-partitions is bounded by $\sum_{q=1}^{p} q!\binom{p}{q}\binom{n}{q} = O(n^p)$.*

Proof. Corollary 2.3.9 shows that all cone-separable partitions with specific q nonempty parts can be constructed from q values selected from the $\phi(A^i)$'s (the representation of Corollary 2.3.9 shows that not all selections will produce cone-separable partitions). It follows that the number of cone-separable partitions with specific q nonempty parts is bounded by $q!\binom{n}{q}$. Allowing the number of nonempty parts to range, and selecting the particular set of nonempty parts, we conclude that the number of cone-separable p-partitions is bounded by $\sum_{q=1}^{p} q!\binom{p}{q}\binom{n}{q} = O(n^p)$. □

The assumption in Theorems 2.4.6 and 2.4.5 that A contains no zero vector cannot be relaxed as cone-separable partitions may assign arbitrarily indices that correspond to zero vectors.

Theorem 2.4.5 shows that when $d = 1$ and A has no zero elements, the number of cone-separable partitions depends on whether or not A has both positive and negative elements. In particular, for $p > 2$, the number of A-cone-separable partitions depends on the elements of a (generic) matrix A and not just on its size.

Enumerating All Cone-Separable Partitions when $d \le 2$ and A has no Zero Vectors

To enumerate the cone-separable partitions when $d = 1$, one first determines if $\mathcal{N}_+ \equiv \{i \in \mathcal{N} : A^i > 0\}$ and $\mathcal{N}_- \equiv \{i \in \mathcal{N} : A^i < 0\}$ are nonempty. If one of them is empty, the cone-separable p-partitions are available by specifying the single part that equals the nonempty $\mathcal{N}_-/\mathcal{N}_+$. If both $\mathcal{N}_-$ and $\mathcal{N}_+$ are nonempty, the $p(p-1)$ cone-separable partitions are available by specifying the two parts that equal $\mathcal{N}_-$ and $\mathcal{N}_+$.

When $d = 2$, the cone-separable partitions with q nonempty parts can be enumerated by the following method: first select q distinct values from $\{\phi(A^1), \ldots, \phi(A^n)\}$ and order these values, say $0 \le \phi_1 < \cdots < \phi_q < 360°$. For each such selection, determine the sets

$$\sigma_t \equiv \{i \in \mathcal{N} : \phi_t \le \phi(A^i) < \min\{\phi_t + 180°, \phi_{t+1}\}\} \quad \text{for} \quad t = 1, \ldots, q-1$$

and

$$\sigma_q \equiv \{i \in \mathcal{N} : \phi_q \le \phi(A^i) < \min\{\phi_q + 180°, 360°\} \text{ or } 0 \le \phi(A^i) < \min\{\phi_q - 180°, \phi_1\}\}.$$

Corollary 2.3.9 shows that whenever this process generates a partition, it is cone-separable. Further, all cone-separable partitions will be determined in this way. Clearly, the generated q-tuples are pairwise disjoint and they form a partition if and only if they cover all the columns of A. A test for the latter is to verify that no $\phi(A^i)$ lies in

$$[\cup_{t=1}^{q-1}(\phi_t + 180°, \phi_{t+1})] \cup (\phi_q + 180°, 360°) \cup (\phi_q - 180°, \phi_1)$$

(where $(\cdot,\cdot)$ stands for open intervals). The amount of effort of generating and testing all $O(n^q)$ tuples is then $O(n^{q+1})$, and the total number of arithmetic operations that will cover all $q = 1, \ldots, p$ is $O(n^{p+1})$. □

We shall apply the notation for perturbing the columns of a matrix (introduced in Section 2.2) to the matrix $[0, A]$ with the first zero vector indexed as its 0-column; in particular, as $M_d^0 = 0 \in \mathbb{R}^d$, we have that $[0, A](\epsilon) = [0, A(\epsilon)]$. Also, for distinct indices $i_1, \ldots, i_d \in \mathcal{N}$, the arguments of Section 2.2 show that the determinant

$$D_{(0,i_1,\ldots,i_d)}(\epsilon) \equiv \det[0, \widetilde{A}(\epsilon)^{i_1}, \ldots, \widetilde{A}(\epsilon)^{i_d}]$$

is a nonzero polynomial of degree d in ϵ, assuring that its sign is invariant for all sufficiently small $\epsilon > 0$. We define the *generic sign* of A at $(i_1, \ldots, i_d)$, denoted $\chi_A(i_1, \ldots, i_d)$, as the common value of $\text{sign}_{A(\epsilon)}(0, i_1, \ldots, i_d)$ for all sufficiently small positive ϵ (as the referenced set has only d indices, no confusion should occur with the corresponding notation used in Section 2.2 which refers to $d+1$ indices). We recall that the efficient algorithm of Section 2.2 applies to computing the new generic signs.

When relevant, we shall use prefix to refer to the matrix of partitioned vectors when referring to cone-separability, e.g., we shall refer to A-cone-separable or $A(\epsilon)$-cone-separable partitions.

Lemma 2.4.7. *Let $A \in \mathbb{R}^{d\times n}$. For all sufficiently small $\epsilon > 0$: (i) $[0, A(\epsilon)]$ is generic, (ii) for every $(d-1)$-set $I \subseteq \mathcal{N}$, the sets $I^-_{A(\epsilon)}$ and $I^+_{A(\epsilon)}$ are independent of ϵ, and (iii) the set of $A(\epsilon)$-cone-separable p-partitions is invariant to ϵ. Further, if $[0, A]$ is generic, "$\epsilon > 0$" can be replaced by "$\epsilon \geq 0$".*

Proof. The proof of the lemma is identical to the one of Lemma 2.2.8, except that the references to Lemmas 2.2.2 and 2.2.3 and Theorem 2.2.4 are replaced with references to Lemmas 2.4.2 and 2.4.3 and Theorem 2.4.4. □

Let $A \in \mathbb{R}^{d\times n}$. A p-partition of $\mathcal{N}$ is called *A-limit-cone-separable* if it is $A(\epsilon)$-cone-separable for all sufficiently small $\epsilon > 0$. Denote by $\Pi^p_{A-lim-cone-sep}$ the set of A-limit-cone-separable p-partitions. Lemma 2.4.7 shows that for all sufficiently small $\epsilon > 0$, the set of $A(\epsilon)$-cone-separable partitions is the same and equals $\Pi^p_{A-lim-cone-sep}$. The following examples show that the new property does not imply cone-separability, and vice versa.

Example 2.4.3. Let $A = \binom{1\ 2}{1\ 2} \subset \mathbb{R}^{2\times 2}$ and consider the partition $\pi = (\{1\},\{2\})$. As cone $\{\binom{1}{1}\}$ = cone $\{\binom{2}{2}\}$, π is not A-cone-separable. But, for sufficiently small $\epsilon > 0$, the vector $C^\epsilon = \binom{2+\epsilon}{-2}$ satisfies $(C^\epsilon)^T\binom{1+\epsilon}{1+\epsilon} = \epsilon + \epsilon^2 > 0$ and $(C^\epsilon)^T\binom{2+\epsilon}{2+4\epsilon} = -2\epsilon + 2\epsilon^2 < 0$, implying that π is $A(\epsilon)$-cone-separable; so, π is A-limit-cone-separable. □

Example 2.4.4. Let $n = 2, d = 1$ and $A = (0, 0)$. In this case, every partition is A-cone-separable, in particular, this is true for $\pi = (\{1\},\{2\})$. But, for each $\epsilon > 0$, the only $A(\epsilon)$-cone-separable partitions are $(\{1,2\},\emptyset)$ and $(\emptyset,\{1,2\})$ and π is not one of these partitions. So, π is not A-limit-cone-separable. □

It is natural to speculate that given a cone-separable partition π, the indices corresponding to zero vectors can be reassigned so that the partition becomes limit-cone-separable. The forthcoming Theorem 2.4.14 and Example 2.4.6 address this issue.

The next result provides a polynomial bound on the number of limit-cone-separable partitions. It is followed by a description of a method for their enumeration.

Theorem 2.4.8. *Let* $A \in \mathbb{R}^{d\times n}$ *and* $p \geq 2$. *Then* $|\Pi^p_{A-lim-cone-sep}| \leq [2^d\binom{n}{d-1}]^{\binom{p}{2}} = O\left(n^{(d-1)\binom{p}{2}}\right)$; *if* $d = 2$, *the bound can be tightened to* $|\Pi^p_{A-lim-cone-sep}| \leq \sum_{q=1}^{p} q!\binom{n}{q} = O(n^p)$.

Proof. Let $\Pi \equiv \Pi^p_{A-lim-cone-sep}$. By Lemma 2.4.7 it is possible to select $\epsilon > 0$ so that $A(\epsilon)$ is generic and Π equals the set of $A(\epsilon)$-cone-separable p-partitions. The bounds on $|\Pi|$ is now immediate from Theorems 2.4.4 and 2.4.6. □

Theorem 2.3.7 determines a necessary and sufficient condition for the single 1-partition to be cone-separable. Also, when $d = 1$, Theorem 2.4.5 implies that the number of limit-cone-separable partitions is bounded by p^2; the order of this bound is $O(1)$ which coincides with the main bound of Theorem 2.4.8.

Enumerating All A-Limit-Cone-Separable Partitions when $d > 2$

Consider $A \in \mathbb{R}^{d\times n}$. If $n \leq d$, then for each $\epsilon > 0$ $A(\epsilon)$ is generic; therefore the set of $A(\epsilon)$-cone-separable p-partitions is the set of all partitions, of which there are $p^n \leq p^d$. It follows that this is the set of A-limit-cone-separable partitions and it can be generated using a fixed number of arithmetic operations. Also, if $p = 1$, $\Pi^p_{A-lim-cone-sep}$ is nonempty and contains the single partition $(\mathcal{N})$ if and only if $\mathcal{N}$ and $\{0\}$ are $A(\epsilon)$-cone-separable for sufficiently small $\epsilon > 0$, that is, if the parametric system $C^T A(\epsilon)$ (with C as the d-variable vector) is feasible for every $\epsilon > 0$ that is sufficiently small. Henceforth we assume that $n > d$ and $p \geq 2$. By Lemma 2.4.7, for sufficiently small positive ϵ, for each $(d-1)$-set $I \subseteq \mathcal{N}$, $I^-_{A(\epsilon)}$ and $I^+_{A(\epsilon)}$ are independent of ϵ. For a $(d-1)$-set $I = \{i_1, \ldots, i_{d-1}\} \subseteq \mathcal{N}$ with $i_1 < \cdots < i_{d-1}$, $I^-_{A(\epsilon)}$ and $I^+_{A(\epsilon)}$ for sufficiently small positive ϵ are available from the generic signs of A at $(0, i_1, \ldots, i_{d-1}, i)$ for each $i \in \mathcal{N}\backslash I$. A method described in Section 2.2 shows how each generic sign can be computed with $O(d^4)$ arithmetic operations; so, all $\binom{n}{d}$ generic signs can be computed with $O\left(\binom{n}{d}d^4\right) = O(n^d)$ arithmetic operations (permutations that put $\widetilde{A}^i$ into the right location may be applied).

Next, from Lemmas 2.4.7 and 2.4.3, $\Pi^p_{A-lim-cone-sep}$ equals the common set of $A(\epsilon)$-cone-separable partitions for sufficiently small positive ϵ; for such ϵ the set of $A(\epsilon)$-cone-separable 2- and p-partitions, $p > 2$, can be determined from the generic signs by the earlier method for enumerating the cone-separable partitions corresponding to generic matrices. As the method will only use the generic signs, the complexity bounds will be $O(n^{(d-1)p^2})$, the same as for a fixed generic matrix; further, this bound subsumes the work for computing the generic signs. □

The next lemma records relations between the set of limit-cone-separable and cone-separable partitions.

Lemma 2.4.9. *Let* $A \in \mathbb{R}^{d\times n}$.

(a) If A has no zero vector, then every A-cone-separable partition is A-separable and A-limit-cone-separable.
(b) If $[0, A]$ is generic, then a partition is A-cone-separable if and only if it is A-limit-cone-separable.

Further, let $N' \equiv \{i \in \mathcal{N} : A^i \neq 0\}$, *let* $A' \equiv A^{N'}$ *and consider the columns of* A' *indexed by the elements in* $\mathcal{N}'$.

(c) If π is an A-cone-separable partition, then the partition $\pi' = (\pi_1 \cap \mathcal{N}', \ldots, \pi_p \cap \mathcal{N}')$ of $\mathcal{N}'$ is A'-limit-cone-separable and A'-cone-separable.

Proof. If A has no zero vector, Lemma 2.3.1 shows that a cone-separable partition is separable and the fact that strict inequalities are preserved under small perturbations implies that a cone-separable partition is limit-cone-separable. This proves part (a). Next, part (b) is immediate from the final conclusion of Lemma 2.4.7 that applies to the case where $[0, A]$ is generic. Next, to verify part (c) observe that A' has no zero vector and that the constructed partition π' inherits cone-separability from π; the conclusion that π' is limit-cone-separable now follows from part (a). □

Theorem 2.4.8 and Lemma 2.4.9 allow us to extend the second conclusion of Theorem 2.4.4 for $d > 2$ from matrices A with $[0, A]$ generic to matrices A that contain no zero vector (for $d \leq 2$, Theorems 2.4.5 and 2.4.6 already handle such matrices).

Corollary 2.4.10. *Suppose* $A \in \mathbb{R}^{d\times n}$ *has no zero vector. Then the number of cone-separable partitions is bounded by* $[2^d \binom{n}{d-1}]^{\binom{p}{2}} = O\left(n^{(d-1)\binom{p}{2}}\right)$.

Proof. As A has no zero vector, Lemma 2.4.9 (part (a)) assures that every A-cone-separable partition is A-limit-cone-separable; the asserted bound then follows from Theorem 2.4.8. □

The assumption in Corollary 2.4.10 that A contains no zero vector cannot be relaxed as cone-separable partitions may assign arbitrarily indices that correspond to zero vectors.

Enumerating All A-Cone-Separable Partitions when A has no Zero Vectors

Consider $A \in \mathbb{R}^{d\times n}$. The case where $d \leq 2$ has already been considered, so assume that $d > 2$. Start by determining the set of all A-limit-cone-separable partitions. This can be accomplished by the methods described earlier in the current section using $O(n^{(d-1)p^2})$ arithmetic operations. Lemma 2.4.9 (part (a)) assures that each A-cone-separable partition will be produced, but Example 2.4.3 demonstrates that other partitions may be generated as well. Still, a test for A-cone-separability is available (in Section 2.3). Specifically, testing for cone-separability of a partition requires a feasibility test of each of the $O(n^{(d-1)\binom{p}{2}})$ generated limit-cone-separable partitions by testing the feasibility of $\binom{p}{2}$ linear systems where each has d variables and up to n constraints (see (2.3.2)). □

Enumerating All A-Cone-Separable Partitions when A has Zero Vectors

Consider $A \in \mathbb{R}^{d\times n}$ which has $n_0 > 1$ zero vectors. Let $\mathcal{N}' \equiv \{i \in \mathcal{N} : A^i \neq 0\}$ and let $A' \equiv A^{\mathcal{N}'}$. First, we enumerate all cone-separable partitions of script N' using the method given for the no-zero-vector case (distinguishing between the cases where $d = 1$, $d = 2$ and $d > 2$). Next, assign the n_0 zero vectors in $N \setminus \mathcal{N}'$ arbitrarily among the p parts. This procedure will clearly generate A-cone-separable partitions and Lemma 2.4.9 (part (c)) assures that it will generate all of them. The method is clearly polynomial. □

The next result shows that every vertex of a single-size partition polytope has a representation as a vector associated with a limit-cone-separable partition. We recall that Corollary 2.3.11 shows that every partition whose associated matrix is a vertex is cone-separable.

Theorem 2.4.11. *Let $A \in \mathbb{R}^{d\times n}$ and v be a vertex of $P_A^{\widehat{\Pi^p}}$. Then $v = A_\pi$ for some A-limit-cone-separable p-partition π.*

Proof. The proof follows the arguments proving Theorem 2.2.12 except that $\widehat{\Pi^p}$ replaces Π^Γ, Lemma 2.4.7 replaces Lemma 2.2.8, Corollary 2.3.11 replaces Theorem 2.2.1 and "-separable" and "-limit-cone-separable" are replaced by "-cone-separable" and "-limit-cone-separable". □

Here again, the number of matrices in the set $\{A_\pi : \pi \in \widehat{\Pi^p}\}$ is exponential in n, even for fixed d, p. But, Theorems 2.4.8 and 2.4.11 yield the following polynomial bound on the number of vertices of this polytope.

Corollary 2.4.12. *Let $A \in \mathbb{R}^{d\times n}$. Then the number of vertices of $P_A^{\widehat{\Pi^p}}$ is bounded by $[2^d\binom{n}{d-1}]^{\binom{p}{2}} = O\left(n^{(d-1)\binom{p}{2}}\right)$; further, if $d = 2$, the bound can be tightened to $\sum_{q=1}^{p} q!\binom{n}{q} = O(n^p)$.*

Proof. The bound is immediate from Theorems 2.4.8 and 2.4.11. □

For $d = 1$, the comment following Theorem 2.4.8 combines with Theorem 2.4.11 to show that the number of vertices of $P_A^{\widehat{\Pi^p}}$ when $d = 1$ is bounded by p^2; the order of this bound is $O(1)$ which coincides with the main bound of Corollary 2.4.12.

The next example demonstrates that the inverse of the implication established in Theorem 2.4.11 need not hold.

Example 2.4.5. Consider Example 2.3.3 (which is also discussed in Example 2.1.2). It exhibits a cone-separable partition π for which A_π is not a vertex of the corresponding single-size partition polytope. Observing that the matrix A for that example has $[0, A]$ generic, the A-cone-separable partition π of the example is A-limit-cone-separable (Lemma 2.4.9, part (b)). □

Theorem 2.4.11 is next used to establish the optimality of limit-cone-separable partitions.

Theorem 2.4.13. *Let $f(\cdot)$ be an (edge-)quasi-convex function on the single-size partition polytope $P_A^{\widehat{\Pi^p}}$. Then there exists an optimal partition π which is A-limit-cone-separable and has A_π as a vertex of $P_A^{\widehat{\Pi^p}}$.*

Proof. The proof is immediate from Theorem 2.4.11 and the arguments proving Theorem 2.2.14 (and Theorem 2.2.1). □

Solving Single-Size Partition Problems with $f(\cdot)$ (Edge-) Quasi-Convex by Enumerating Limit-Cone-Separable Partitions

The method for solving constrained-shape partition problems with $f(\cdot)$ (edge-)quasi-convex by enumerating limit-cone-separable partitions adapts word-by-word except that it will be based on the generation of limit-cone-separable partitions, instead of limit-separable partitions. For $d > 2$ we get the modified time complexity bound of $O\left(n^{(d-1)p^2}\right)$ arithmetic operations and queries to the f-oracle (but no tests on partition-shapes are needed). Of course, the basis of the algorithm is Theorem 2.4.13 rather than Theorem 2.2.14. For $d = 2$ and $d = 1$, one gets the corresponding improved bounds. □

The enumeration of limit-cone-separable partitions and associated matrices is next used to enumerate the vertices of single-size partition polytopes. Recall that an alternative method that accomplishes this goal was provided in Section 1.2, based on the simple structure of the set of edge-directions of these polytopes.

Enumerating the Vertices of Single-Size Partition Polytopes Using Limit-Cone-Separable Partitions

The above algorithm for solving single-size partition problems is based on enumerating limit-cone-separable partitions along with the associated matrices. As the set of associated matrices, say U, is guaranteed to include the set V of vertices of $P_A^{\widehat{\Pi^p}}$, the second paragraph of the Algorithm for Enumerating the Vertices of Constrained-Shape Partition Polytopes of Section 2.2 (and the following paragraph) shows how the method can be enhanced with an efficient test for matrices in U to be in V. □

Recall from Section 1.2 that a method for enumerating vertices of a polytope can be used to enumerate facets. Also, characterization of partitions whose associated matrix is a vertex of a corresponding bounded-shape partition polytope is provided in Theorems 1.3.1, 1.3.10 and 1.4.3. These can replace the test that is used in the above algorithm.

The remark following Example 2.4.4 raises the question of whether given a cone-separable partition π, the indices corresponding to zero vectors can be reassigned so that the resulting partition becomes limit-cone-separable.

Theorem 2.4.14. *Let $A \in \mathbb{R}^{d\times n}$ and π be an A-cone-separable 2-partition.*

Then there exists an A-limit-cone-separable 2-partition σ which coincides with π on the assignment of all indices i with $A^i \neq 0$ and has $A_\sigma = A_\pi$.

Proof. Let $J \equiv \{i \in \mathcal{N} : A^i = 0\}$. The cone-separability of π assures that for some vector $C \in \mathbb{R}^d$,

$$C^T A^t > 0 > C^T A^s \text{ for each } t \in \pi_1 \setminus J \text{ and } s \in \pi_2 \setminus J. \tag{2.4.3}$$

By Lemma 3.1.9 of Vol. I with $\mathcal{L} = \{0\}$, there is a vector $H \in \mathbb{R}^d$ satisfying

$$H^T M_d^i \neq 0 \text{ for each } i \in J. \tag{2.4.4}$$

For sufficiently small positive δ, (2.4.3) is satisfied with $C+\delta H$ replacing C and (2.4.4) is satisfied with $C+\delta H$ replacing H. By possibly replacing C with $C+\delta H$ for such δ, we may and will assume that (2.4.3) is satisfied and, in addition, $C^T M_d^i \neq 0$ for each index $i \in J$. Let $J_+ \equiv \{i \in J : C^T M_d^i > 0\}$ and $J_- \equiv \{i \in J : C^T M_d^i < 0\}$; we then have that (J_+, J_-) partitions J and for sufficiently small positive ϵ

$$C^T A(\epsilon)^t > 0 > C^T A(\epsilon)^s \text{ for each } t \in (\pi_1 \setminus J) \cup J_+ \text{ and } s \in (\pi_2 \setminus J) \cup J_-,$$

assuring that $\sigma \equiv ((\pi_1 \setminus J) \cup J_+, (\pi_2 \setminus J) \cup J_-)$ is $A(\epsilon)$-cone-separable 2-partition. Further, $\sigma_r \cap (\mathcal{N} \setminus J) = \pi_r \cap (\mathcal{N} \setminus J)$ for $r = 1, 2$ and $A_\pi = A_\sigma = (\sum_{i \in \pi_1 \setminus J} A^i, \sum_{i \in \pi_2 \setminus J} A^i)$. □

The final conclusion of Theorem 2.4.14 can be proved directly. Specifically, consider the single-size partition polytope P corresponding to $p = 2$. As π is a cone-separable 2-partition, Theorem 2.3.12 assures that A_π is a vertex of P and therefore Theorem 2.4.11 assures that there is an A-limit-cone-separable partition σ with $A_\sigma = A_\pi$.

The following example shows that neither of the two conclusions of Theorem 2.4.14 extends to single-size problems with $p > 2$.

Example 2.4.6. Let $n = 9, d = 3$ and

$$A = \begin{bmatrix} 0 & 1 & 1 & 1 & 1 & 1 & 1 & 1 & 1 \\ 0 & -2 & -2 & -1.99 & 100 & 2 & 2 & 1.99 & -100 \\ 0 & -1.99 & 100 & 2 & 2 & 1.99 & -100 & -2 & -2 \end{bmatrix}.$$

Consider the 4-partition $\pi = (\{1,2,3\}, \{4,5\}, \{6,7\}, \{8,9\})$ which has

$$A_\pi = \begin{bmatrix} 2 & 2 & 2 & 2 \\ -4 & 98.1 & 4 & -98.1 \\ 98.1 & 4 & -98.1 & -4 \end{bmatrix}.$$

The partition π is A-cone-separable as the vectors $C^{12} = (-1.999, -1, 0)^T$, $C^{13} = (0, -1, 0)^T$, $C^{14} = (1.999, 0, 1)^T$, $C^{23} = (-1.999, 0, 1)^T$, $C^{24} =$

$(0,0,1)^T$, $C^{34} = (-1.999,1,0)^T$ constitute the corresponding separating vectors.

Next let σ be a partition that either coincides with π or is obtained from π by reassigning the zero vector and we will show that σ is not A-limit-cone-separable. Let B be the matrix obtained from A by replacing the first zero vector by $(1,1,1)^T$, that is,

$$B = \begin{bmatrix} 1 & 1 & 1 & 1 & 1 & 1 & 1 & 1 & 1 \\ 1 & -2 & -2 & -1.99 & 100 & 2 & 2 & 1.99 & -100 \\ 1 & -1.99 & 100 & 2 & 2 & 1.99 & -100 & -2 & -2 \end{bmatrix}.$$

Observe that for each $\epsilon > 0$, $B(\epsilon)$ is obtainable from $A(\epsilon)$ by multiplying the first column by $\frac{1+\epsilon}{\epsilon}$; as such transformation of the underlying matrix preserves cone-separability, it suffices to show that σ is not B-limit-cone-separable, that is, it is not $B(\epsilon)$-cone-separable for all sufficiently small positive ϵ. It is easily verified that $[0, B]$ is generic. By Lemma 2.4.7(iii), the invariance of the set of $B(\epsilon)$-cone-separable p-partitions on ϵ also includes the case $\epsilon = 0$. Therefore it suffices to show that the corresponding partitions are not B-cone-separable.

Observe that all columns of B lie on the plane $H = \{x \in \mathbb{R}^3 : x_1 = 1\}$; Figure 2.4.1 illustrates the projection of these columns on H. The figure demonstrates that there is no separable partition $\sigma = (\sigma_1, \sigma_2, \sigma_3, \sigma_4)$ with $\sigma_r \supseteq \{2r, 2r+1\}$ for $r = 1,2,3,4$; precisely, the convex hull of the part containing B^1 overlaps with another convex hull. Thus, σ is not B-separable and therefore not B-cone-separable by Lemma 2.3.1. It follows that σ is not an A-limit-cone-separable partition, and thus $A_\sigma \neq A_\pi$ for every partition $\sigma \neq \pi$. □

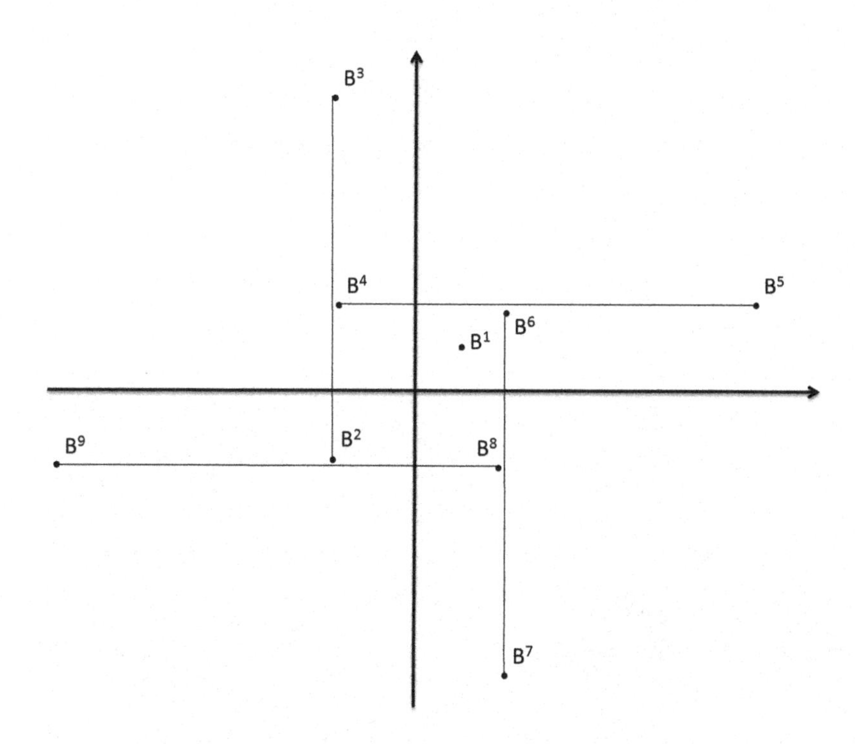

Figure 2.4.1 A cone-separable partition where no reassignment of zero vectors will convert to a limit-cone-separable one.

Chapter 3

Partitions over Multi-Parameter Spaces: Combinatorial Structure

In this chapter, multi-parameter partition problems other than the sum-partition problems are studied. As the polyhedral approach of Chapters 1 and 2 no longer applies, the "combinatorial approach" as introduced in Chapters 6 and 7 in Vol. I will be extended to the multi-parameter environment to solve these more general partition problems. The general plan follows Vol. I: to identify partition properties which define classes of partitions satisfying these properties respectively, to bound and to enumerate these partition-property classes, and to prove for certain types of objective functions, some corresponding partition-property classes contain optimal partitions. When such a class is of polynomial size, then the brute-force method of enumerating all members of this class, comparing their objective values and selecting a best partition works.

Partition-property classes are also specified by restrictions on the shape of partitions, for example, single-shape, bounded-shape, constrained-shape, single-size.... Instead of stating the results for every such class (having the same partition property), Lemma 1.2.1 of Vol. I allows us to state it just for the smallest such class (usually "single-shape") which automatically extends to the other larger classes (having the same partition property) containing it.

Finally, we remind the reader that even if a multi-parameter property Q introduced here reduces to a single-parameter property Q' introduced in Chapter 6 of Vol. I, we cannot automatically take the results of Q' as the special case of Q when $d = 1$. This is because Q' is defined on index-partition while Q is on vector-partition. Cases where this inheritance is allowed will be addressed in the text.

3.1 Properties of Partitions

For a finite set $\Omega \in \mathbb{R}^d$, let $\operatorname{int}\Omega$ denote the interior of $\operatorname{conv}\Omega$. Given finite subsets Ω^1 and Ω^2 of $\mathbb{R}^d$, we say that Ω^1 *penetrates* Ω^2, written $\Omega^1 \to \Omega^2$, if $\Omega^1 \cap \operatorname{conv}\Omega^2 \neq \emptyset$. We say that Ω^1 strictly penetrates Ω^2 if $\Omega^1 \cap \operatorname{int}\Omega^2 \neq \emptyset$.

Lemma 3.1.1. *If Ω^1 and Ω^2 are finite sets in $\mathbb{R}^d$ such that the vectors in their union are in general position, then Ω^1 penetrates Ω^2 if and only if Ω^1 strictly penetrates Ω^2.*

Proof. As the points of Ω^2 are in general position, a face of dimension $0 < k < d$ of $\operatorname{conv}\Omega^2$ contains exactly k points; further, as the points of $\Omega^1 \cup \Omega^2$ are in general position, no such face can contain a point of Ω^1. As the relative interiors of the nonempty faces of $\operatorname{conv}\Omega^2$ partition $\operatorname{conv}\Omega^2$ (Proposition 3.1.6 of Vol. I), we conclude that if $\Omega^1 \cap (\operatorname{conv}\Omega^2) \neq \emptyset$; any point in the intersection must then lie in $\operatorname{int}\Omega^2$. This proves that penetration implies strict penetration. As the opposite implication is trite, the proof of the lemma is complete. □

We note that some implications about penetration that are true in $\mathbb{R}^1$ do not hold in $\mathbb{R}^d$ with $d > 1$. For example, $\Omega^1 \to \Omega^2$ and $\Omega^2 \not\to \Omega^1$ no longer imply $\operatorname{conv}\Omega^1 \subset \operatorname{conv}\Omega^2$ (see Figure 3.1.1(a)), while $\Omega^1 \not\to \Omega^2$ and $\Omega^2 \not\to \Omega^1$ no longer imply $(\operatorname{conv}\Omega^1) \cap (\operatorname{conv}\Omega^2) = \emptyset$ (see Figure 3.1.1(b)).

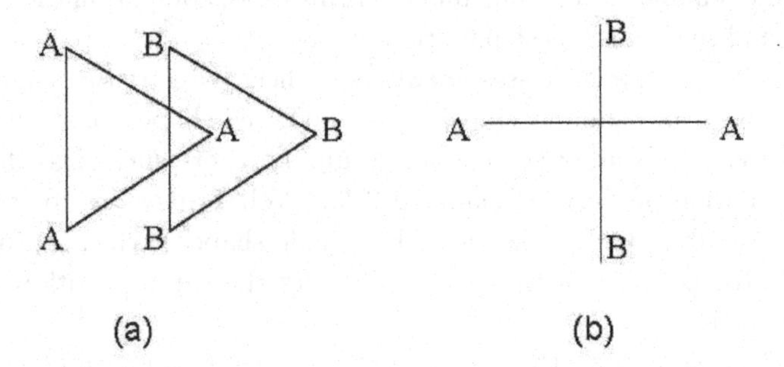

Figure 3.1.1 Implications of penetration not in $\mathbb{R}^d$ for $d > 1$

We next introduce properties of partitions of vectors in $\mathbb{R}^d$, with $d > 1$. The fact that some implications of penetration in $\mathbb{R}^1$ do not hold in $\mathbb{R}^d$ for $d > 1$ implies that some properties of partitions of points in $\mathbb{R}^1$ that were defined in terms of penetration have more than one counterpart for partitions of points in $\mathbb{R}^d$.

A *sphere* in $\mathbb{R}^d$ is defined as a set of the form $S = \{x \in \mathbb{R}^d : \|x - a\| \leq R\}$ for some $a \in \mathbb{R}^d$ and $R > 0$, where $\| \ \|$ stands for the Euclidean norm. The interior and boundary of such a sphere is given by $\operatorname{int} S = \{x \in \mathbb{R}^d : \|x - a\| < R\}$ and $\operatorname{bd} S = \{x \in \mathbb{R}^d : \|x - a\| = R\}$.

Consider a partition $\pi = (\pi_1, \ldots, \pi_p)$ of a finite set of vectors in $\mathbb{R}^d$. We shall consider the following properties of such partitions. The first three were introduced in Chapter 2.

- Separable (S) (also referred to as "disjoint" in the literature): For all distinct $r, s = 1, \ldots, p$, π_r and π_s are separable (see the definition of separability in Section 2.1 – by Corollary 2.1.3, an equivalent condition is that $(\operatorname{conv} \pi_r) \cap (\operatorname{conv} \pi_s) = \emptyset$).
- Almost-separable (AS): For all distinct $r, s = 1, \ldots, p$, π_r and π_s are almost-separable (see the definition of almost-separability in Section 2.1 – by Corollary 2.1.3, an equivalent condition is that $(\operatorname{conv} \pi_r) \cap (\operatorname{conv} \pi_s)$ contains at most a single point, and if the intersection contains a point, it is a vertex of both $\operatorname{conv} \pi_r$ and $\operatorname{conv} \pi_s$).
- Cone-separable (C_nS): For all distinct $r, s = 1, \ldots, p$, π_r and π_s are cone-separable (see the definition of cone separability in Section 2.3 – by Lemma 2.3.5, an equivalent condition is that $(\operatorname{cone} \pi_r) \cap (\operatorname{cone} \pi_s) = \{0\}$.
- Sphere-separable (SS): For all distinct $r, s = 1, \ldots, p$, there exists a sphere $S \subset \mathbb{R}^d$ such that either $(U, W) = (\pi_r, \pi_s)$ or $(U, W) = (\pi_s, \pi_r)$, satisfies $S \supseteq U$ and $S \cap W = \emptyset$.
- Convex-separable (C_vS): For all distinct $r, s = 1, \ldots, p$, there exists a convex set $S \subset \mathbb{R}^d$ such that either $(U, W) = (\pi_r, \pi_s)$ or $(U, W) = (\pi_s, \pi_r)$, satisfies $S \supseteq U$ and $S \cap W = \emptyset$. (Evidently, without loss of generality, it is possible to assume that $\dim S = d$.)
- Almost-sphere-separable (ASS): For all distinct $r, s = 1, \ldots, p$, there exists a sphere $S \subset \mathbb{R}^d$ such that either $(U, W) = (\pi_r, \pi_s)$ or $(U, W) = (\pi_s, \pi_r)$, satisfies $S \supseteq U$ and $|S \cap W| \leq 1$ and if x is a single point in $S \cap W$, then $U \cap (\operatorname{bd} S) = \{x\}$.
- Nonpenetrating (NP): For all distinct $r, s = 1, \ldots, p$, $\pi_r \not\to \pi_s$.
- Noncrossing (NC): For all distinct $r, s = 1, \ldots, p$, either $(\operatorname{conv} \pi_r) \cap (\operatorname{conv} \pi_s) = \emptyset$ or one convex hull is contained in the other with no member of the larger part penetrating the smaller part.
- Acyclic (AC): There do not exist $q \geq 2$ parts of π, say $\pi_{i_1}, \ldots, \pi_{i_q}$, such that $\pi_{i_1} \to \pi_{i_2} \to \ldots \to \pi_{i_q} \to \pi_{i_1}$.
- Monopolistic (M_p): One part has all elements. Note that if empty parts

are prohibited, there is only one partition satisfying M_p and its size is 1, if empty parts are allowed it means that $p-1$ parts are empty.

- Nearly monopolistic (N_eM_p): At most one part of π has more than a single point.
- Nearly cone-separable (N_eC_nS): For all distinct $r, s = 1, \ldots, p$, π_r and π_s are either cone-separable or at least one of them is a singleton.

We note that S (called disjoint) and C_nS were first considered in Barnes, Hoffman and Rothblum [7], C_vS (called "noncrossing") in Boros and Hwang [10], AS (called "separable") in Hwang, Onn and Rothblum [53], SS, NP (called "nested") and AC (called "connected") in Boros and Hammer [9], NC (for $d=1$) in Kreweras [65] and N_eM_p in Pfersky, Rudolf and Woeginger [80]. We also note that two subsets Ω^1 and Ω^2 of $\mathbb{R}^d$ are separable if and only if there exists a (closed) half-space S such that $S \supseteq \Omega^1$ and $S \cap \Omega^2 = \emptyset$, and Ω^1 and Ω^2 are cone-separable if and only if there exists a (closed) cone S of dimension d such that $S \supseteq \Omega^1$ and $S \cap \Omega^2 = \{0\}$.

Two subsets Ω^1 and Ω^2 of $\mathbb{R}^d$ are *weakly separable* if there exists a nonzero d-vector C such that

$$C^T u^1 \geq C^T u^2 \text{ for all } u^1 \in \Omega^1 \text{ and } u^2 \in \Omega^2. \tag{3.1.1}$$

Evidently, this condition is equivalent to the existence of a half-space S with $S \supseteq \Omega^1$ and $(\text{int } S) \cap \Omega^2 = \emptyset$. We say that two sets Ω^1 and Ω^2 in $\mathbb{R}^d$ are *weakly cone-separable* if there exists a nonzero d-vector C such that

$$C^T u^1 \geq 0 \geq C^T u^2 \text{ for all } u^1 \in \Omega^1 \text{ and } u^2 \in \Omega^2. \tag{3.1.2}$$

Evidently, this condition is equivalent to the existence of a closed cone S with $S \supseteq \Omega^1$ and $(\text{int } S) \cap \Omega^2 \subseteq \{0\}$. We can then define the following weak version of separability

- Weakly Separable (WS): For all distinct $r, s = 1, \ldots, p$, π_r and π_s are weakly separable.
- Weakly cone-separable (WC_nS): For all distinct $r, s = 1, \ldots, p$, π_r and π_s are weakly cone-separable.

Weak versions of SS and C_vS can then be defined by relaxing the requirement that $S \cap W$ equals $\emptyset$ or $\{0\}$, respectively, to $(\text{int } S) \cap W = \emptyset$.

- Weakly sphere-separable (WSS): For all distinct $r, s = 1, \ldots, p$, there exists a sphere $S \subset \mathbb{R}^d$ such that either $(U, W) = (\pi_r, \pi_s)$ or $(U, W) = (\pi_s, \pi_r)$, satisfies $S \supseteq U$ and $(\text{int } S) \cap W = \emptyset$.

- Weakly convex-separable (WC_vS): For all distinct $r, s = 1, \ldots, p$, there exists a convex set S with dim $S = d$ such that either $(U, W) = (\pi_r, \pi_s)$ or $(U, W) = (\pi_s, \pi_r)$, satisfies $S \supseteq U$ and $(\text{int } S) \cap W = \emptyset$.

The definitions of properties NP, NC and AC include requirements that prohibit penetration. We can obtain *weaker* versions of these properties by replacing "penetration" with "strict penetration". We then get the following new "weak" properties:

- Weakly Nonpenetrating (WNP)
- Weakly Noncrossing (WNC)
- Weakly Acyclic (WAC).

Finally, property M_p is defined independently of the notion of penetration, thus does not have a weak version.

We refer to the properties S, C_nS, SS, C_vS, NP, NC, AC and $(N_e)M_p$ as *strong*, to AS, ASS, WS, WC_nS, WC_vS, WSS, WNP, WNC, WAC and N_eC_nS as *weak properties*. Note that a partition can satisfy a strong property only if all equivalent elements are assigned to the same part. Therefore, when a result concerns strong properties only, we can treat the set of distinct elements as the set to be partitioned. In particular, when empty parts are not allowed in a p-partition problem, we always assume n (the number of distinct elements) $\geq p$ to avoid trivial discussion.

The 1-dim partition property "nestedness" can be extended to $\mathbb{R}^d$ by the following requirement: A partition is *nested* if for any two parts, the convex hull of one of them is contained in the convex hull of the other. However, nested partitions are not very interesting in $\mathbb{R}^d$ for $d > 1$ since the number of nonempty parts is dependent on the geometry of A. For example, if A is the set of vertices of a polytope in $\mathbb{R}^d$, then the only nested p-partition is the partition with $p - 1$ empty parts. Thus, we will not consider nested partitions for general d. It is noted that for $d = 1$ in Vol. I, there always exists a nested partition for any shape not containing two singletons.

When $d = 1$, both S and NP reduce to consecutiveness, while N_eC_nS, NC, SS, AC and C_vS all reduce to noncrossingness. Further, when $d = 1$ partitions in C_nS can have at most one part containing positive elements, at most one part containing negative elements, some parts each consisting of only the 0-element (but may have multiple copies). For a set of points with k 0-elements, this set is not size(p)-regular for $p > k + 2$ and not shape-regular for almost all shapes. Hence, C_nS was not studied in Vol. I.

For $d = 1$, when there are enough 0-elements, C_nS reduces to bi-extremal. But, recall that N_eM_p does not reduce to extremal since the singletons can be anywhere not necessarily the smallest elements.

Theorem 3.1.2. *The relations among the properties of partitions that we introduced are represented in Figure 3.1.2. We draw the relations among the strong properties and the relations among the weak properties separately except that C_nS is inserted into the latter figure since all its relations are to weak properties. The implication links from the strong graph to the weak graph are: $S \Rightarrow AS$, $SS \Rightarrow ASS$ and $Q \Rightarrow WQ$ for $Q \in \{S, NP, NC, SS, AC, C_vS\}$.*

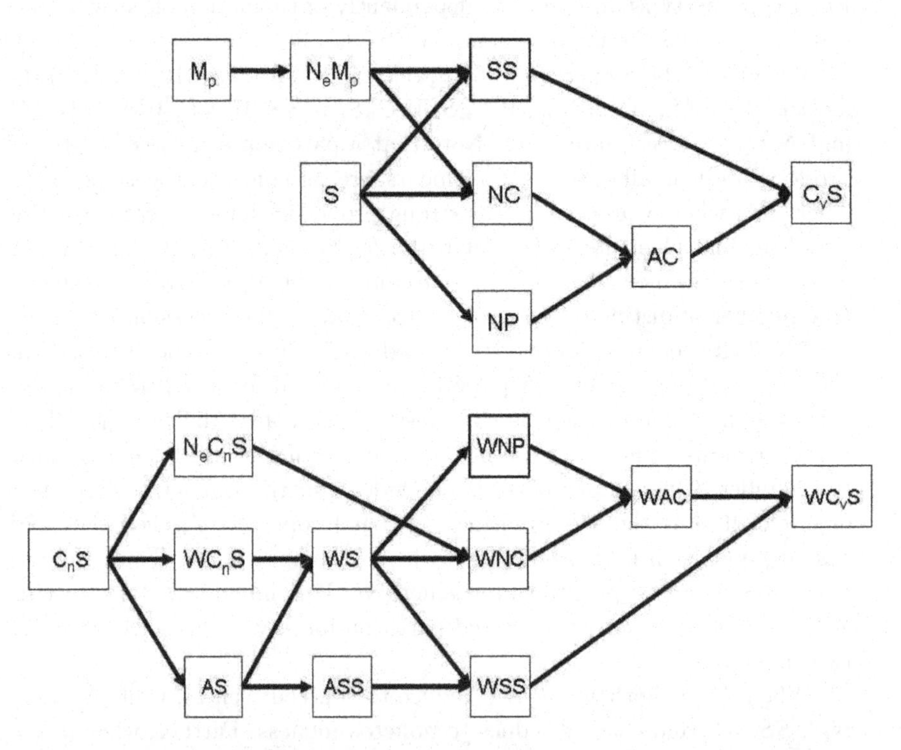

Figure 3.1.2 Implications among properties of multi-parameter partitions

Proof. Most implications and non-implications are pretty straightforward. We only verify the less obvious statements. For non-implication, we use figures where common capital letters indicate points that are in the same part.

(i) $S \Rightarrow SS$. It suffices to show that if two finite sets are separable, then they are sphere-separable. So, let Ω^1 and Ω^2 be two separable finite sets in $\mathbb{R}^d$ and let $C \in \mathbb{R}^d$ be a separating vector satisfying

$$C^T x < C^T y \text{ for every } x \in \Omega^1 \text{ and } y \in \Omega^2.$$

It then follows from the finiteness of Ω^1 and Ω^2 that for some sufficiently small positive ϵ

$$\epsilon\|y\|^2 - 2C^T y < \epsilon\|x\|^2 - 2C^T x \text{ for every } x \in \Omega^1 \text{ and } y \in \Omega^2,$$

or equivalently, $\|y - \frac{C}{\epsilon}\|^2 < \|x - \frac{C}{\epsilon}\|^2$. Letting γ be any positive number satisfying

$$\|y - \frac{C}{\epsilon}\| < \gamma < \|x - \frac{C}{\epsilon}\| \text{ for every } x \in \Omega^1 \text{ and } y \in \Omega^2,$$

we have that the sphere $S \equiv \{z \in \mathbb{R}^d : \|z - \frac{C}{\epsilon}\| < \gamma\}$ separates between Ω^1 and Ω^2.

(ii) $AS \Rightarrow ASS$. The above proof that $S \Rightarrow SS$ proves that $AS \Rightarrow ASS$, except that one has to qualify throughout that $x \neq y$.

(iii) $SS \not\Rightarrow AC$. See Figure 3.1.3.

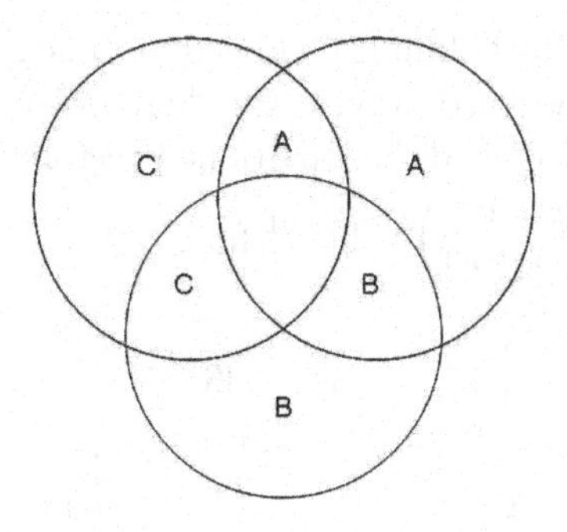

Figure 3.1.3 $SS \not\Rightarrow AC$ in Theorem 3.1.2

(iv) $NP \not\Rightarrow SS$. See Figure 3.1.4.

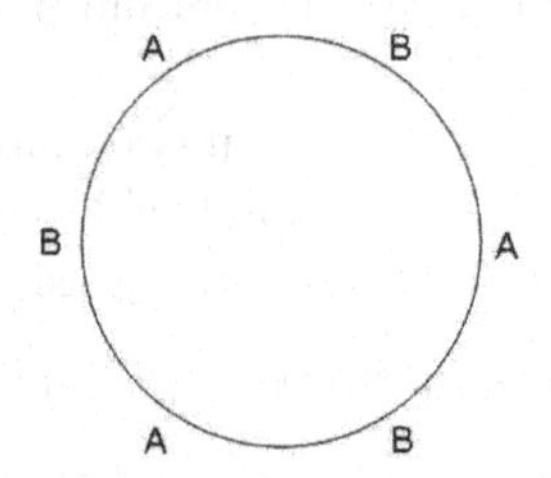

Figure 3.1.4 $NP \not\Rightarrow SS$ in Theorem 3.1.2

(v) $WS \not\Rightarrow WSS$. See Figure 3.1.5.

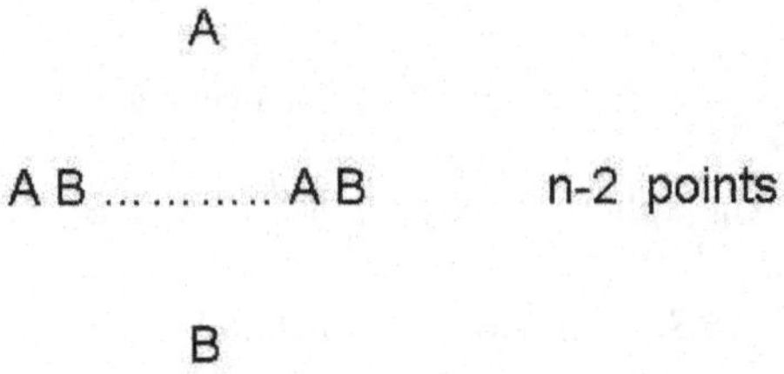

A

A B .... A B n-2 points

B

Figure 3.1.5 $WS \not\Rightarrow WSS$ in Theorem 3.1.2

(vi) $M_p \not\Rightarrow C_nS$. Suppose $d = p = 2$ and $A = \{(0,1),(0,-1)\}$. Then $\pi = (A, \emptyset)$ is monopolistic but not cone-separable. □

Golany, Hwang and Rothblum [41] proved an (unexpected) relation between S and SS. Here we also provide the "almost" version of the result. To state these results, we embed the partitioned vectors in $\mathbb{R}^{d+1}$ by defining

$$\widehat{A}^i \equiv \begin{pmatrix} A^i \\ \|A^i\|^2 \end{pmatrix} \in \mathbb{R}^{d+1} \text{ for } i = 1,\ldots,n \tag{3.1.3}$$

and

$$\widehat{A} \equiv \left[\widehat{A}^1,\ldots,\widehat{A}^n\right] \in \mathbb{R}^{(d+1)\times n}. \tag{3.1.4}$$

Next, for each p-partition $\pi = (\pi_1,\ldots,\pi_p)$, let $\widehat{\pi}$ be the p-partition for the problem with partitioned vectors $\widehat{A}^1,\ldots,\widehat{A}^n$ with $\widehat{\pi}_j = \{\widehat{A}^i : i \in \pi_j\}$ for $j = 1,\ldots,p$; in particular,

$$\widehat{A}_{\widehat{\pi}} = \left[\sum_{i\in\pi_1} \widehat{A}^i,\ldots,\sum_{i\in\pi_p} \widehat{A}^i\right] \in \mathbb{R}^{(d+1)\times p} \tag{3.1.5}$$

Theorem 3.1.3. *A partition π is (weak/almost) sphere-separable if and only if $\widehat{\pi}$ is (weak/almost) separable for data where the partitioned vectors are $\widehat{A}^1, \ldots, \widehat{A}^n$.*

Proof. We first observe that for every $x, y, u \in \mathbb{R}^d$,

$$\|x+u\| < \|y+u\| \text{ if and only if } \|x\|^2 + 2u^T x < \|y\|^2 + 2u^T y\,. \quad (3.1.6)$$

Assume first that $\widehat{\pi}$ is a separable partition (for the problem where the partitioned vectors are $\widehat{A}^1, \ldots, \widehat{A}^n$). Consider any pair of parts of π, say π_1 and π_2. It then follows that there exists a vector $b \in \mathbb{R}^d$ and scalar β such that

$$(b^T, \beta)\begin{pmatrix} A^i \\ \|A^i\|^2 \end{pmatrix} < (b^T, \beta)\begin{pmatrix} A^t \\ \|A^t\|^2 \end{pmatrix} \quad \text{for each } i \in \pi_1 \text{ and } t \in \pi_2. \quad (3.1.7)$$

Now, if $\beta = 0$, then (3.1.7) implies that π_1 and π_2 are separable (as they satisfy the definition of separability from Section 2.1, with b as the separating vector); as $S \Rightarrow SS$ (Theorem 3.1.2), we conclude that π_1 and π_2 are sphere-separable. Alternatively, if $\beta > 0$, (3.1.7) can be rewritten as

$$\left(\frac{b}{\beta}\right)^T A^i + \|A^i\|^2 < \left(\frac{b}{\beta}\right)^T A^t + \|A^t\|^2 \quad \text{for each } i \in \pi_1 \text{ and } t \in \pi_2, \quad (3.1.8)$$

or equivalently (using (3.1.6) with $x = A^i$, $y = A^t$ and $u = \frac{b}{2\beta}$)

$$\left\| A^i + \frac{b}{2\beta} \right\| < \left\| A^t + \frac{b}{2\beta} \right\| \quad \text{for each } i \in \pi_1 \text{ and } t \in \pi_2. \quad (3.1.9)$$

It follows that with γ as any positive scalar satisfying

$$\left\| A^i + \frac{b}{2\beta} \right\| < \gamma < \left\| A^t + \frac{b}{2\beta} \right\| \quad \text{for each } i \in \pi_1 \text{ and } t \in \pi_2, \quad (3.1.10)$$

the sphere

$$S = \left\{ x \in \mathbb{R}^d : \left\| x - \left(-\frac{b}{2\beta} \right) \right\| \leq \gamma \right\} \quad (3.1.11)$$

verifies the sphere separability of π_1 and π_2. In the remaining case where $\beta < 0$, the inverse of the inequalities of (3.1.8) and (3.1.9) hold. With γ as any positive scalar for which the inverse inequalities of (3.1.10) hold, π_1 and π_2 are separable by the sphere given in (3.1.11).

Next assume that π is a sphere-separable partition and consider any pair of parts of π, say π_1 and π_2. Let $S = \{x \in \mathbb{R}^d : \|x - a\| \leq \gamma\}$ be a separating sphere between π_1 and π_2. By perturbing γ and possibly exchanging the roles of π_1 and π_2, we have that

$$\|A^i - a\| < \gamma < \|A^t - a\| \quad \text{for each } i \in \pi_1 \text{ and } t \in \pi_2. \quad (3.1.12)$$

It now follows from (3.1.6) that

$$-2a^T A^i + \|A^i\|^2 < -2a^T A^t + \|A^t\|^2 \quad \text{for each } i \in \pi_1 \text{ and } t \in \pi_2. \quad (3.1.13)$$

So, (3.1.7) holds with $b = -2a$ and $\beta = 1$, demonstrating that $\widehat{\pi}_1$ and $\widehat{\pi}_2$ are separable.

The weak version of our theorem follows from the above argument with strict inequalities replaced by weak inequalities in (3.1.6)–(3.1.13). We next consider the equivalence of the "almost" versions. Again, we consider any pair of parts of π, say π_1 and π_2, and will demonstrate that they are almost-sphere-separable if and only if $\widehat{\pi}_1$ and $\widehat{\pi}_2$ are almost-separable. In view of the first part of the proof, the only case of interest is when $|\pi_1 \cap \pi_2| \neq \emptyset$ and x is a unique common vector occurring in π_1 and π_2. Of course, $\widehat{x}$ is then a unique common vector of $\widehat{\pi}_1$ and $\widehat{\pi}_2$. We next observe that if $\widehat{\pi}_1$ and $\widehat{\pi}_2$ are almost-separable, there exists a vector $b \in \mathbb{R}^d$ and scalar β such that

$$(b^T, \beta)\begin{pmatrix} A^i \\ \|A^i\|^2 \end{pmatrix} < (b^T, \beta)\begin{pmatrix} A^t \\ \|A^t\|^2 \end{pmatrix} \text{ for each } i \in \pi_1 \setminus \{x\} \text{ and } t \in \pi_2 \setminus \{x\}. \quad (3.1.14)$$

Now, if $\beta = 0$ we have that (3.1.14) assures that π_1 and π_2 are almost-separable, hence almost-sphere-separable (see Theorem 3.1.2). Alternatively, if $\beta > 0$, the earlier arguments about (sphere) separability show that the sphere defined by (3.1.11) has $\pi_1 \subseteq S$, $\pi_2 \cap S = \{x\}$ and $\pi_1 \cap (\text{bd } S) = \{x\}$, proving that π_1 and π_2 are almost-sphere-separable. In the remaining case where $\beta < 0$, we get the same conclusions as for the case $\beta > 0$ except for an exchange of π_1 and π_2.

We finally assume that π_1 and π_2 are almost-sphere-separable with $S = \{x \in \mathbb{R}^d : \|x - a\| \leq \gamma\}$ as a corresponding separating sphere. By possibly exchanging the roles of π_1 and π_2, we then have that $\pi_1 \subseteq S$, $\pi_2 \cap S = \{x\}$ and $\pi_1 \cap (\text{bd } S) = \{x\}$. Following the arguments that showed that sphere separability of π implies separability of $\widehat{\pi}$, we conclude that the linear functional determined by $\binom{-2a}{1}$ (on $\mathbb{R}^{d+1}$) attains at $\widehat{A}^u$ a unique maximum over $\widehat{\pi}_1$ and a unique minimum over $\widehat{\pi}_2$, proving that π_1 and π_2 are almost-sphere-separable. □

We observe that the proofs of the equivalences of Theorem 3.1.3 are constructive and show how to convert separating vectors that verify (almost-) separability into spheres that verify (almost-) sphere separability, and vice versa.

Lemma 3.1.4. *Suppose Ω^1 and Ω^2 are subsets of $\mathbb{R}^d$.*

(a) *Suppose Ω^1 and Ω^2 are weakly separable and C is a nonzero d-vector which satisfies (3.1.1). Let $\sup_{u\in\Omega^1} C^T u \leq \gamma \leq \inf_{u\in\Omega^2} C^T u$ and $H \equiv \{x \in \mathbb{R}^d : C^T x = \gamma\}$. Then Ω^1 and Ω^2 are (almost, weakly) separable if and only if $\Omega^1 \cap H$ and $\Omega^2 \cap H$ are (almost, weakly) separable.*

(b) *Suppose Ω^1 and Ω^2 are weakly cone-separable and C is a nonzero d-vector which satisfies (3.1.2). Let $H \equiv \{x \in \mathbb{R}^d : C^T x = 0\}$. Then Ω^1 and Ω^2 are cone-separable if and only if $\Omega^1 \cap H$ and $\Omega^2 \cap H$ are cone-separable.*

Proof. We first establish part (a). Necessity of the condition for Ω^1 and Ω^2 to be (almost-) separable is trite. To verify sufficiency of the condition for almost-separability, suppose $\Omega^1 \cap H$ and $\Omega^2 \cap H$ are almost-separable and D is a corresponding separating vector, that is,

$$D^T u^1 > D^T u^2 \quad \text{for all } u^1 \in \Omega^1 \cap H \text{ and } u^2 \in \Omega^2 \cap H \text{ with } u^1 \neq u^2. \tag{3.1.15}$$

As

$$C^T u^1 > C^T u^2 \quad \text{for all } u^1 \in \Omega^1 \setminus H \text{ and } u^2 \in \Omega^2 \setminus H \tag{3.1.16}$$

we have that for a sufficiently small positive ϵ,

$$(C + \epsilon D)^T u^1 > (C + \epsilon D)^T u^2 \text{ for all } u^1 \in \Omega^1 \text{ and } u^2 \in \Omega^2 \text{ with } u^1 \neq u^2, \tag{3.1.17}$$

proving the almost-separability of Ω^1 and Ω^2. The same argument proves the sufficiency of the condition for separability, except for the drop of "$u^1 \neq u^2$" in (3.1.15) and (3.1.17).

The proof of part (b) follows from the (modified) arguments that establish part (a), with "$>$" replaced by "$> 0 >$" and with the exclusion of $u^1 = 0$ and $u^2 = 0$. □

We emphasize that the sets in Lemma 3.1.4 are allowed to be empty. The next result identifies situations where the weak (almost-) properties imply the corresponding strong properties.

Lemma 3.1.5. *If the A^i's are in general position, then $WS \Rightarrow S$, $WC_vS \Rightarrow C_vS$, $WNP \Rightarrow NP$, $WNC \Rightarrow NC$ and $WAC \Rightarrow AC$. If $\binom{A^1}{\|A^1\|^2}, \ldots, \binom{A^n}{\|A^n\|^2}$ are in general position, then $WSS \Rightarrow SS$. If $0, A^1, \ldots, A^n$ are in general position, then $WC_nS \Rightarrow C_nS$. If the A^i's are distinct, then $AS = S$ and $ASS = SS$.*

Proof. Assume that the A^i's are in general position. We first consider the first three implications. For that purpose, consider a partition π and an arbitrary pair of parts of π, say π_r and π_s. It suffices to show that for each such pair, if the 2-partition (π_r, π_s) satisfies the corresponding weak property, then it satisfies the strong property.

First assume that (π_r, π_s) are weakly separable and we will show that it is separable. Indeed, let H be a hyperplane that weakly separates π_r and π_s. As the A^i's are in general position there are at most d A^i's on H and Lemma 2.2.2 assures that the intersection of $(\pi_r \cap H, \pi_s \cap H)$ is a separable 2-partition. It then follows from Lemma 3.1.4 that (π_r, π_s) is a separable 2-partition.

Next assume that (π_r, π_s) is weakly convex-separable. Let S is a convex set of dimension d which contains one of the two parts and whose interior has an empty intersection with the other part, say the first part is π_r and the second one is π_s. Without loss of generality it can be assumed that S is a polytope (in particular, $\operatorname{conv} \pi_r$ will do when the dimension of this set is d). Express S as the intersection of half-spaces; points of π_s that are in S are on its boundary, that is, on a boundary hyperplane of one of the half-spaces. Next apply to each hyperplane the perturbation argument of the proof of Lemma 3.1.4, yielding a perturbed polytope S that still contains π_r and that contains no point of π_s in its interior.

Next, the three implications about the strong and weak versions of NP, NC and AC, are immediate from Lemma 3.1.1 which implies that when the A^i's are in general position, penetration among parts of a partition implies strict penetration. Next, the implication $WSS \Rightarrow SS$ when $\binom{A^1}{\|A^1\|^2}, \ldots, \binom{A^n}{\|A^n\|^2}$ are in general position, is immediate from Theorem 3.1.3 and the first conclusion of the current theorem. Next, the case of cone separability follows from arguments that are similar to those that were used for separability, relying on the second part of Lemma 3.1.4 (instead of the first part) and on Lemma 2.4.2 (instead of Lemma 2.2.2). Finally, the part of the theorem concerning the weak properties is trivial. □

The following two examples demonstrate the assertion that the A^i's in general position is not sufficient for $WSS \Rightarrow SS$ and $WC_nS \Rightarrow C_nS$.

Example 3.1.1. Let $d = 2$, $n = 4$ and $A^1 = \binom{1}{1}$, $A^2 = \binom{1}{-1}$, $A^3 = \binom{-1}{-1}$ and $A^4 = \binom{-1}{1}$. The partition $\pi = (\pi_1, \pi_2)$ with $\pi_1 = \{A^1, A^2, A^3\}$ and $\pi_2 = \{A^4\}$ is WSS, with $S = \{x \in \mathbb{R}^2 : \|x\|^2 \leq 2\}$ as a separating sphere between π_1 and π_2; but π is not SS as any sphere that contains π_1 contains π_2. Here, A^1, A^2, A^3, A^4 are in general position, but $\widehat{A}^1, \widehat{A}^2, \widehat{A}^3, \widehat{A}^4$ are not (as they are contained in the hyperplane $\{z \in \mathbb{R}^3 : z_3 = 2\}$). □

Example 3.1.2. Let $d = 2$, $n = 2$ and $A^u = \binom{u}{0}$ for $u = 1, 2$. The partition $\pi = (\pi_1, \pi_2)$ with $\pi_1 = \{A^1\}$ and $\pi_2 = \{A^2\}$ is WC_nS, with $C = \binom{0}{1}$ as a weakly separating vector between π_1 and π_2; but π is not C_nS as a vector $C \in \mathbb{R}^2$ satisfies $C^T A^1 > 0$ if and only if $C^T A^2 > 0$. Here, A^1, A^2 are in general position, but $0, A^1, A^2$ are not. □

When the A^i's are not in general position, the hope is that we can transform a weakly separable partition to an (almost-) separable partition. Note that if the number of distinct points is at least p, then a separable p-partition always exists. Such a separable partition can be obtained by looking at the convex hull of the vectors; this polytope has a vertex and assigning its copies as the only points in π_1 keeps π_1 nonempty and separable from the set of all other vectors. Iteratively, looking at a vertex of the convex hull of all remaining vectors and assigning its copies as the only points in the next part until $p - 1$ parts have been set. Having all remaining vectors at the last part will complete the construction. If the number of distinct points is less than p, but $n \geq p$, we can repeat the above construction but with the assignment of a single copy of a vertex at each stage, and the process will result in an almost-separable partitions.

The construction in the above paragraph demonstrates that a transformation from a weakly separable partition to an (almost-) separable partition is trivial if we do not put any restrictions on the transformation. Since the ultimate goal of dealing with (almost-) separable partitions is to obtain an optimal partition with the corresponding property, the restriction should be concerned with preserving optimality. On the other hand, we do not want to tie the transformation to a particular objective function. Thus, we will choose a restriction which is broadly consistent with either preserving optimality or being used as the statistics $s(\cdot)$ over partitions in proving sortability. The restriction that we shall impose is that after the transformation, $\operatorname{conv} \pi'_j \subseteq \operatorname{conv} \pi_j$ for $j = 1, \ldots, p$. Note that this inclusion guarantees that volume, diameter, smallest radius of an enclosing sphere of $\operatorname{conv} \pi'_j$ does not exceed the corresponding characteristics of $\operatorname{conv} \pi_j$. Unfortunately, we have useful transformations of this type only for $d = 2$ (as suggested by Lemma 3.1.4).

Theorem 3.1.6. *Let $d = 2$ and let π be a weakly separable p-partition. If empty parts are allowed, then there exists a separable p-partition π' with*

$$\text{conv } \pi'_j \subseteq \text{conv } \pi_j \quad \text{for } j = 1, ..., p, \text{ with strict inequality holding for at least one } j \text{ when } \pi \neq \pi'. \tag{3.1.18}$$

Proof. Consider two parts π_r and π_s of π which are not separable. If all vectors in $\pi_r \cup \pi_s$ coincide, let $\pi'_r = \pi_r \cup \pi_s$ and $\pi'_s = \emptyset$; these sets are separable and satisfy (3.1.18) for $j = r, s$. So, assume that $\pi_r \cup \pi_s$ has at least two distinct elements. As π is weakly separable, π_r and π_s are weakly separable. Using terminology of Chapter 2, there exists a line L that weakly separates π_r and π_s, and (using Lemma 3.1.4) it can be assumed that L contains (at least) two distinct points of $\pi_r \cup \pi_s$; in particular, π_r is included in the union of $L_r \equiv L \cap \pi_r$ and one of the two open half-spaces determined by L, say L^-, while π_s is included in the union of $L_s \equiv L \cap \pi_s$ and the other open half-space, say L^+.

Next observe that the distinct points of $L_r \cup L_s = (\pi_r \cup \pi_s) \cap L$ can be ordered as $\ell_1, ..., \ell_m$, $m \geq 2$, such that $\text{conv}\,\{\ell_1, \ell_t\}$ is (weakly) increasing in $t \in \{1, \ldots, m\}$ and ℓ_1 is a common vertex of these intervals (each of the ℓ_i's may correspond to multiple copies of the partitioned vectors). By possibly switching the indices r and s, we may assume that $\ell_1 \in \pi_r$. Let $x = \max\{t : \ell_t \in \pi_r\}$. As π_r and π_s are not separable, $\{\ell_1, \ldots, \ell_x\}$ must contain a point that is present in π_s (for otherwise L_r and L_s are separable and Lemma 3.1.5 would imply that π_r and π_s are separable). Let $y = \min\{t : \ell_t \in \pi_s\}$. In particular, $y \leq x$ and ℓ_y is a vertex of $\text{conv}\, L_s$. As $\text{conv}\, L_s$ is an extreme set of $\text{conv}\,\pi_s$, Theorem 3.1.1 of Vol. I assures that ℓ_y is a vertex of $\text{conv}\,\pi_s$. Now, let $\pi'_r = \mathcal{L}^- \cup \{\ell_1, \ldots, \ell_x\}$ and $\pi'_s = L^+ \cup \{l_{x+1} \ldots \ell_m\}$ (the latter could be empty). Then $\text{conv}\,\pi'_r = \text{conv}\,\pi_r$, $\pi'_s \subseteq \pi_s \setminus \{\ell_y\}$ and $\text{conv}\,\pi'_s \subset \text{conv}\,\pi_s$ (the last inclusion is strict as $\ell_y \in \text{conv}\,\pi_s$ but $\ell_y \notin \text{conv}\,\pi'_s$). Further, $\pi'_r \cap L$ and $\pi'_s \cap L$ are separable and L (continues to) weakly separate π'_r and π'_s; consequently, Lemma 3.1.4 assures that π'_r and π'_s are separable. Evidently, for $j = r, s$, $\text{conv}\,\pi'_j \subseteq \text{conv}\,\pi_j$ assures that weak and regular separability of pairs of parts of π is preserved when π_r and π_s are replaced, respectively, by π'_r and π'_s. Hence, the above construction can be executed iteratively on all pairs of parts, with each pair considered at most once. The conclusion of this process will produce a partition with the asserted properties. $\square$

The next example shows that when empty parts are prohibited, it is impossible to change a non-separable partition into another while preserving (3.1.18), let alone to gain separability.

Example 3.1.3. Let $d = 2$, $n = 3$, $p = 2$, $A = \begin{pmatrix} -1 & 0 & 1 \\ 0 & 0 & 0 \end{pmatrix}$ and $\pi = (\{\binom{-1}{0}, \binom{1}{0}\}, \{\binom{0}{0}\})$. Then π is not separable and there is no partition $\pi' \neq \pi$ with nonempty parts that satisfies (3.1.18), let alone one that is separable.

The next example shows that the conclusion of Theorem 3.1.6 cannot be extended to $d = 3$ whether or not empty sets are allowed.

Example 3.1.4. Let $d = 3$, $n = 8$, $p = 2$ and let

$$A = \begin{pmatrix} 0 & 0 & 0 & 0 & 0 & 0 & 1 & -1 \\ 1 & 1 & -2 & -1 & -1 & 2 & 0 & 0 \\ 1 & -1 & 0 & 1 & -1 & 0 & 0 & 0 \end{pmatrix}$$

so, $A^1, \ldots, A^6$ are the vertices of two triangles in the plane $x_1 = 0$ which form a star of David (as is illustrated in Figure 3.1.6) and A^7 and A^8 are on the two sides of that plane. Consider the 2-partition π with $\pi_1 = \{A^1, A^2, A^3, A^7\}$ and $\pi_2 = \{A^4, A^5, A^6, A^8\}$. As no column of A is in the convex hull of the other 7 columns, moving any column from π_1 or π_2 to the other part cannot result in a new part whose convex hull is contained in the convex hull of the original part. Thus, there is no partition $\pi' \neq \pi$ that satisfies (3.1.18), let alone one that is separable (and π is not separable). $\square$

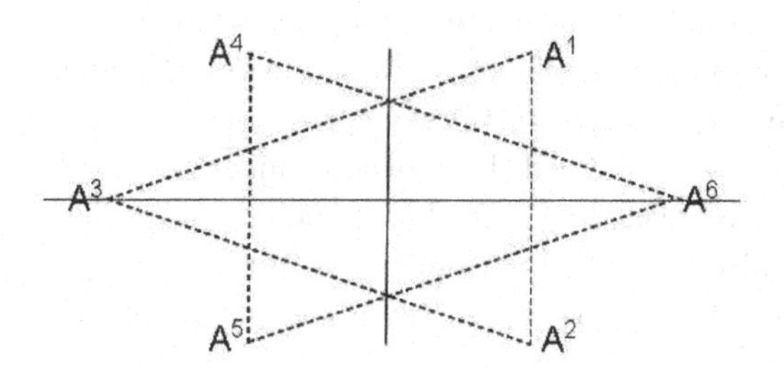

Figure 3.1.6 The projection of $A^1, \ldots, A^6$ on the plane $x_1 = 0$ for Example 3.1.4

Lemma 3.1.7. *Assume that all the A^i's lie on the boundary of a sphere. Then a partition is (almost-) sphere-separable if and only if it is (almost-) separable.*

Proof. As scaling and transposition preserve $(A)SS$ and $(A)S$, we may and will assume that the A^i's lie on the boundary unit ball $U \equiv \{x : \|x\| = 1\}$, that is, $\|A^i\| = 1$ for each i. The definitions of $(A)SS$ assure that $(A)S \Rightarrow (A)SS$ (without restriction on the location of the A^i's). To see that $ASS \Rightarrow AS$ consider an almost-sphere-separable partition $\pi = (\pi_1, \ldots, \pi_p)$ and distinct $r, s = 1, \ldots, p$. Suppose $S = \{x \in \mathbb{R}^d : \|x - a\| \leq R\} \subseteq \mathbb{R}^d$ with $S \supseteq \pi_r$ and $S \cap \pi_s$ including at most a single element and if x is such a point, then $\pi_r \cap (\text{int } S) = \{x\}$. Either π_r or π_s is empty or contains (possibly multiple copies of) a single point is trite; so assume that this is not the case which implies that $a \neq 0$. Next observe that with $\alpha \equiv 1 - R^2 + \|a\|^2$ and $\propto$ represents $<, >$ or $=$,

$$U \cap \{x \in \mathbb{R}^d : \|x - a\| \propto R\} = U \cap \{x \in \mathbb{R}^d : a^T x \geq \alpha\}.$$

As $\pi_r \cup \pi_s \subseteq U$, we conclude that $\pi_r \subseteq V \equiv \{x \in \mathbb{R}^d : a^T x \geq \alpha\}$ and $\pi_s \cap V$ consists of at most a single element. Moreover, any point in $\pi_s \cap V$ is contained in $\pi_s \cap S$ and the assumptions about such a point assure that it is also in π_r and that it uniquely maximizes/minimizes the linear function $a^T x$ defined by a over π_s/π_r, assuring that such a point is a vertex of $\operatorname{conv} \pi_r$ and $\operatorname{conv} \pi_s$. This proves that π_r and π_s are almost-separable. The argument about $SS \Rightarrow S$ follows similarly. □

Lemma 3.1.7 does not have a "weak" version – when all the A^i's lie on the boundary of a sphere, weak sphere separability does not imply weak separability; see the proof of Theorem 3.1.2.

In Sections 2.2 and 2.4 we considered "lim-separable" and "cone-lim-separable" partitions and enumerations of these classes of partitions were presented as a device to enumerate vertices of corresponding partition polytopes. These properties were defined through some invariance under certain perturbations of the partitioned vectors. We do not consider these properties here as their definitions depend not only on A, but also on the selection of directions along which the partitioned vectors are to be perturbed.

3.2 Counting and Enumerating Partition Classes of Single-Size

In this section we bound the number of Q-partitions for each property Q that was introduced in Section 3.1. Recall that in Chapter 6 of Vol. I we define $\#_Q(n,p)$ as the number of single-parameter p-partitions on n distinct elements (not allowing empty parts); as we focused on index-partitions, the number was independent of the parameters associated with the indices. Here, we consider vector-partitions and the counts may depend on the partitioned vectors. For each $A \in \mathbb{R}^{d \times n}$, let $\#_Q^A(p)$ be the number of p-partitions of the columns of A that satisfy Q, allowing empty parts. The emphasis in the current section is to examine whether $\#_Q(n,p,d) \equiv \max_{A \in \mathbb{R}^{d\times n}} \#_Q^A(p)$ grows exponentially or polynomially in n, with $p \geq 2$ and d fixed; in the latter case, an enumeration algorithm will be given.

Counting of a polynomial class is usually done by providing an upper bound of the form $O(n^m)$ for some positive m.

Every labeled p-partition corresponds to at most $p!$ unlabeled partitions (the option for identical parts implies that the bound needs not always be tight). Thus, the number of labeled p-partitions is bounded by $p!$ times the number of unlabeled p-partitions (for index-partitions there are no identical parts, so the bound is realized), a multiplier that is independent of n. Consequently, the $O(\cdot)$ order of the classes of labeled and unlabeled partitions with a given property coincide. In the current section we find it convenient to count the labeled classes. We also continue to allow empty parts unless specified otherwise.

When columns of A are not distinct, we refer to a vector $x \in \mathbb{R}^d$ as a *multiple point* (of A) if it appears more than once among the columns of A; the *multiplicity* of a multiple point x is the number of times x appears among the columns of A.

Note that a partition of a multiset A can satisfy a strong partition property only if all copies of a multiple point go to the same part. Therefore, we might as well assume that we are partitioning the subset A' of A which consists of a single copy of each point, multiple or not. To simplify writing, we will rename A' as A, which now stands for a set of n distinct vectors whenever dealing with strong partition properties. This interpretation is not only true in this chapter but also true in Chapter 2 when counting separable partitions and cone-separable partitions (in the latter case, a 0-vector can remain to be multiple), although we did not state it there explicitly.

As in Chapter 2, we shall use prefix "A-" to index that a partition property refers to partitions of the columns of the matrix A. Also, for $A \in \mathbb{R}^{d\times n}$ and $E \in \mathbb{R}^{d\times u}$, a p-partition π is $(A|E)$-*separable* if every pair of parts of π are separable by a hyperplane that contains the columns of E.

In Section 2.2, a direct bound of $\#_S^A(p)$ was given for A generic. When A is not generic, then a bound of the number of limit-separable partitions is given which can serve as a bound for $\#_S^A(p)$. The same strategy is used to bound $\#_{CS}^A(p)$. In this section, we give an exact count of the number of $(A|E)$-separable partitions, when $[A, E]$ is generic, and then use it to tighten the bounds of $\#_S^A(p)$ and $\#_{CS}^A(p)$ by specifying E.

Recall that Corollary 2.2.11 gave the following bound of A-separable partitions,

$$\#_S^A(p) \leq \left[2^{d+1}\binom{n}{d}\right]^{\binom{p}{2}} = O(n^{d\binom{p}{2}}). \tag{3.2.1}$$

We now give a better bound. Harding [45, Theorem 1] gave an exact count of the number of A-separable unlabeled 2-partitions when A is generic (it does not improve on the order of the bound of Lemma 2.2.3). We present Harding's result in a more general context which corrects Harding [45, Theorem 2]; see [60].

For $n, d \geq 1$, define the (n, d)-*Harding number*, denoted $H(n, d)$, by

$$H(n,d) \equiv \sum_{j=0}^{d} \binom{n-1}{j} \tag{3.2.2}$$

(where we use the standard convention that $\binom{n}{0} = 1$ for each $n \geq 0$ and $\binom{n}{k} = 0$ if $k > n$). Using a standard binomial identity that for $n, d \geq 2$

$$\begin{aligned} H(n,d) &= \sum_{j=0}^{d} \binom{n-1}{j} = \sum_{j=0}^{d} \left[\binom{n-2}{j} + \binom{n-2}{j-1}\right] \\ &= H(n-1,d) + H(n-1,d-1), \end{aligned} \tag{3.2.3}$$

in particular, we have that for each d, $H(n, d)$ is increasing in n and in d.

Lemma 3.2.1. $H(n, d) < 2^d\binom{n}{d}$.

Proof. Obviously, this is true for $n = d = 1$. Suppose this lemma holds for any (n, d) with $n + d < k$, $n \geq d \geq 1$ and $k \geq 2$. We prove it for k. From (3.2.3),

$$\begin{aligned} H(n,d) &= H(n-1,d) + H(n-1,d-1) \\ &< (2^d)\binom{n-1}{d} + (2^{d-1})\binom{n-1}{d-1} \\ &= (2^{d-1})\binom{n-1}{d} + (2^{d-1})\left[\binom{n-1}{d} + \binom{n-1}{d-1}\right] \\ &= (2^{d-1})\binom{n-1}{d} + (2^{d-1})\binom{n}{d} \\ &< (2^d)\binom{n}{d}. \end{aligned}$$

□

Theorem 3.2.2. *Suppose $A \in \mathbb{R}^{d\times n}$ and $E \in \mathbb{R}^{d\times u}$, where $0 \leq u \leq d$ and $[A, E]$ is generic. Further, if $u = d$, then assume that A is not on one side of the hyperplane uniquely determined by E. With empty parts allowed, the number of $(A|E)$-separable 2-partitions is $2H(n, d-u)$ (and it is $2H(n, d-u) - 2$ if empty parts are prohibited).*

Proof. Fix $d \geq 1$. The proof follows by a double induction on u and n. If $n = 2$ and $0 \leq u < d$, then there are precisely two partitions. As $[A, E]$ is generic, there is a hyperplane containing the columns of E and excluding both columns of A. Thus, the two partitions are $(A|E)$-separable and the number of such partitions is $2 = 2H(2, d-u)$. Next consider the case where $u = d$. As $[A, E]$ is generic, there is a unique $(d-1)$-plane that contains the d columns of E, contains no point of A, and splits A into two nonempty parts (since A is not on one side of the plane). So, the unique $(d-1)$-plane containing the columns of E defines precisely two $(A|E)$-separable partition and the number of such partitions is $2 = 2H(n, 0) = 2H(n, d-d)$.

Next assume that $0 \leq u' < d$ and $n' \geq 3$ are integers for which the conclusion of our theorem holds whenever $(-u, n)$ is lexicographically less than $(-u', n')$ and we will prove it when $(-u, n) = (-u', n')$. For finite sets $X, Y \subseteq \mathbb{R}^d$, let $\Pi_S(X|Y)$ be the set of 2-partitions of X that are separable by a $(d-1)$-plane that contains Y. Our goal is to show that $|\Pi_S(A|E)| = 2H(n, d-u)$.

Select any point $v \in A$. Evidently, if $\pi = (\pi_1, \pi_2) \in \Pi_S(A|E)$ is separable by a $(d-1)$-plane G_π that contains E, then $\sigma = (\pi_1 \setminus \{v\}, \pi_2 \setminus \{v\})$ is separable by G_π and it is therefore in $\Pi_S(A \setminus \{v\}|E)$. Next consider a 2-partition $\sigma = (\sigma_1, \sigma_2) \in \Pi_S(A \setminus \{v\}|E)$ which is separable by a $(d-1)$-plane G_σ that contains E. If v is on the side of G_σ that contains σ_1, then $(\sigma_1 \cup \{v\}, \sigma_2) \in \Pi_S(A|E)$. Similarly, if v is on the side of G_σ that contains σ_2, then $(\sigma_1, \sigma_2 \cup \{v\}) \in \Pi_S(A|E)$. It remains to consider the case where v is on G_σ itself. As the points in $E \cup \{v\}$ are in general position, G_σ can be perturbed in two ways to generate two hyperplanes G_σ^1 and G_σ^2, each containing E and separating σ_1 and σ_2 with v on the side of G_σ^1 that contains σ_1 and on the side of G_σ^2 that contains σ_2. It follows that all partitions of $\Pi_S(A \setminus \{v\}|E)$ generate all partitions of $\Pi_S(A|E)$ by adding v to one of their parts. Of course, no partition in $\Pi_S(A|E)$ can be generated by different partition of $\Pi_S(A \setminus \{v\}|E)$.

We have seen that each partition in $\sigma \in \Pi_S(A \setminus \{v\}|E)$ generates at least one partition in $\Pi_S(A|E)$; it generates two if and only if there exist two $(d-1)$-planes that contain E and separate σ_1 and σ_2, one having v on the side that contains σ_1 and the other having v on the side that contains σ_2. We have already observed that this situation occurs whenever there is a $(d-1)$-plane that contains $E \cup \{v\}$ and separates σ_1 and σ_2. We next show that this condition is not just sufficient, but also necessary. So, assume that $\sigma = (\sigma_1, \sigma_2) \in \Pi(A \setminus \{v\}|E)$ generates two partitions of $\Pi(A|E)$. It then follows that there exist two nonzero vectors $C^1, C^2 \in \mathbb{R}^d$ and scalars

$\alpha^1, \alpha^2 \in \mathbb{R}$ such that for $i = 1, 2$

$(C^i)^T x < \alpha^i, (C^i)^T y > \alpha^i$ and $(C^i)^T z = \alpha^i$ for all $x \in \sigma_1, y \in \sigma_2$ and $z \in E$

and

$$(C^1)^T v < \alpha^1 \text{ and } (C^2)^T v > \alpha^2.$$

Let $\gamma \equiv \dfrac{(C^2)^T v - \alpha^2}{[\alpha^1 - (C^1)^T v] + [(C^2)^T v - \alpha^2]}$, $C \equiv \gamma C^1 + (1-\gamma)C^2$ and $\alpha \equiv \gamma\alpha^1 + (1-\gamma)\alpha^2$. Then $0 < \gamma < 1$,

$C^T x < \alpha, C^T y > \alpha$ and $C^T z = \alpha$ for all $x \in \sigma_1, y \in \sigma_2$ and $z \in E \cup \{v\}$

and $C \neq 0$ (as either σ_1 or σ_2 are nonempty). So, the $(d-1)$-plane $G = \{x \in \mathbb{R}^d : C^T x = \alpha\}$ separates σ and contains $E \cup \{v\}$. We conclude that

$$\begin{aligned}\#_S(A|E) &= 2\#_S(A \setminus \{v\}|E) + 2\#_S(A \setminus \{v\}|E \cup v) \\ &= 2H(n-1, d-u) + 2H(n-1, d-u-1) = 2H(n, d-u),\end{aligned}$$

where $\#_S(A \setminus \{v\}|E) = H(n-1, d-u)$ and $\#_S(A \setminus \{v\}|E \cup v) = H(n-1, d-u-1)$ follow from the induction assumption (as the vectors in the respective unions of the two pairs of sets are in general position) and the last equality follows from (3.2.3). □

The inductive proof of Theorem 3.2.2 yields a recursive algorithm with complexity bound of $O(n)H(n, d-u)$; see Hwang and Rothblum [60] for details. (To understand the role of the extra power of n, consider the case where $u = d$. After determining the unique hyperplane that contains the columns of E, one has to evaluate n determinants in order to split the columns of A between the two sides of that hyperplane.) We do not formalize this algorithm as it does not lend itself naturally to parallel execution on $\binom{n}{d}$ processors as does the algorithm that follows Lemma 2.2.3 (which apply to the enumeration of separable vector partitions and have the same overall complexity bound). In fact, all references to enumeration methods that are based on enumeration of separable 2-partitions are to be interpreted as using the method that follows Lemma 2.2.3.

The next example demonstrates that the requirement in Theorem 3.2.2 that $[A, E]$ is generic cannot be weakened by requiring that A and E are disjoint and separately generic (the example demonstrates that Harding [45, Theorem 2] is false as stated, because it misses the "joint generic" requirement).

Example 3.2.1. Let $d = 2$, $n = 3$, $E = \{\binom{0}{0}\}$ and for $w = 1, 2$, let $A^w = \{\binom{w}{1}, \binom{2}{2}, \binom{1}{2}\}$. Now, the number of $(A^w|E)$-separable 2-partitions is

$6 = 2H(3, 2-1)$ when $w = 2$ and is $4 \neq 2H(3, 2-1)$ when $w = 1$. It is easy to verify that the vectors of $A^2 \cup E$ are in general position whereas the vectors of $A^1 \cup E$ are not. □

Theorem 3.2.2 and Lemma 2.2.5 establish, respectively, a precise count of the number of separable partitions for $p = 2$ and for $d = 1$ which applies to generic matrices and depends only on the number of columns of the underlying matrix A and the dimension d. The following example demonstrates that this invariance of the number of A-separable p-partitions under a generic matrix $A \in \mathbb{R}^{d\times n}$ does not extend to $d = 2$ and $p = 3$, that is, the number of A-separable partitions needs not be a function only of p and the size of A.

Example 3.2.2. We shall consider two sets of points (matrices), both with $d = 2$, $n = 5$ and $p = 3$. The matrix for the first example is given by

$$A(1) = \begin{pmatrix} -1 & 0 & 1 & 1 & -1 \\ 1 & 2 & 1 & -1 & -1 \end{pmatrix}$$

while the matrix for the second example is given by

$$A(2) = \begin{pmatrix} -1 & 1 & 0 & -0.1 & 0.1 \\ -1 & -1 & 1 & 0 & 0 \end{pmatrix};$$

the columns of the two matrices are illustrated in Figure 3.2.1.

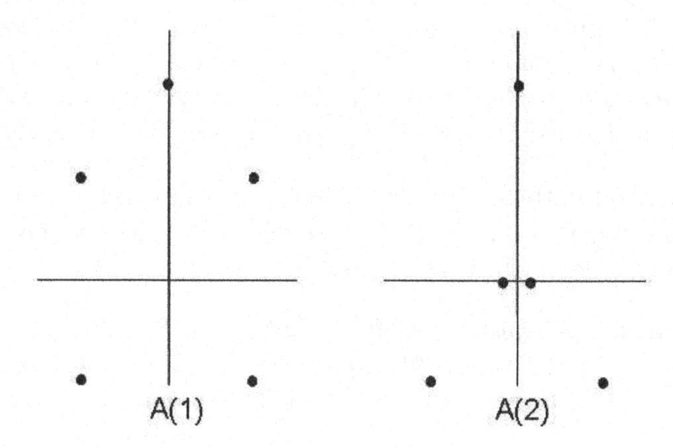

Figure 3.2.1 The partitioned vectors for Example 3.2.2

The following table counts for the number of separable unlabeled partitions (with nonempty parts) for the two examples, classified by their unlabeled shape (the figure in parentheses indicates the total number of partitions in each category).

unlabeled shape	$A(1)$-separable partitions	$A(2)$-separable partitions
$\{3,1,1\}$	10(out of 10)	7(out of 10)
$\{2,2,1\}$	10(out of 15)	14(out of 15)
total	20(out of 25)	21(out of 25)

The counting for the first example (corresponding to $A(1)$) is simple, using symmetry. A more delicate counting is required for the second example (corresponding to $A(2)$). Here, the counting of the 10 partitions with shape $\{3,1,1\}$ is by examining the 10 options for the triplet (which determine the unlabeled partition) and observing that $A(2)$-separable partitions are obtained for the 7 options that exclude the triplet being $\{1,2,3\}$,$\{1,3,5\}$ and $\{2,3,4\}$. The counting of the partitions with shape $\{2,2,1\}$ is by examining the 3 options for the singleton, each yielding 3 options for splitting the remaining 4 points; the only non-separable partition is obtained when the singleton $\{3\}$ is matched with the two pairs $\{2,4\}$ and $\{1,5\}$. The table shows that there are 20 $A(1)$-separable partitions and 21 $A(2)$-separable partitions.

If one is interested in labeled partitions, we get a similar inequality as each unlabeled partition corresponds to 6 labeled partitions. Also, we get a similar inequality if empty parts are allowed as the number of partitions that have at least one empty part is the same under $A(1)$ and $A(2)$ and is given by the Harding formula for 2-partitions (see Theorem 3.2.2). □

Hwang and Rothblum [60], by imitating the approach used in proving Theorem 2.2.4, gave the following generalization. Here we give a much more detailed proof (the construction).

Theorem 3.2.3. *Suppose $A \in \mathbb{R}^{d\times n}$ and $E \in \mathbb{R}^{d\times u}$, where $0 \le u \le d$. Then the set of $(A|E)$-separable p-partitions equals the set of p-partitions associated with lists of $\binom{p}{2}$ $(A|E)$-separable 2-partitions.*

Proof. If $A \cap (\text{aff } E) \neq \emptyset$, then every hyperplane containing the columns of E contains one of A's columns, implying that there are no $(A|E)$-separable partitions. So, in this case the assertions of the theorem are trite. Henceforth we shall assume that $A \cap (\text{aff } \overline{E}) = \emptyset$.

Consider a p-partition $\pi = (\pi_1, \ldots, \pi_p)$ associated with a list of $\binom{p}{2}$ $(A|E)$-separable 2-partitions whose elements are $\{\pi^{r,s} = (\pi_r^{r,s}, \pi_s^{r,s}) : 1 \le r < s \le p\}$. For each $1 \le r < s \le p$, $\pi_r \subseteq \pi_r^{r,s}$ and $\pi_s \subseteq \pi_s^{r,s}$ and therefore the $(A|E)$-separability of $\pi^{r,s} = (\pi_r^{r,s}, \pi_s^{r,s})$ implies that (π_r, π_s) is also $(A|E)$-separable.

Conversely, let $\pi = (\pi_1, \ldots, \pi_p)$ be an $(A|E)$-separable p-partition. We first prove that π has the desired representation under the assumption that $E \neq \emptyset$ (that is, $u > 1$). Consider any pair $1 \le r < s \le p$. Then there exists a vector $C^{rs} \in \mathbb{R}^d$ and a constant γ^{rs} such that

$$(C^{rs})^T A^u > \gamma^{rs} > (C^{rs})^T A^v \text{ for each } u \in \pi_r \text{ and } v \in \pi_s$$

and

$$(C^{rs})^T x = \gamma^{rs} \text{ for each } x \in \text{aff } E.$$

As aff E is a translation of the nonempty linear set tng E, we conclude from Lemma 3.1.9 of Vol. I that there are $H \in \mathbb{R}^d$ and $\mu \in \mathbb{R}$ such that $H^T A^i \neq \mu$ for each $i \in \mathcal{N}$, and $H^T x = \mu$ for each $x \in \text{aff } E$. For ϵ positive and sufficiently small, the replacement of (C^{rs}, γ^{rs}) by $(C^{rs} + \epsilon H, \gamma^{rs} + \epsilon\mu)$ preserves the above two displayed properties, while guaranteeing that $(C^{rs})^T A^i \neq \gamma^{rs}$ for all $i \in \mathcal{N}$. Let $\pi^{r,s} \equiv (\pi_r^{r,s}, \pi_s^{r,s})$ be the 2-partition with $\pi_r^{r,s} \equiv \{i \in \mathcal{N} : (C^{rs})^T A^i > \gamma^{rs}\}$ and $\pi_s^{r,s} \equiv \{i \in \mathcal{N} : (C^{rs})^T A^i < \gamma^{rs}\}$. Then π^{rs} is an $(A|E)$-separable 2-partition of $\mathcal{N}$, $\pi_r \subseteq \pi_r^{r,s}$ and $\pi_s \subseteq \pi_s^{r,s}$. Now, let $\pi' = (\pi'_1, \ldots, \pi'_p)$ be the p-tuple associated with the list of $\binom{p}{2}$ separable 2-partitions whose elements are the $\pi^{r,s}$'s. Then the sets of π' are pairwise disjoint and we have

$$\pi_j \subseteq \left(\cap_{t=j+1}^{p} \pi_j^{j,t}\right) \bigcap \left(\cap_{t=1}^{j-1} \pi_j^{t,j}\right) = \pi'_j \quad \text{for } j = 1, \ldots, p.$$

Since $\mathcal{N} = \cup_{i=1}^{p} \pi_i \subseteq \cup_{i=1}^{p} \pi'_i \subseteq \mathcal{N}$, it follows that $\pi = \pi'$, that is, π is the p-partition associated with the constructed list of $\binom{p}{2}$ $(A|E)$-separable 2-partitions. It remains to consider the case where E is empty (that is, $u = 0$). In this case the requirement $(C^{rs})^T x = \gamma^{rs}$ for each $x \in \text{aff } E$ is vacuous and one can perturb (C^{rs}, γ^{rs}) by $(C^{rs}, \gamma^{rs} + \epsilon)$ to preserve the first property of (C^{rs}, γ^{rs}) while guaranteeing $(C^{rs})^T A^i \neq \gamma^{rs}$ for each $i \in \mathcal{N}$. The remaining arguments used when E is not empty apply unchanged. □

Theorem 3.2.4. *Suppose* $A \in \mathbb{R}^{d \times n}$, $p \ge 2$. *Then* $\#_S^A(p) \le [\#_S^A(2)]^{\binom{p}{2}} \le [2H(n,d)]^{\binom{p}{2}}$*; further, the bound can be tightened to* $O(n^{6p-12})$ *when* $d = 2$ *and* $p \ge 3$, *and* $\#_S^A(p) = p! \sum_{j=1}^{p} \binom{n-1}{j-1}$ *when* $d = 1$.

Proof. As identical columns cannot be separated, we may assume for the purpose of counting separable partitions that A has only n columns, which are distinct. For general d, $\#_S^A(p) \leq [\#_S^A(2)]^{\binom{p}{2}}$ follows from Theorem 2.2.4 and the remaining inequalities follow from Theorem 3.2.2. Next the bounds for $d = 1$ and $d = 2$ follow from Theorem 6.2.1 of Vol. I (recall that separability reduces to consecutiveness) and Theorem 2.2.7, respectively.□

Combining the above results, we obtain

Theorem 3.2.5. *$\#_S(n,p,d) \leq [\#_S(n,2,d)]^{\binom{p}{2}} = [2H(n,d)]^{\binom{p}{2}} \leq [2^{d+1}\binom{n}{d}]^{\binom{p}{2}} = O(n^{d\binom{p}{2}})$ for $d \geq 2$, $\#_S(n,p,1) = O(n^{p-1})$ and $\#_S(n,p,2) = O(n^{6p-12})$.*

Proof. Theorem 3.2.2 shows that for a matrix $A \in \mathbb{R}^{d\times n}$ whose columns are in general position $\#_S^A(2) = 2H(n,d)$, hence, $\#_S(n,2,d) = 2H(n,d)$. By Theorem 3.2.4, we further conclude that $\#_S(n,2,d) \leq 2H(n,d)$ from $\#_S^A(2) \leq 2H(n,d)$ for all $A \in \mathbb{R}^{d\times n}$. Thus, $\#_S(n,2,d) = 2H(n,d)$. □

We next extend the above results to sphere-separable partitions.

Theorem 3.2.6. *Suppose $A \in \mathbb{R}^{d\times n}$, $p \geq 2$. Then $\#_{SS}^A(p) \leq [\#_{SS}^A(2)]^{\binom{p}{2}} \leq [2H(n,d+1)]^{\binom{p}{2}} \leq [2^{d+1}\binom{n}{d+1}]^{\binom{p}{2}}$, and if $d = 1$, then $\#_{SS}^A(p) = \sum_{q=1}^{p-1}\binom{p}{q}q!\binom{n}{p-q-1}\binom{n}{p-q}\frac{1}{nq!}$.*

Proof. As identical columns cannot be separated by sphere-separability, we may assume for the purpose of counting sphere-separable partitions that A has distinct columns. Let $\widehat{A}$ be defined by (3.1.3) and (3.1.4). It then follows from Theorems 3.1.3 and 3.2.4 that for general d,

$$\#_{SS}^A(p) = \#_S^{\widehat{A}}(p) \leq [\#_S^{\widehat{A}}(2)]^{\binom{p}{2}} = [\#_{SS}^A(2)]^{\binom{p}{2}}$$

and

$$\#_{SS}^A(2) = \#_S^{\widehat{A}}(2) \leq 2H(n,d+1) \leq 2^{d+1}\binom{n}{d+1}.$$

The bound for $d = 1$ follows from Theorem 6.2.3 of Vol. I and the formula given in Section 1.3 of Vol. I for transforming the number of p-partitions from not allowing empty parts to the corresponding number of allowing.□

Given $A \in \mathbb{R}^{d\times n}$, Theorem 3.1.3 also shows that any enumeration method for separable partitions immediately yields a method for enumerating the A-sphere-separable partitions (with the complexity bound obtained by substituting $d+1$ for d).

Theorems 3.1.3 and 3.2.2 imply that if the columns of $\widehat{A}$ are in general position, then $\#_S^{\widehat{A}}(2) = 2H(n, d+1)$. The following two examples demonstrate that it is possible that both A and $\widehat{A}$ are generic, but A being generic does not imply $\widehat{A}$ is.

Example 3.2.3. Given $d \geq 1$, let $A = [0, I, 2e] \in \mathbb{R}^{d\times(d+2)}$ with 0 as the 0-vector in $\mathbb{R}^d$, I as the $d \times d$ identity matrix and $e = (1, \ldots, 1)^T \in \mathbb{R}^d$. To see that the columns of A are in general position, consider the matrix B obtained from A by adding two rows to it – one at the bottom consisting of the squares of the norms of the columns of A, that is, $[0, e^T, 4d] \in \mathbb{R}^{1\times(d+2)}$, and the other at the top consisting of 1's. It is easily verified that the determinant of B is $2d \neq 0$; so, B is nonsingular and therefore the columns of $\widehat{A}$ are in general position. □

Example 3.2.4. Given $d \geq 1$, let $A \in \mathbb{R}^{d\times(d+2)}$ be any matrix whose columns are in general position and have unit (Euclidean) norm. It follows that the last row of $\widehat{A}$ has all 1's and therefore adding a row of 1's to $\widehat{A}$ results in a matrix B with two identical rows (of 1's). It follows that B is singular. Consequently, the columns of $\widehat{A}$ are not in general position. □

Theorem 3.2.7. $\#ss(n,p,d) \leq [\#ss(n,2,d)]^{\binom{p}{2}} \leq [2H(n,d+1)]^{\binom{p}{2}} \leq [2^{d+1}\binom{n}{d+1}]^{\binom{p}{2}} = O\left(n^{(d+1)\binom{p}{2}}\right)$ *for* $d \geq 2$, *and* $\#ss(n,p,1) \leq \frac{p!}{n}\binom{n}{p-1}\binom{n}{p} = O\left(n^{2p-2}\right)$.

Proof. The inequalities $\#ss(n,p,d) \leq [\#ss(n,2,d)]^{\binom{p}{2}} \leq [2H(n,d+1)]^{\binom{p}{2}} \leq [2^{d+1}\binom{n}{d+1}]^{\binom{p}{2}} = O\left(n^{(d+1)\binom{p}{2}}\right)$ follow from Theorem 3.2.6. □

We next turn our attention to cone-separability.

Theorem 3.2.8. *Let* $A \in \mathbb{R}^{d\times n}$ *have* $[0, A]$ *generic. Then the number of* A*-cone-separable 2-partitions is* $2H(n, d-1)$.

Proof. The result follows from Theorem 3.2.2 where E is the 0-vector in $\mathbb{R}^{d+1}$ and from the observation that when $[0, A]$ is generic, A has no 0-vector and thus A-cone-separability and $(A|0)$-separability coincide. □

The next result shows that the Harding numbers provide bounds on the number of A-cone-separable 2-partitions when the underlying matrix has no 0-vector.

Lemma 3.2.9. *Suppose* $A \in \mathbb{R}^{d\times n}$ *has* n *distinct columns and no 0-vectors. Then* $\#_{C_n S}^{A}(2) \leq 2H(n, d-1)$.

Proof. By Lemma 2.4.9, the set of A-cone-separable 2-partitions is contained in the set of "A-limit-cone-separable" 2-partitions which equals the set of cone-separable 2-partitions with respect to a perturbation $A(\epsilon)$ of A that has $[0, A(\epsilon)]$ generic; it now follows from Theorem 3.2.8 that $\#^{A}_{C_nS}(2) \leq \#^{A(\epsilon)}_{C_nS}(2) = 2H(n, d-1)$. □

Lemma 3.2.10. *Suppose $A \in \mathbb{R}^{d \times n}$ has n_0 0-vectors and $n - n_0$ distinct nonzero columns. Then*

$$\#^{A}_{C_nS}(2) \leq 2(n_0+1)H(n-n_0, d-1) \leq 2H(n, d); \tag{3.2.4}$$

further, if 0 and the distinct columns of A are in general position, then the first two inequalities hold with equality.

Proof. Let A' be a submatrix of A consisting of the nonzero columns. We observe that all A-cone-separable 2-partitions can be constructed from the A'-cone-separable 2-partitions and splitting the n_0 copies of the 0 vector between the two parts. As there are n_0+1 ways to split the n_0 0 vectors, we get that

$$\#^{A}_{C_nS}(2) = (n_0+1)\#^{A'}_{C_nS}(2) \leq 2(n_0+1)H(n-n_0, d-1),$$

the last inequality following from Lemma 3.2.9.

We next prove the second inequality of (3.2.4). We prove it by induction on n_0. The inequality is trivial (with equality) for $n_0 = 0$. So, assume it holds for $n_0 \geq 0$ and consider n_0+1 replacing n_0. We then have from the induction assumption (ignoring a factor of 2 throughout) that

$$\begin{aligned}[(n_0+1)+1]H(n-n_0-1, d-1) &\leq H(n-1, d) + H(n-n_0-1, d-1) \\ &\leq H(n-1, d) + H(n-1, d-1) \\ &= H(n, d),\end{aligned}$$

the second inequality following from the monotonicity of $H(n, d)$ in n and last equality following from (3.2.3). The third inequality in (3.2.4) is immediate from the monotonicity of $H(n, d)$ in n. Finally, assume that $[0, A']$ is generic, implying that A has no 0-vector and $n_0 = 0$. Then the above arguments and Theorem 3.2.8 show that $\#^{A}_{C_nS}(2) = \#^{A'}_{C_nS}(2) = 2H(n-n_0, d-1)$, proving that the first two inequalities of (3.2.4) hold with equality. □

When A has no 0-vectors, the first bound in (3.2.4) has order $O(n^{d-1})$. But, when 0-vectors are present that bound can have order $O(n^d)$, for example, when $n_0 = \frac{n}{2}$.

We next bound the number of cone-separable p-partitions (Hwang, Lee, Liu and Rothblum [51] considered the case where the distinct columns of A are in general position).

Theorem 3.2.11. *Suppose $A \in \mathbb{R}^{d\times n}$ has no 0-vectors and n distinct nonzero columns with $p \geq 2$. Then $\#^A_{C_nS}(p) \leq [\#^A_{C_nS}(2)]^{\binom{p}{2}} \leq [2H(n,d-1)]^{\binom{p}{2}}$ and if $d=2$, the bound can be sharpened to $\#^A_{C_nS}(p) \leq p!\binom{n}{p}$. Further, if A has $n_0 > 0$ 0-vectors, and $n-n_0$ nonzero distinct columns, then $\#^A_{C_nS}(p) \leq [2H(n-n_0,d-1)]^{\binom{p}{2}}\binom{n_0+p-1}{p-1}$ and if $d=2$ the bound can be sharpened to $\#^A_{C_nS}(p) \leq p!\binom{n}{p}\binom{n_0+p-1}{p-1}$. For $d=1$ and $p\geq 2$, $\#^A_{C_nS}(p) \leq p(p-1)\binom{n_0+p-1}{p-1}$.*

Proof. The no 0-vector case follows from Theorem 2.4.4 and Lemma 3.2.9 for general d, and from Theorem 2.4.6 for $d=2$. When A has $n_0 \geq 1$ 0-vectors, then $\binom{n_0+p-1}{p-1}$ is the number of ways of partitioning the 0-vectors into p labeled parts (see the second paragraph of this section). Let A' denote the subset of the nonzero elements in A. In Section 2.4 we give a method to enumerate the set of cone-separable partitions of A by first enumerate the set of cone-separable partitions of A', and then transplant this set of partitions to partitions of A by mixing in the partitions of the 0-vectors. The bounds in Theorem 3.2.11 records this procedure.

For $d=1$ and $p\geq 2$, by Theorem 2.4.5, there are $p(p-1)$ cone-separable partitions for the nonzero elements. Further, there are $\binom{n_0+p-1}{p-1}$ ways of distribution the n_0 zero elements to the p parts. Consequently, the bound holds. □

Consider the assumptions of Theorem 3.2.11 with general d and the presence of 0-vectors. An alternative approach to bound $\#^A_S(p)$ is to use the bound $\#^A_S(2) \leq 2H(n-n_0,d-1)(n_0+1)$ (Lemma 3.2.10) and Theorem 2.4.4 to get a bound $[2H(n-n_0,d-1)(n_0+1)]^{\binom{p}{2}} = O(n^{d\binom{p}{2}})$. Then the exponent of the order $d\binom{p}{2}$ of this bound is larger than the exponent of n of the bound in Theorem 3.2.11 which is $(d-1)\binom{p}{2}+p-1$ (under all p).

Theorem 2.4.6 shows that when A has a single row, all positive elements must be in the same part and all negative elements must be in the same part and the zero vectors can be split arbitrarily. So, an A-cone-separable partition can have at most two parts that include nonzero vectors and it is possible that no A-cone-separable partition exists.

Corollary 3.2.12. $\#_{C_nS}(n,p,d) \leq [H(n,d-1)]^{\binom{p}{2}}\binom{n+p-1}{p-1} =$

$O\left(n^{(d-1)\binom{p}{2}+p-1}\right)$ *and this bound can be sharpened when* $d = 2$ *to* $\#_{C_nS}(n,p,2) \leq p!\binom{n}{p}\binom{n+p-1}{p-1} = O\left(n^{2p-1}\right)$ *and when* $d = 1$, *to* $\#_{C_nS}(n,p,1) \leq p(p-1)\binom{n+p-1}{p-1} = O(n^{(p-1)})$.

Proof. The bounds are immediate from Theorem 3.2.11. □

The next two theorems establish relations between bounds on the number of separable and cone-separable partitions. Recall that $\widetilde{A}$ is the matrix obtained from A by appending a first coordinate 1 to each of its columns.

Theorem 3.2.13. *Let* $A \in \mathbb{R}^{d\times n}$ *and* $p \geq 1$. *Then* $\#^{A}_{C_nS}(p) \leq \#^{A}_{S}(p) \leq \#^{\widetilde{A}}_{C_nS}(p)$.

Proof. The first inequality follows from Theorem 3.2.11. To establish the second inequality assume that π is an A-separable p-partition. It then follows that for each $r, s = 1,\ldots,p$ there exists a vector $C^{rs} \in \mathbb{R}^d$ and a scalar α^{rs} such that if $i \in \pi_r$ and $u \in \pi_s$ then

$$(C^{rs})^T A^i > \alpha^{rs} > (C^{rs})^T A^u;$$

with $\widetilde{C} = \binom{-\alpha^{rs}}{C^{rs}}$ we then have that

$$(\widetilde{C}^{rs})^T \widetilde{A}^i = -\alpha^{rs} + (C^{rs})^T A^i > 0 > -\alpha^{rs} + (C^{rs})^T A^u = (\widetilde{C}^{rs})^T \widetilde{A}^u.$$

Thus, the $\widetilde{A}$-partition $\pi' = (\pi'_1,\ldots,\pi'_p)$ with $\pi'_j = \{\widetilde{A}^i : i \in \pi_j\}$ for $j = 1,\ldots,p$ is cone-separable. As the map $\pi \to \pi'$ is one-to-one, we have that $\#^{A}_{S}(p) \leq \#^{\widetilde{A}}_{C_nS}(p)$. □

Theorem 3.2.14. $\#_{C_nS}(n,p,d) \leq \#_S(n,p,d) \leq \#_{C_nS}(n,p,d+1) \leq O\left(n^{d\binom{p}{2}+p-1}\right)$.

Proof. The inequalities follow immediately from Theorems 3.2.13 and 3.2.12. □

The inequality $\#_S(n,p,d) \leq O\left(n^{d\binom{p}{2}}\right)$ of Theorem 3.2.14 is consistent with the conclusion of Theorem 3.2.5.

We next consider almost-separable partitions, starting with 2-partitions. The results appear in Hwang and Rothblum [60].

Lemma 3.2.15. *Suppose* $A \in \mathbb{R}^{d\times n}$ *has* $\bar{n}$ *distinct columns. Then*

$$\#^{A}_{AS}(2) \leq 2H(\bar{n},d) + (n-\bar{n})[2H(\bar{n}-1,d-1)] \leq 2H(n,d); \qquad (3.2.5)$$

further, if the distinct columns of A are in general position, then the first inequality holds with equality.

Proof. Let $v^1, \ldots, v^{\bar{n}}$ be the distinct columns of A and let $A' = [v^1, \ldots, v^{\bar{n}}] \in \mathbb{R}^{d \times \bar{n}}$. Also, for $v \in \{v^1, \ldots, v^{\bar{n}}\}$, let $\nu^A(v)$ be the number of occurrences of v in A. We call v a multi-vector in A if $\nu^v(A) > 1$. We first observe that every A'-separable partition is A'-almost-separable, the numbers of A- and A'-separable partitions coincide and, by Theorem 3.2.4, $2H(\bar{n}, d)$ bounds these numbers; further, by Theorem 3.2.2, the bound is tight when $v^1, \ldots, v^{\bar{n}}$ are in general position. We next observe that each almost-separable 2-partition of $A^1, \ldots, A^n$ which is not separable is uniquely determined by a triplet consisting of:
(i) a vector $v \in \{v^1, \ldots, v^{\bar{n}}\}$,
(ii) a separable partition of $\{v^1, \ldots, v^{\bar{n}}\} \setminus \{v\}$ for which there exists a separating hyperplane that contains v, and
(iii) a partition of the $\nu^A(v)$ copies of v into two nonempty sets; further, the correspondence of triplets to almost-separable partitions that are not separable is one-to-one. So, in order to count for all A-almost-separable 2-partition that are not separable, it suffices to count the corresponding triplets. For each $v \in \{v^1, \ldots, v^{\bar{n}}\}$, the number of separable partitions of $\{v^1, \ldots, v^{\bar{n}}\} \setminus \{v\}$ for which there exists a separating hyperplane that contains v is the number of cone-separable partitions corresponding to the matrix whose columns are $\{v^1 - v, \ldots, v^{\bar{n}} - v\} \setminus \{0\}$. As this matrix has no 0-vector, Lemma 3.2.9 shows that the corresponding number of cone-separable 2-partitions is bounded by $2H(\bar{n} - 1, d - 1)$ and Lemma 3.2.8 shows that the bound is tight whenever $v^1 - v, \ldots, v^{\bar{n}} - v$ are in general position, or equivalently, $v^1, \ldots, v^{\bar{n}}$ are in general position. Also, the number of ways to split ν identical points into two nonempty sets is $\nu - 1$. As $\sum_{i=1}^{\bar{n}} \nu^A(v^i) = n$, we conclude that the number of triplets corresponding to the A-almost-separable partitions which are not A-separable (with empty sets allowed) is bounded by

$$\sum_{i=1}^{\bar{n}} 2H(\bar{n} - 1, d - 1)[\nu^A(v^i) - 1] = 2H(\bar{n} - 1, d - 1)(n - \bar{n})$$

and therefore

$$\#_{AS}^{A}(2) \leq 2H(\bar{n}, d) + 2(n - \bar{n})H(\bar{n} - 1, d - 1).$$

The above arguments further show that the bound is tight whenever the columns of A are in general position.

We verify the second inequality of (3.2.5) by showing that for given $d \geq 1$ and $n \geq 1$,

$$H(n, d) \geq H(k, d) + (n - k)H(k - 1, d - 1) \quad \text{for} \quad 1 \leq k \leq n. \tag{3.2.6}$$

Evidently, (3.2.6) is trite when $k = n$. Next, assume that (3.2.6) holds for k and consider $k-1$ replacing k. We recall from (3.2.3) that $H(k,d) = H(k-1,d) + H(k-1,d-1)$. Thus, using the induction assumption, we have that

$$\begin{aligned} H(n,d) &\geq H(k,d) + (n-k)H(k-1,d-1) \\ &= [H(k-1,d) + H(k-1,d-1)] + (n-k)H(k-1,d-1) \\ &= H(k-1,d) + [n-(k-1)]H(k-1,d-1) \\ &\geq H(k-1,d) + [n-(k-1)]H(k-2,d-1), \end{aligned}$$

where the last inequality follows from the fact that for each d, $H(n,d)$ is increasing in n (see (3.2.3)). □

Given $A \in \mathbb{R}^{d \times n}$, the proof of Lemma 3.2.15 offers the following method for enumerating the set of almost-separable 2-partitions. First, produce all separable 2-partitions (say, using the method described in Section 2.2). Next, for each multiple point v generate all separable partitions of the remaining columns of A (that exclude all copies of v) where the separating hyperplane contains v. This task can be accomplished by determining all cone-separable (index) 2-partitions of the $A^1 - v, \ldots, A^n - v$ (a method for enumerating all cone-separable 2-partitions when the underlying matrix does not contain a 0-vector is described in Section 2.4). Next augment each generated partition with the $\nu^A(v) - 1$ splits of the $\nu^A(v)$ copies of v into two nonempty sets.

Corollary 3.2.16. $\#_{AS}(n,2,d) = \#_S(n,2,d) = 2H(n,d)$.

Proof. Theorem 3.2.5 shows that $\#_S(n,2,d) = 2H(n,d)$ and Lemma 3.2.15 implies that $\#_{AS}(n,2,d) \leq 2H(n,d)$. Finally, the inequality $\#_{AS}(n,2,d) \geq \#_S(n,2,d)$ is trite as separability implies almost-separability. □

We next show how the almost-separable 2-partitions can be used to generate all almost-separable p-partitions using a construction that is motivated by the one used to prove Theorems 2.2.4 and 2.4.4. For that purpose we need some definitions that will allow us to execute algebra on finite multi-subsets of $\mathbb{R}^d$. Given a finite multi-subset U of $\mathbb{R}^d$, the *multiplicity* of a vector $x \in \mathbb{R}^d$, denoted $\nu^U(x)$, is the number of occurrences of x in U; in particular, if B is a matrix with d rows, we use the notation $\nu^B(x)$ with the interpretation that B stands for the multiset consisting of the columns of B (counting for multiple occurrences). Also, if U and V

are two finite multi-subsets of $\mathbb{R}^d$, then $U \cap V$ denotes the multiset with $\nu^{U\cap V}(x) = \min\{\nu^U(x), \nu^V(x)\}$ for each $x \in \mathbb{R}^d$. Further, we write $U \subseteq V$ if $U \cap V = U$.

Let $A \in \mathbb{R}^{d\times n}$ and $p \geq 2$ be given. We shall consider $\binom{p}{2}$-tuples of almost-separable 2-partitions of $\mathcal{N}$, indexed by $\{(r,s) : 1 \leq r < s \leq p\}$; we refer to such an indexed tuple as a *list of* $\binom{p}{2}$ *almost-separable 2-partitions*, denote it by $\mathcal{L}$ and denote its elements (which are almost-separable 2-partitions) by $\pi^{r,s} = (\pi_r^{r,s}, \pi_s^{r,s})$ where $1 \leq r < s \leq p$. With such a list, say $\mathcal{L} = \{\pi^{r,s} : 1 \leq r < s \leq p\}$, we associate a p-tuple $\pi = (\pi_1, \ldots, \pi_p)$ of multi-subsets of $\mathcal{N}$ by a two-step construction (which is motivated in the next paragraph). First, (using multiset-intersection) set

$$\pi'_j \equiv \left(\cap_{t=j+1}^{p} \pi_j^{j,t}\right) \bigcap \left(\cap_{t=1}^{j-1} \pi_j^{t,j}\right) \quad \text{for } j = 1, \ldots, p. \tag{3.2.7}$$

We next produce each multiset π_j from the multiset π'_j by (possibly) reducing some of the $\nu^{\pi'_j}(x)$'s. Formally, for each $x \in \{A^1, \ldots, A^n\}$, let $j^{\pi'}(x) \equiv \max\{j = 1, \ldots, p : x \in \pi'_j\}$. If $j \neq j^{\pi'}(x)$, we let $\nu^{\pi_j}(x) = \nu^{\pi'_j}(x)$. Alternatively, if $j = j^{\pi'}(x)$ we set $\nu^{\pi_j}(x) = \min\{\nu^{\pi'_j}(x), [\nu^A(x) - \sum_{t=1}^{j-1} \nu^{\pi'_t}(x)]_+\}$ (where ζ_+ for $\zeta \in \mathbb{R}$ stands for $\max\{\zeta, 0\}$). For a vector $x \in \{A^1, \ldots, A^n\}$, the number $\sum_{j=1}^{p} \nu^{\pi_j}(x)$ may be larger or smaller than $\nu^A(x)$ (in fact, the sum may be 0, indicating that x does not appear in either of the multisets $\pi_1, \ldots, \pi_p$). If it happens that for each $x \in \{A^1, \ldots, A^n\}$, $\sum_{j=1}^{p} \nu^{\pi_j}(x) = \nu^A(x)$, then $\pi = (\pi_1, \ldots, \pi_p)$ is a p-partition which will be called *the p-partition associated with $\mathcal{L}$*.

We next motivate the construction of the above paragraph. The goal is to facilitate the generation of each almost-separable partition (formally accomplished in the proof of Theorem 3.2.17 below). Given an almost-separable partition σ, each pair of parts constitutes an almost-separable 2-partition of a subset of $A^1, \ldots, A^n$. The idea is then to extend the two parts of each pair to a partition of $A^1, \ldots, A^n$ so that for each j the intersection of the $p-1$ extended parts containing σ_j coincides with σ_j. A key problem is the handling of situations where multiple points are present and copies of a vector x are shared by more than 2 parts of σ; say 12 copies of x are present and σ_1, σ_2 and σ_3 contain, respectively, 2,3 and 7 copies. The construction of the partitions of $A^1, \ldots, A^n$ sends the access copies of x to the part with the larger index. So, we get partitions $\pi^{1,2}, \pi^{1,3}, \pi^{2,3}$ of $A^1, \ldots, A^n$ such that $\pi_1^{1,2}$ has 2 copies of x, $\pi_2^{1,2}$ has 10 copies, $\pi_1^{1,3}$ has 2 copies, $\pi_3^{1,3}$ has 10 copies, $\pi_2^{2,3}$ has 3 copies and $\pi_3^{2,3}$ has 9 copies. Next, the intersection formula (3.2.7) produces π'_1,π'_2,π'_3 having, respectively, 2,3,9 copies of x. In general, we get $\nu^{\pi'_j}(x) = \nu^{\sigma_j}(x)$ for $j \neq j^* \equiv j^{\pi'}(x)$ and $\nu^{\pi'_{j^*}}(x) \geq \nu^{\sigma_{j^*}}(x)$. In our example,

$j^* = 3$ and the number of occurrences of x in the third part is reduced in the second step from 9 to $\nu^A(x) - [\nu^{\pi'_1}(x) + \nu^{\pi'_2}(x)] = 12 - (2+3) = 7 = \nu^{\sigma_3}(x)$.

Theorem 3.2.17. *Suppose $A \in \mathbb{R}^{d\times n}$ and $p \geq 2$. Then the set of A-almost-separable p-partitions equals the set of p-partitions associated with lists of $\binom{p}{2}$ A-almost-separable 2-partitions; in particular, $\#^A_{AS}(p) \leq [\#^A_{AS}(2)]^{\binom{p}{2}} \leq [2H(n,d)]^{d\binom{p}{2}}$.*

Proof. Consider a p-partition $\pi = (\pi_1, \ldots, \pi_p)$ associated with a list of $\binom{p}{2}$ almost-separable 2-partitions whose elements are $\{\pi^{r,s} = (\pi^{r,s}_r, \pi^{r,s}_s) : 1 \leq r < s \leq p\}$ and let $1 \leq r < s \leq p$. As $\pi^{r,s}$ is almost-separable and (3.2.7) implies that $\pi_r \subseteq \pi^{r,s}_r$ and $\pi_s \subseteq \pi^{r,s}_s$, we have that π_r and π_s are almost-separable. So, π is almost-separable.

Conversely, let $\sigma = (\sigma_1, \ldots, \sigma_p)$ be an almost-separable p-partition and we will construct a list of $\binom{p}{2}$ almost-separable 2-partitions which has an associated partition and this partition is σ. We start by considering any specific pair $1 \leq r < s \leq p$ and constructing 2-partition $\pi^{r,s} = (\pi^{r,s}_r, \pi^{r,s}_s)$ (of the columns of A) that has the following properties:

(i) $\sigma_r \subseteq \pi^{r,s}_r$ and $\sigma_s \subseteq \pi^{r,s}_s$.

(ii-a) If σ_r and σ_s are separable, then so are $\pi^{r,s}_r$ and $\pi^{r,s}_s$.

(ii-b) If σ_r and σ_s are not separable, then $\pi^{r,s}_r$ and $\pi^{r,s}_s$ are almost-separable but not separable; further, in this case the unique vector x occurring in σ_r and σ_s is the unique vector occurring in both $\pi^{r,s}_r$ and $\pi^{r,s}_s$,

$$\nu^{\pi^{r,s}_r}(x) = \nu^{\sigma_r}(x) \quad \text{and} \quad \nu^{\pi^{r,s}_s}(x) \geq \nu^{\sigma_s}(x). \tag{3.2.8}$$

We first consider the case where σ_r and σ_s are separable. In this case there exists a vector $C^{rs} \in \mathbb{R}^d$ and a constant γ^{rs} such that $(C^{rs})^T y > \gamma^{rs} > (C^{rs})^T z$ whenever $y \in \sigma_r$ and $z \in \sigma_s$. By possibly perturbing γ^{rs} we may and will assume that $(C^{rs})^T A^i \neq \gamma^{rs}$ for each $i \in \mathcal{N}$. Let $\pi^{r,s} \equiv (\pi^{r,s}_r, \pi^{r,s}_s)$ be the 2-partition with $\pi^{r,s}_r$ and $\pi^{r,s}_s$ as, respectively, the multisets $\{y \in \{A^1, \ldots, A^n\} : (C^{rs})^T y > \gamma^{rs}\}$ and $\{z \in \{A^1, \ldots, A^n\} : (C^{rs})^T z < \gamma^{rs}\}$ (counting for multiplicities). Then $\pi^{r,s}$ is separable, $\sigma_r \subseteq \pi^{r,s}_r$ and $\sigma_s \subseteq \pi^{r,s}_s$.

Next assume that σ_r and σ_s are not separable (while being almost-separable). Then there exists a vector $C^{rs} \in \mathbb{R}^d$ and a unique point $x \in \{A^1, \ldots, A^n\}$ that occurs in both σ_r and σ_s such that

$$(C^{rs})^T y > (C^{rs})^T x > (C^{rs})^T z \quad \text{if } y \in \sigma_r, z \in \sigma_s \text{ and } y, z \neq x \tag{3.2.9}$$

By Lemma 3.1.9 of Vol. I, there exists a vector $H \in \mathbb{R}^d$ with $H^T(A^i - x) \neq 0$ for each $i = 1, \ldots, n$ with $A^i \neq x$. It then follows that for $\epsilon > 0$ sufficiently small, $(C^{rs} + \epsilon H)^T(A^i - x) \neq 0$ for all $i = 1, \ldots, n$ with $A^i \neq x$; in particular, ϵ can be selected so that (3.2.9) is preserved when C^{rs} is replaced by $C^{rs} + \epsilon H$. It follows that we may and will assume that C^{rs} and x satisfy (3.2.9) and $\gamma^{r,s} \equiv (C^{rs})^T x \neq (C^{rs})^T A^i$ for each $i = 1, \ldots, n$ with $A^i \neq x$. Next, let $\pi^{r,s} = (\pi_r^{r,s}, \pi_s^{r,s})$ be the partition with $\pi_r^{r,s}$ as the multiset consisting of the union $\{y \in \{A^1, \ldots, A^n\} : (C^{rs})^T y > \gamma^{r,s}\}$ (counting for multiplicities) and $\nu^{\sigma_r}(x)$ copies of x, while $\pi_s^{r,s}$ as the multiset consisting of the union $\{z \in \{A^1, \ldots, A^n\} : (C^{rs})^T z < \gamma^{r,s}\}$ (counting for multiplicities) and $\nu^A(x) - \nu^{\sigma_r}(x) \geq \nu^{\sigma_s}(x)$ copies of x. It is easily verified that $\pi^{r,s}$ satisfies (i) and (ii-b).

Our construction produces a list $\mathcal{L} = \{\pi^{r,s} : 1 \leq r < s \leq p\}$ of $\binom{p}{2}$ A-almost-separable 2-partitions. Let $\pi = (\pi_1, \ldots, \pi_p)$ be the tuple of multiset associated with $\mathcal{L}$ and let $\pi' = (\pi'_1, \ldots, \pi'_p)$ be the intermediary construction. We will next show that π is a partition which coincides with σ. As the elements of the σ_j's and of the π_j's come from $A^1, \ldots, A^n$, it suffices to show that $\nu^{\pi_j}(x) = \nu^{\sigma_j}(x)$ for each $x \in \{A^1, \ldots, A^n\}$. Consider any such vector x. Property (i) of our construction assures (the multiset-inclusion)

$$\sigma_j \subseteq \left(\cap_{t=j+1}^{p} \pi_j^{j,t}\right) \bigcap \left(\cap_{t=1}^{j-1} \pi_j^{t,j}\right) = \pi'_j \quad \text{for } j = 1, \ldots, p, \tag{3.2.10}$$

implying that

$$\nu^{\sigma_j}(x) \leq \nu^{\pi'_j}(x) \quad \text{for } j = 1, \ldots, p. \tag{3.2.11}$$

Let

$$J \equiv \{j = 1, \ldots, p : x \in \sigma_j\} \subseteq \{j = 1, \ldots, p : x \in \pi'_j\} \tag{3.2.12}$$

and let s be the maximal element in J. We next argue that the last inclusion in (3.2.12) holds as equality. Indeed, if $j \notin J$ then $x \notin \sigma_j$. As σ is a partition of the columns of A, there is an index $t \in \{1, \ldots, p\}$ such that $x \in \sigma_t$. Depending on whether $j < t$ or $j > t$, our construction implies that $x \in \pi_t^{j,t}$ and $x \notin \pi_j^{j,t}$, or $x \in \pi_t^{t,j}$ and $x \notin \pi_j^{t,j}$; in either case, (3.2.7) implies that $x \notin \pi'_j$. So, indeed, equality holds in (3.2.12); in particular, $s = \max\{j = 1, \ldots, p : x \in \pi'_j\} = j^{\pi'}(x)$. Now, if $j \in J \setminus \{s\}$ then $j < s$ and x occurs in both σ_j and σ_s, implying that these parts are not separable. It then follows from (3.2.8), (3.2.7) and the construction of π from π' that $\nu^{\sigma_j}(x) = \nu^{\pi_j^{j,s}}(x) \geq \nu^{\pi'_j}(x) = \nu^{\pi_j}(x)$. This inequality combines with (3.2.11) to show that

$$\nu^{\sigma_j}(x) = \nu^{\pi_j}(x) \quad \text{for} \quad j \in J \setminus \{s\}. \tag{3.2.13}$$

We next extend (3.2.13) for $j = s$. From (3.2.11), $\sum_{j\in J} \nu^{\pi'_j}(x) \geq \sum_{j\in J} \nu^{\sigma_j}(x) = \nu^A(x)$ and therefore

$$\nu^{\pi'_s}(x) \geq \nu^A(x) - \sum_{j\in J\setminus\{s\}} \nu^{\pi'_j}(x) = \nu^A(x) - \sum_{j\in J\setminus\{s\}} \nu^{\sigma_j}(x) = \nu^{\sigma_s}(x) \geq 0.$$

Next, the rules of constructing π from π' imply that

$$\nu^{\pi_s}(x) = \min\{\nu^{\pi'_s}(x), [\nu^A(x) - \sum_{j=1}^{s-1} \nu^{\pi'_j}(x)]_+\} = \nu^{\sigma_s}(x).$$

As (3.2.13) was extended to $j = s$, the π_j's coincide (as multisets) with the parts of σ; thus, π is a partition and it coincides with σ. The asserted inequality $\#^A_{AS}(p) \leq [\#^A_{AS}(2)]^{\binom{p}{2}}$ now follows immediately. Finally, the second inequality in the statement of our theorem are immediate from Corollary 3.2.16. □

Theorem 3.2.17 and the construction described in the paragraph preceding this theorem suggest a method for enumerating all almost-separable p-partitions. First enumerate all almost-separable 2-partitions and then use lists of $\binom{p}{2}$ almost-separable 2-partitions to produce p-tuples of multisets. The generated tuples should then be tested for being partitions of the columns of A; those that pass the test constitute the almost-separable p-partitions (duplications may occur).

We recall that Corollary 2.1.6 shows that every partition $\pi = (\pi_1, \ldots, \pi_p)$ for which $A_\pi = (\sum_{i\in\pi_1}, \ldots, \sum_{i\in\pi_p})$ is a vertex of a constrained-shape partition polytope is almost-separable. Hence, an enumeration of all almost-separable partitions along with tests on each generated partition π that determine if A_π is a vertex and if the shape of π is in a prescribed set, yields a method for enumerating the vertices of the underlying constrained-shape partition polytopes. As discussed in Section 2.1, vertex enumeration yields a solution method for corresponding partition problems with objective function $F(\pi) = f(A_\pi)$ where f is (edge-)quasi-convex(on the constrained-shape partition polytope). In Section 2.2 we developed the same method, focusing on limit-separable partitions rather than almost-separable partitions. The approach was based on Theorem 2.2.12 which demonstrates that every vertex of a constrained-shape partition polytope has a representation A_π where π is limit-separable. It is noted that the complexity bounds for generating all limit-separable partitions and almost-separable partitions coincide.

We next derive a precise count on the number of almost-separable partitions when $d = 1$.

Theorem 3.2.18. *Suppose $A \in \mathbb{R}^{1\times n}$. Then $\#^A_{AS}(p) \le p!\binom{n-1}{p-1}$, with equality holding when the columns of A are distinct.*

Proof. As A has a single row, its columns are numbers. Without loss of generality assume that $A^1 \le \cdots \le A^n$. In this case, A-almost-separable vector-partitions are in one-to-one correspondence with consecutive partitions, the number of which is available from Lemma 6.2.1 of Vol. I and then the general formula that translates the number of labeled partitions into unlabeled ones. □

Theorem 3.2.19. $\#_{AS}(n,p,d) \le [\#_{AS}(n,2,d)]^{\binom{p}{2}} = [2H(n,d)]^{\binom{p}{2}} = O\left(n^{d\binom{p}{2}}\right)$*; further,* $\#_{AS}(n,p,1) = p!\binom{n-1}{p-1} = O\left(n^{p-1}\right)$.

Proof. The conclusions under general d follow from Theorem 3.2.17 and Corollary 3.2.16. The proof when $d = 1$ follows from Theorem 3.2.18. □

The bound $O\left(n^{d\binom{p}{2}}\right)$ on $\#_{AS}(n,p,d)$, obtained in Theorem 3.2.19, coincides with the bound on $\#_S(n,p,d)$, obtained in Theorem 3.2.5. Further, Theorem 3.2.5, Corollary 3.2.16 and Theorem 3.2.19 show that $\#_S(n,2,d) = 2H(n,d) = \#_{AS}(n,2,d)$ and $\#_S(n,p,1) = p!\binom{n-1}{p-1} = \#_{AS}(n,p,1)$. We pose an open problem to determine whether or not $\#_{AS}(n,p,d) = \#_S(n,p,d)$ for $p > 2$ and $d > 1$.

Theorem 3.2.20. $\#_{ASS}(n,p,d) \le \#_{AS}(n,p,d+1) \le [2H(n,d+1)]^{\binom{p}{2}} = O\left(n^{(d+1)\binom{p}{2}}\right)$.

Proof. The result is immediate from Theorems 3.1.3 and 3.2.19. □

Theorem 3.1.3 also shows that the enumeration of almost-sphere-separable partitions is possible by using the enumeration method for almost-separable partitions.

Next, we turn our attention to two "nearly" properties.

Theorem 3.2.21. $\#_{N_eC_nS}(n,p,d) = \sum_{q=0}^{p-1}\binom{n}{q}q!\#_{C_nS}(n-q,p-q,d)$.

Proof. Clearly, $\#_{N_eC_nS}(n,p,d)$ achieves maximum when A has n distinct columns. Consider the case where there are exactly q singletons. The q singletons can be chosen from the n elements in $\binom{n}{q}$ ways. Once they are chosen, any permutation of them yields a distinct partition. Finally, the non-singleton parts can be arranged in $\#_{C_nS}(n-q,p-q,d)$ ways. □

Theorem 3.2.22. $\#_{N_e M_p}(n, p, d) = p!\binom{n}{p-1}$ *for* $n > p$.

Proof. Again, it suffices to consider n distinct columns. Each choice of $p-1$ points induces a distinct $N_e M_p$ unlabeled p-partition with each point in the chosen set corresponding to a singleton part. Then there are $p!$ choices for assigning the sets to the parts. This establishes an upper bound, which is clearly tight when the elements are distinct. □

Finally, we deal with some exponential classes.

Theorem 3.2.23. *The number of NP 2-partitions in $\mathbb{R}^2$ with shape $(\alpha n, (1-\alpha)n)$ for some constant α is exponential in n.*

Proof. Consider n points on a circle. Then the partition (π_1, π_2) where π_1 consists of any αn points on the circle satisfies NP, and there are $\binom{n}{\alpha n}$ of them. Note that the convex hull of points on a circle does not cover other points on the circle. □

Corollary 3.2.24. *The number of AC and of $C_v S$ 2-partitions in $\mathbb{R}^2$ with shape $(\alpha n, (1-\alpha)n)$ for some constant α is exponential in n.*

Proof. As $NP \Rightarrow AC \Rightarrow C_v S$ (Theorem 3.1.2), the conclusions of the corollary are immediate from Theorem 3.2.23. □

Theorem 3.2.25. *The number of NC 2-partitions in $\mathbb{R}^2$ with shape $(\alpha n, (1-\alpha)n)$ for some constant α is exponential in n.*

Proof. Consider the example constructed in the proof of Theorem 3.2.23 and augment it with a triangle outside of the circle. □

Corollary 3.2.26. *The number of p-partitions satisfying NP or AC or $C_v S$ or NC is exponential in n.*

Proof. The conclusions about NP and NC follow by adding $p-2$ points to the examples constructed in the proofs of Theorems 3.2.23 and 3.2.25, respectively, where each of the added points represents a part. The conclusions about AC and $C_v S$ follow from the arguments of the proof of Corollary 3.2.24. □

If Q is an exponential class, then of course WQ is an exponential class. Consider the two polynomial classes S and $C_n S$. We can construct point-sets such that the number of WS or $WC_n S$ p-partitions is exponential. For example, in Figure 3.1.6, let half of the points from the middle line be in π_1. These examples exist only when $O(n)$ points lie on a separating hyperplane. If the number of such points are bounded by a constant, then proofs analogous to Theorems 3.2.5 and 3.2.12 show that the number of WS or WC_nS p-partitions is polynomial. Thus, we say WS and WC_nS are in semi-polynomial classes.

property	bound($d>2$)	$d=2$	$d=1$	reference
S, AS	$O(n^{d\binom{p}{2}})$	$O(n^{(6p-12)})$	$O(n^{(p-1)})$	3.2.5, 3.2.19
SS	$O(n^{(d+1)\binom{p}{2}})$	$O(n^{3\binom{p}{2}})$	$O(n^{(2p-2)})$	3.2.7
C_nS, N_eC_nS	$O(n^{(d-1)\binom{p}{2}+p-1})$	$O(n^{(2p-1)})$	$O(n^{(p-1)})$	3.2.12, 3.2.21
M_p	p	p	p	trivial
N_eM_p	$p!\binom{n}{p-1}$	$p!\binom{n}{p-1}$	$p!\binom{n}{p-1}$	3.2.22

3.3 Consistency and Sortability of Particular Partition-Properties

Although heredity, consistency and sortability were all defined in Vol. I (Section 6.4) under index-partition and the assumption of no empty part, their definitions are independent of these issues and hence can be used in this volume. Since heredity is satisfied by all partition properties studied in this book, we will push it into the background and ignore it to simplify statements. But we need to study consistency and sortability. We first explain how to treat empty parts in dealing with consistency and sortability. The rule is extremely simple: Given a set S of parts containing a subset S' of empty parts, whether S satisfies a partition property Q depends on whether $S \setminus S'$ satisfies it. Under this rule, it is easily seen that consistency and sortability of a partition property is not affected by the appearance of empty parts; in particular, results in Sections 6.3 and 6.4 of Vol. I can all be adapted to the allowing empty parts case. We will follow the other chapters to allow empty parts in this section to allow wider applications (empty parts can always be barred by specifying shapes).

In Vol. I, the partition is performed on the index set with distinct indices, hence all partition properties studied in Vol. I are strong. On the other hand, to satisfy any strong property, 1-dim or otherwise, all identical elements must go to the same part (for cone separability, the 0-element is an exception). Thus in discussing a strong property under element-partition, we might as well combine all copies of the same point into a single copy and perform partition on the new set A' consisting of n' distinct elements. Then the analysis is the same as that of the index partition, i.e., we can use the index-partition results for $d = 1$ from Vol. I, i.e., to use the $d = 1$ result to initiate a deduction (on d) proof. On the other hand, a weak partition property is not defined under the index partition, so the $d = 1$ case needs to be worked out too.

We start by studying the minimum consistency indices of these properties. We shall use the notation $(W)Q$ to denote Q and WQ, and $N_e(Q)$ to denote Q and N_eQ. We also refer the reader to Section 6.4 of Vol. I for definitions of consistency, regularity, sortability and related terms.

Theorem 3.3.1. *The minimum consistency index is 2 for $(W)C_nS$, $(W)S$, $(W)NP$, $(W)SS$, $(W)NC$, AS, $(W)C_vS$ and $(N_e)M_p$, and ∞ for $(W)AC$. All the above properties are hereditary.*

Proof. The definitions of $(W)C_nS$, $(W)S$, $(W)NP$, $(W)SS$, $(W)NC$, AS and $(W)C_vS$ use binary relations, hence they are 2-consistent. The proof for $(N_e)M_p$ is trivial. Next, Figure 3.3.1 demonstrates that $W(AC)$ is not 2-consistent.

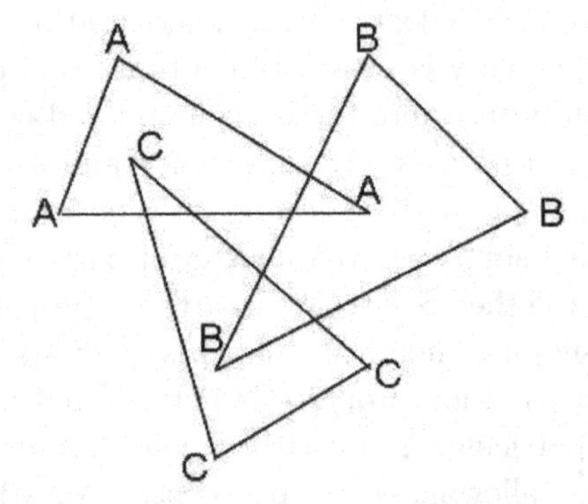

Figure 3.3.1 As shown, any two of parts A, B and C are acyclic but $A \to B \to C \to A$. Thus, $(W)AC$ is not 2-consistent.

This example can be extended to a $(k+1)$-cycle of parts where each part penetrates the next part but no other. The resulting $(k+1)$-partition does not satisfy $W(AC)$, but each of its k-subpartitions does. So, $(W)AC$ is not k-consistent for any finite k. □

Next, some basic relations between sortabilities of related properties.

Lemma 3.3.2. *If Q is not (ℓ, k, t)-sortable in $\mathbb{R}^d$, then it is not so in $\mathbb{R}^{d'}$ for $d' > d$. In particular, let Q' be a multi-parameter property which can be reduced to a single parameter property Q for $d = 1$. Then Q is not (ℓ, k, t)-sortable implies that Q' and WQ' are not (ℓ, k, t)-sortable.*

Proof. A counter-example in $\mathbb{R}^d$ is a counter-example in $\mathbb{R}^{d'}$. □

We next record a general result about t-regularity (see p. 263, Vol. I for definition).

Lemma 3.3.3. *Let Q' and Q be two multi-parameter properties with $Q' \Rightarrow Q$. If Q' is t-regular, then also Q is.*

Proof. If there exists a partition π in a family Π, then $Q' \Rightarrow Q$ assures that π also satisfies Q. □

The next two lemmas and their proofs parallel Lemmas 6.4.12 and 6.5.5 of Vol. I, respectively.

Lemma 3.3.4. *Let Q and Q' be k-consistent, hereditary multi-parameter properties of partitions with $Q' \Rightarrow Q$. If Q' is (sort-specific, k, t)-sortable, then Q is (sort-specific, k, t)-sortable.*

Proof. The proof of Lemma 6.4.12 of Vol. I applies. □

Lemma 3.3.5. *Let $Q \in \{(W)S,\ M_p,\ (W)NP,\ (W)SS,\ (W)C_vS,\ (W)NC,\ (W)AC,\ AS,\ ASS\}$ and let $k' > k \geq 2$, with k at least as large as the minimum consistency index of Q. If Q is (strongly, k', t)-sortable, then Q is (strongly, k, t)-sortable.*

Proof. All the properties are invariant when a new part consisting of an isolated element is added. Consequently, the arguments of the proof of Lemma 6.5.5 of Vol. I apply. □

Lemma 3.3.6. *All properties mentioned in Section 3.1 are shape-regular, except for $(N_e)C_nS$ and $(N_e)M_p$ which are not shape-regular but are support-regular.*

Proof. The shape-regularity of AS follows from Theorem 2.1.8 (and the existence of p distinct vectors which assures that the corresponding partition polytope is nonempty). As S and AS coincide when the columns of A are distinct, we have that S is also shape-regular. Then the asserted regularities of all other properties except for $(W, N_e)C_nS$ and $(N_e)M_p$ are implied by Lemma 3.3.3 and Theorem 3.1.2. Next, the regularity conclusions about $(N_e)M_p$ are easily verified.

To see that C_nS is support-regular, recall the underlying assumption that every pair of nonzero columns of A are linearly independent. As 0-vectors can be assigned arbitrarily, we assume that the partition is on the nonzero subset of A. We next argue that one can always find a hyperplane containing 0 that splits the points into two nonempty parts. Indeed, as $\text{rank}[A^1, A^2] = 2$, there exists a vector $C \in \mathbb{R}^d$, $C^T A^1 = 1$ and $C^T A^2 = -1$. For each i, $(C + \sum_{i=3}^n \epsilon^i A^i)^T A^i$ is a nonzero polynomial in ϵ; consequently, for sufficiently small positive ϵ, the replacement of C with $C + \sum_{u=3}^n \epsilon^u A^u$ will have $C^T A^1 > 0 > C^T A^2$ and $C^T A^i \neq 0$ for every i, establishing the assertion. Recursive construction of additional $k-2$ half-hyperplanes will induce a cone-separable k-partition. For WC_nS, since any set of points in the same separating hyperplane can be split to parts arbitrarily, the above argument for C_nS can be adapted to prove that WC_nS is support-regular. The above arguments also apply to N_eC_nS after assigning singletons.

To see that $(N_e)C_nS$ is not shape-regular, let $d = 2$ and let the columns of A consist of $n \geq 4$ points in the plane that are uniformly spread on the unit circle. Then there is no cone-separable partition with shape $(n - 1, 1)$. It is noted that for this example, the columns of $[0, A]$ are in general position. □

Although not as natural as in $\mathbb{R}^1$, there are still ways to order points in $\mathbb{R}^d$. One way is to order the points lexicographically with vector $x \in \mathbb{R}^d$ represented by $(x_1, \ldots, x_d)$; we refer to this ordering as the *coordinates-ordering*. A partition that is consecutive in this ordering is called *coordinates-consecutive*.

Lemma 3.3.7. *A coordinates-consecutive partition is almost-separable (separable if points are distinct).*

Proof. Since both properties are defined by behavior of 2-partitions, it suffices to prove the lemma for $p = 2$. The lemma is trivially true for $d = 1$, as both properties reduce to consecutiveness. We prove the general d case by induction. So, assume that the lemma holds when the dimension of the vectors is less than d and let $\pi = (\pi_1, \pi_2)$ be a coordinates-consecutive 2-partition. The case where all points share the same first coordinate is reducible to the case where the dimension is $d - 1$. In the alternative case, there exists some k such that π_1 consists of all points x with $x_1 < k$ and a set $\sigma_1 \subset H \equiv \{x \in \mathbb{R}^d : x_1 = k\}$, π_2 consists of all points x with $x_1 > k$ and a set $\sigma_2 \subset H$ and σ_1 and σ_2 are coordinates-consecutive. By induction hypothesis, σ_1 and σ_2 form an almost-separable partition, therefore Lemma 3.1.4 implies that (π_1, π_2) is almost-separable. Finally, the parenthetical result in the statement of our lemma is immediate from the main result of the current lemma and the final conclusion of Lemma 3.1.5. □

Note that the converse of Lemma 3.3.7 is not true, except for $d = 1$, as coordinates-consecutiveness only induces a special class of weakly separable partitions. Therefore, it will be treated as a specific S-sorting.

For a k-consistent, hereditary partition property, Lemma 6.4.11 of Vol. I lists various implications among different specifications of sortability and summarize them in Table 6.4.1 of Vol. I. Since these implications also hold for multi-parameter partitions, we reproduce Table 6.4.1 here for easy references.

Table 6.4.1 of Vol. I: Sortability-implications

(st,k,op) $\Rightarrow^*$	(st,k,supp) $\Rightarrow^*$	(st,k,shape) $\Rightarrow$	(s-s,k,shape) $\Rightarrow$	(s-s,k, supp) $\Rightarrow$	(s-s,k,op)
$\Downarrow$	$\Downarrow$	$\Downarrow$	$\Downarrow$	$\Downarrow$	$\Downarrow$
(p-s,k,op) $\Rightarrow^*$	(p-s,k,supp) $\Rightarrow^*$	(p-s,k,shape) $\Rightarrow$	(w,k,shape) $\Rightarrow$	(w,k,supp) $\Rightarrow$	(w,k,op)

When stating sortability for a strong property we assume that A or $[0, A]$, respectively, satisfies the condition of t-regularity (Lemma 3.3.6).

Theorem 3.3.8. *$(W)S$ and AS are (sort-specific, k,t)-sortable under coordinates-consecutive ordering for all $k \geq 2$ and all t.*

Proof. By Table 6.4.1, it suffices to prove for t=shape. Consider a special kind of coordinates-ordering, called *index-coordinates-ordering* (ICO), such that parts with smaller indices get smaller points (in the sense of lexicographic order given in the definition of coordinates-ordering). Define the statistics $s(\cdot)$ as the lexicographic order on A_π when viewed as the dp-vector

$$\left(\sum_{i\in\pi_1} A_1^i, \ldots, \sum_{i\in\pi_p} A_1^i, \sum_{i\in\pi_1} A_2^i, \ldots, \sum_{i\in\pi_p} A_2^i, \ldots, \sum_{i\in\pi_1} A_d^i, \ldots, \sum_{i\in\pi_p} A_d^i \right).$$

Then it is easily seen that each ICO-(ℓ, k, t)-sorting induces a decrease of $s(\cdot)$. □

Corollary 3.3.9. *$(W)NP$, $(W)SS$, $(W)NC$ and $(W)C_vS$ are all (sort-specific, k, shape)-sortable for all $k \geq 2$.*

Proof. The proof follows from Theorem 3.3.8 and Lemma 3.3.4. □

Note that we can actually apply coordinates-ordering-sorting to the whole partition π (the so-called p-sorting) to achieve separability at once. But Theorem 3.3.8 provides us flexibility in sorting in those environments where the size of sorting may be restricted.

Theorem 3.3.10. *$(N_e)M_p$ is (strongly, k, open)-sortable for all $k \geq 2$.*

Proof. For N_eM_p, set $s(\pi)$ to be the number of parts containing more than one element. For M_p, set $s(\pi)$ to be the number of nonempty parts. It is easily seen that $s(\pi)$ decreases, respectively, under N_eM_p-sortings or M_p-sortings. □

In the rest of this section, we are going to prove some negative results using counter-examples in $\mathbb{R}^2$ (see Lemma 3.3.2). Since the point-sets in these examples are all generic, an example against Q is also against WQ and AQ (if it is defined).

Theorem 3.3.11. *$(W)C_nS$ is not (part-specific, k, shape)-sortable for all $k \geq 3$.*

Proof. Consider a single-shape partition problem with $k \geq 3$, $p = k+1$ and shape $(1,\ldots,1,2)$ on a set of $n = k+2$ points which are the vertices of a regular n-gon. Let the origin be the center of the regular n-gon. Let Π be a family of partitions consisting of all partitions with the two elements of the doubleton part π_{k+1} flanking a single element (to create non-cone-separability). If this element is in π_j (its only element), we call the partition a type-j partition. Clearly, Π does not satisfy C_nS. Let π be a type-j partition. Then any set of k parts of π does not satisfy C_nS except for two sets - one missing π_j and the other missing π_{k+1}. Consider a set K of k parts in π (of type-j) not satisfying C_nS, i.e., the missing part π_i in K is neither π_j nor π_{k+1}. Sort π into a type-i partition by placing the two elements of π_{k+1} at the two ends of the $k+1$ consecutive points in K. Clearly, this is a C_nS-sorting and the new partition is in Π. Since j and i are arbitrary, we have established that Π is a (sort-specific, k, shape)-invariant family without satisfying C_nS.

□

Note that if $k = 2$, then the specified sorting in the above argument does not exist.

Hwang, Lee, Liu and Rothblum [51] first proved the following result. Here, we adopt the simpler proof by Chang and Guo [22].

Theorem 3.3.12. *$(W)S$ and AS are not (strongly, k, shape)-sortable for*

$k \geq 2$.

Proof. By Theorem 3.3.5, it suffices to prove for $k = 2$.

Consider nine points equally spaced on a circle partitioned into three triangles (see Figure 3.3.2(a)).

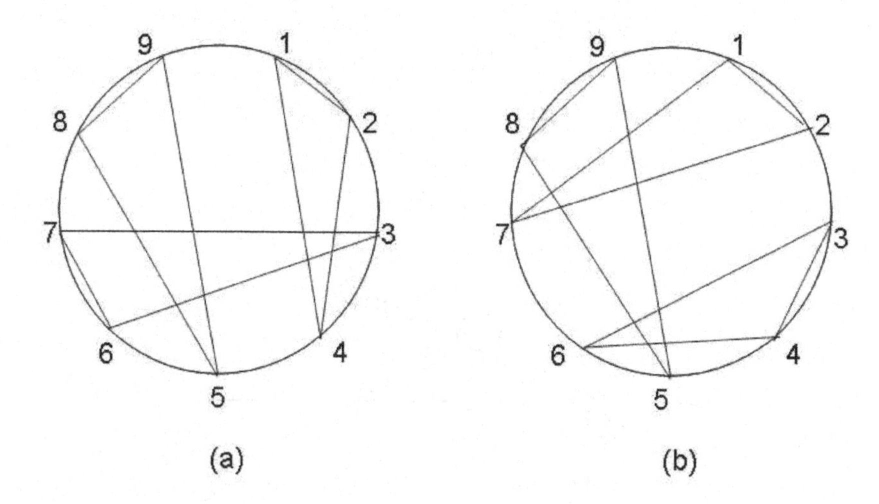

Figure 3.3.2 A (weakly, 2, shape)-invariant family not satisfying S

S-sort the two intersecting triangles $\Delta 124$ and $\Delta 367$ into $\Delta 127$ and $\Delta 346$ as shown in Figure 3.3.2(b). Then, the two figures differ only in a rotation. Hence, we can construct a (weakly, 2, shape)-invariant family with nine such figures which form a cycle with nine rotations, but none of the nine figures satisfies S. This implies S is not (strongly, k, shape)-sortable for $k \geq 2$, and so do $(W)S$ and AS. □

Theorem 3.3.13. *$(W)C_nS$ is not (strongly, k, shape)-sortable for all $k \geq 2$.*

Proof. In Figure 3.3.2, let the center be the common vertex of the cones. Then the existence of two nonseparable triangles implies that they are not cone-separable. Further, it is easily verified that the S-sorting used is also a C_nS-sorting. Consequently, Figure 3.3.2 is also a counter-example against C_nS. □

Chang and Guo [22] also proved the following result.

Theorem 3.3.14. *$(W)NP$ is not (strongly, k, shape)-sortable for all $k \geq 2$.*

Proof. Again, by Lemma 3.3.5 it suffices to prove for $k = 2$. Consider nine points with coordinates $(1,3)$, $(3,2)$, $(2,0)$, $(0,0)$, $(-1,2)$, $(0,1)$, $(2,1)$, $(0.5, 2.5)$ and the 9th point is in the quadrilateral enclosed by the four lines $= [4,7]$, $[4,2]$, $[3,5]$, $[3,6]$ where $[x,y]$ is the line connecting points x and y (see Figure 3.3.3).

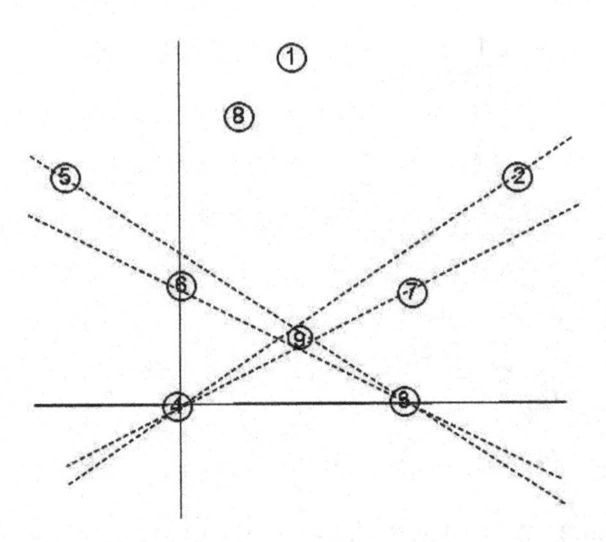

Figure 3.3.3 A configuration of 9 points for NP

Let the shape be (3,3,3). We will represent a partition by giving the three triangles. Then the seven partitions (248, 179, 356) $\rightarrow$ (248, 679, 135) $\rightarrow$ (123, 679, 458) $\rightarrow$ (123, 489, 567) $\rightarrow$ (367, 489, 125) $\rightarrow$ (367, 129, 458) $\rightarrow$ (478, 129, 356) $\rightarrow$ (248, 179, 356), where "$\rightarrow$" implies NP-sort, constitute a (weakly, 2, shape)-invariant family for NP which contains no NP-partition. □

Theorem 3.3.15. *$(W)NC$, $(W)SS$ and $(W)C_vS$ are not (strongly, k, shape)-sortable for all $k \geq 2$.*

Proof. NC, SS and C_vS reduce to NC in $\mathbb{R}^1$. As NC is not (strongly, k, shape)-sortable for all $k \geq 2$, the conclusions of our theorem follow. □

Sortability for the various multi-parameter properties are summarized in the next six tables which follow the format of the tables of Section 6.5 of Vol. I; in particular, the set of k's for which a property is (ℓ, k, t)-sortable is given in the (ℓ, t) cell, along with the set of $\bar{k}$'s for which the property is not (ℓ, k, t)-sortable.

Table 3.3.1. Sortability of $(W)S$ and AS

	Open	*Support*	*Shape*
strongly	$\bar{k} \geq 2$	$\bar{k} \geq 2$	$\bar{k} \geq 2$
part-specific			
sort-specific	$k \geq 2$	$k \geq 2$	$k \geq 2$
weakly	$k \geq 2$	$k \geq 2$	$k \geq 2$

Table 3.3.2. Sortability of $(W)NC$

	Open	*Support*	*Shape*
strongly	$\bar{k} \geq 2$	$\bar{k} \geq 2$	$\bar{k} \geq 2$
part-specific			
sort-specific	$k \geq 2$	$k \geq 2$	$k \geq 2$
weakly	$k \geq 2$	$k \geq 2$	$k \geq 2$

Table 3.3.3. Sortability of $(W)NP$, $(W)SS$ and $(W)C_vS$

	Open	*Support*	*Shape*
strongly	$\bar{k} \geq 2$	$\bar{k} \geq 2$	$\bar{k} \geq 2$
part-specific			
sort-specific	$k \geq 2$	$k \geq 2$	$k \geq 2$
weakly	$k \geq 2$	$k \geq 2$	$k \geq 2$

Table 3.3.4. Sortability of $(W, N_e)C_nS$
(the shape column applies only to WC_nS)

	Open	*Support*	*Shape*
strongly	$\bar{k} \geq 2$	$\bar{k} \geq 2$	$\bar{k} \geq 2$
part-specific	$\bar{k} \geq 3, k = 2$	$\bar{k} \geq 3, k = 2$	$\bar{k} \geq 3, k = 2$
sort-specific			
weakly			

Table 3.3.5. Sortability of $(N_e)M_p$

	Open	*Support*	*Shape*
strongly	$k \geq 2$	$k \geq 2$	$k \geq 2$
part-specific	$k \geq 2$	$k \geq 2$	$k \geq 2$
sort-specific	$k \geq 2$	$k \geq 2$	$k \geq 2$
weakly	$k \geq 2$	$k \geq 2$	$k \geq 2$

Table 3.3.6. Sortability of $(W)AC$

	Open	*Support*	*Shape*
strongly	$\bar{k} \geq 2$	$\bar{k} \geq 2$	$\bar{k} \geq 2$
part-specific	$\bar{k} \geq 2$	$\bar{k} \geq 2$	$\bar{k} \geq 2$
sort-specific	$\bar{k} \geq 2$	$\bar{k} \geq 2$	$\bar{k} \geq 2$
weakly	$\bar{k} \geq 2$	$\bar{k} \geq 2$	$\bar{k} \geq 2$

The following two S-sortings will be heavily used in Section 4.2. One type of S-sorting, which we will call *join and split* (abbreviated as JS-sorting), will play a major role in the study of partition problems with objective functions that depend on the perimeter of the parts. To avoid trivial cases, we assume that the number of distinct points is at least p. Let π_r and π_s be a pair of parts (of a partition π) which is either not separable or contains an empty part. A JS-sorting is then the 2-partition (π'_r, π'_s) with one set consisting of a vertex $\{v\}$ of $\text{conv}(\pi_r \cup \pi_s)(= (\text{conv } \pi_r) \vee (\text{conv } \pi_s))$

and the other set consisting of $(\pi_r \cup \pi_s) \setminus \{v\}$ (of course, a vertex of $\text{conv}(\pi_r \cup \pi_s)$ is a column of A).

Formally, the *perimeter* of a convex polytope $P \subseteq \mathbb{R}^2$, denoted peri(P), is defined as follows: (i) if $\dim P = 2$ (that is, P is a convex polygon), then peri(P) equals the sum of the lengths of P's edges, (ii) if $\dim P = 1$, then peri(P) equals twice the length of P, and (iii) if $\dim P = 0$ or P is empty, then peri(P) $= 0$. The *perimeter* of a finite set $\Omega \subseteq \mathbb{R}^2$, denoted peri($\Omega$), is defined as peri(conv Ω). A motivation for the definition of the perimeter of a line is the fact that it assures continuity of the perimeter of finite sets under perturbations of their points.

Lemma 3.3.16. *Let S and T be two polytopes in $\mathbb{R}^2$ where $\emptyset \neq S \subset T$. Then* peri(S) $<$ peri(T).

Proof. The inequality peri(S) $<$ peri(T) is trite if $\dim(T) = 1$ or if $\dim(S) = 0$. To consider the remaining cases we observe that given a convex polytope in $\mathbb{R}^2$, a cut that removes a triangle having two edges on the boundary and leaving a nonempty set strictly reduces the perimeter (due to the triangle inequality). Now, if $\dim(T) = \dim(S) = 2$, S can be obtained from T through a sequence of such cuts, implying that peri(S) $<$ peri(T). Finally, if $\dim(T) = 2$ and $\dim(S) = 1$, then a sequence of cuts can reduce T to a triangle that has S as one of its edges; in particular, peri(T) is larger than the perimeter of the triangle. But (again by the triangle inequality), the perimeter of the triangle is larger than twice the length of any one of its edges, implying that it is larger than peri(S). □

We shall consider the lattice of polytopes in $\mathbb{R}^2$, where the *meet* of two polytopes S and T, denoted $S \wedge T$, is their intersection and their join, denoted $S \vee T$, is $\text{conv}(S \cup T)$ (which is the convex hull of the union of the set of vertices of S and T). The next lemma will be useful in the study of partition problems with $d = 2$ and objective functions that involve parts' perimeters. It will be convenient to use the notation Δxyz for a triangle with vertices x, y, z.

Lemma 3.3.17. *Let S and T be two polytopes in $\mathbb{R}^2$ having nonempty intersection. Then*

$$\text{peri}(\mathrm{S} \wedge \mathrm{T}) + \text{peri}(\mathrm{S} \vee \mathrm{T}) \leq \text{peri}(\mathrm{S}) + \text{peri}(\mathrm{T}). \tag{3.3.1}$$

Proof. We first consider the case where $\dim(S \cap T) = 2$. Select a point o in the interior of $S \cap T$ and let $v^1, \ldots, v^q$ be the vertices of $S \vee T$; each of these vertices is a vertex of either S or T or both. For $j = 1, \ldots, q$, let u^j, s^j and

t^j be the (unique) intersection of the ray that emanates at o and includes v^j with bd $(\mathrm{S} \cap \mathrm{T})$, bd S and bd T; note that $\{v^j, u^j\} = \{s^j, t^j\}$ and $v^j = u^j$ is possible (the latter happens when S and T intersect at a point which is a vertex of both and in this case $s^j = t^j$). We verify the asserted perimeter inequality by considering the segments of the corresponding boundaries with each triangle $\Delta_j \equiv \Delta_{o,v^j,v^{j+1}}$ (with $v^{q+1} = v^1$). So, let $\mathrm{peri_j(X)}$ denote the length of the intersection of the boundary of $X \in \{S, T, S \vee T, S \wedge T\}$ with Δ_j. If $v^j = t^j$ and $v^{j+1} = t^{j+1}$, then $u^j = s^j$ and $u^{j+1} = s^{j+1}$, $\mathrm{peri_j(S \vee T) = peri_j(T)}$, $\mathrm{peri_j(S \wedge T) = peri_j(S)}$ and

$$\mathrm{peri_j(S \wedge T) + peri_j(S \vee T) = peri_j(S) + peri_j(T)}.$$

A symmetric argument applies when $v^j = s^j$ and $v^{j+1} = s^{j+1}$. In the remaining case v^j and v^{j+1} are not both in S, nor are they both in T, say $v^j = s^j \in S \setminus T$ and $v^{j+1} = t^{j+1} \in T \setminus S$. It follows that the intersection of the boundaries of S and T with Δ_j are, respectively, piecewise-linear paths – one emanating at $v^j \neq u^j$ and ending at $u^{j+1} \in S \cap T$ while the other emanating at $u^j \in S \cap T$ and ending at $v^{j+1} \neq u^{j+1}$. These two paths cross at a single point m^j. Using the notation "length(a $\rightarrow$ b)" to denote the length of a corresponding path between a and b, we have that

$$\begin{aligned}
\mathrm{peri_j(S \wedge T)} &= \mathrm{length(u^j \rightarrow m^j) + length(m^j \rightarrow u^{j+1})},\\
\mathrm{peri_j(S \vee T)} &= \|v^j - v^{j+1}\|,\\
\mathrm{peri_j(S)} &= \mathrm{length(v^j \rightarrow m^j) + length(m^j \rightarrow u^{j+1})},\\
\mathrm{peri_j(T)} &= \mathrm{length(u^j \rightarrow m^j) + length(m^j \rightarrow v^{j+1})}.
\end{aligned}$$

As a line is the shortest path between two points we further have that $\|v^j - v^{j+1}\| \leq \mathrm{length(v^j \rightarrow m^j) + length(m^j \rightarrow v^{j+1})}$ and therefore

$$\begin{aligned}
&\mathrm{peri_j(S \wedge T) + peri_j(S \vee T)}\\
&= \mathrm{length(u^j \rightarrow m^j) + length(m^j \rightarrow u^{j+1}) + \|v^j - v^{j+1}\|}\\
&\leq \mathrm{length(u^j \rightarrow m^j) + length(m^j \rightarrow u^{j+1})}\\
&\quad \mathrm{+length(v^j \rightarrow m^j) + length(m^j \rightarrow v^{j+1})}\\
&= \mathrm{peri_j(S) + peri_j(T)}.
\end{aligned}$$

We use a perturbation argument to verify the case where $\dim(S \cap T) < 2$. Let V_X denote the set of vertices of a polytope X. Also, let ω be any particular point in $S \cap T$ (which is assumed to nonempty) and for $\epsilon > 0$ let $W_\epsilon \equiv \{\omega + \binom{\epsilon}{0}, \omega + \binom{-\epsilon}{0}, \omega + \binom{0}{\epsilon}, \omega + \binom{0}{-\epsilon}\}$, $S_\epsilon \equiv \mathrm{conv}(V_S \cup W_\epsilon)$ and $T_\epsilon \equiv \mathrm{conv}(V_T \cup W_\epsilon)$. For each $\epsilon > 0$ we then have that $\dim(S_\epsilon \cap T_\epsilon) = 2$, $S_\epsilon \vee T_\epsilon = \mathrm{conv}(V_S \cup V_T \cup W_\epsilon) = \mathrm{conv}(V_{S \cup T} \cup W_\epsilon)$, $S_\epsilon \wedge T_\epsilon = \mathrm{conv}(V_{S \cap T} \cup W_\epsilon)$ and (with the definition of the perimeter of a line segment as twice its length) $\lim_{\epsilon \downarrow 0} \text{peri}(X_\epsilon) = \text{peri}(X)$ for $X \in \{S, T, S \vee T, S \wedge T\}$. The conclusion of the lemma next follows from the application of (3.3.1) with S_ϵ and T_ϵ replacing S and T and letting $\epsilon \downarrow 0$. □

Corollary 3.3.18. *Let Ω^1 and Ω^2 be two finite sets in $\mathbb{R}^2$ which are not separable. Then*

$$\text{peri}(\Omega^1 \cup \Omega^2) + \text{peri}[(conv\ \Omega^1) \cap (conv\ \Omega^2)] \leq \text{peri}(\Omega^1) + \text{peri}(\Omega^2). \quad (3.3.2)$$

Proof. As Ω^1 and Ω^2 are not separable, their convex hulls intersect. The conclusions of the corollary are then immediate from Lemma 3.3.17 with $S = \text{conv}\ \Omega^1$ and $T = \text{conv}\ \Omega^2$ (noting that $[(\text{conv}\ \Omega^1) \vee (\text{conv}\ \Omega^2)] = \text{conv}(\Omega^1 \cup \Omega^2)$. □

We are now ready to prove a critical inequality about perimeters.

Lemma 3.3.19. *Let $d = 2$. Suppose π is a partition with nonseparable parts π_r and π_s. If (π'_r, π'_s) is the outcome of JS-sorting (π_r, π_s), then π'_r and π'_s are nonempty, separable and satisfy*

$$\text{peri}(\pi'_r) + \text{peri}(\pi'_s) < \text{peri}(\pi_r) + \text{peri}(\pi_s). \quad (3.3.3)$$

Proof. Assume that π_r and π_s are not separable. Let v be the vertex of $\text{conv}(\pi_r \cup \pi_s)$ that is used in the JS-sorting that generates (π'_r, π'_s). Without loss of generality, assume that $\pi'_r = \{v\}$. A vertex of a convex set is the unique maximizer of a linear functional over the set. Thus, there exists a vector $C \in \mathbb{R}^2$ with $C^T v > C^T x$ for each $x \in [\text{conv}(\pi_r \cup \pi_s) \setminus \{v\}]$, implying that $\pi'_r = \{v\} \neq \emptyset$ and $\pi'_s = (\pi_r \cup \pi_s) \setminus \{v\} \neq \emptyset$ are separable. Next observe that as $\text{conv}\ \pi'_s = \text{conv}[(\pi_r \cup \pi_s) \setminus \{v\}] \subset \text{conv}(\pi_r \cup \pi_s)$, Lemma 3.3.16 implies that $\text{peri}(\pi'_s) < \text{peri}(\pi_r \cup \pi_s)$. Also, trivially, $\text{peri}(\pi'_r) = 0 \leq \text{peri}[(\text{conv}\ \pi_r) \wedge (\text{conv}\ \pi_s)]$. Finally, as π_r and π_s are not separable, their convex hulls are not disjoint and Lemma 3.3.17 applies to them. We conclude from that lemma and the above inequalities that

$$\begin{aligned} \text{peri}(\pi_r) + \text{peri}(\pi_s) &\geq \text{peri}[(\text{conv}\ \pi_r) \wedge (\text{conv}\ \pi_s)] + \text{peri}[(\text{conv}\ \pi_r) \vee (\text{conv}\ \pi_s)] \\ &= \text{peri}[(\text{conv}\ \pi_r) \wedge (\text{conv}\ \pi_s)] + \text{peri}(\pi_r \cup \pi_s) \\ &> 0 + \text{peri}(\pi_r \cup \pi_s) \setminus \{v\} = \text{peri}(\pi'_r) + \text{peri}(\pi'_s). \end{aligned}$$

□

Corollary 3.3.20. *For $d = 2$, S is (sort-specific, 2, support)-sortable with "specific" referring to JS-sorting.*

Proof. The stated result is immediate from Lemma 3.3.19, with $\sum_{j=1}^{p} \text{peri}(\pi_j)$ as the corresponding partition-statistics. □

The validity of Corollary 3.3.20 does not depend on the definition of the perimeter of a line segment. In particular, if the perimeter of a line is taken as its length, one can prove the corollary using the statistics $\sum_{j=1}^{p} w_j \text{peri}(\pi_j)$ with $w_j = 1$ if $\dim \pi_j = 1$ and $w_j = 2$ otherwise.

We now define another S-sorting called CRW-sorting, first proposed by Capoyleas, Rote and Woeginger [14], which covers the following three cases:

(i) Either π_r contains π_s or vice versa. Then CWR-sorting is the same as JS-sorting.
(ii) π_r and π_s are weakly separable. Then CRW-sorting is the sorting used in Theorem 3.1.6.
(iii) π_r and π_s have a 2-dim intersection not belonging to the above two cases.

We define CRW-sorting for the 2-dim intersection case which is none of the two special subcases. Given two polytopes S and T in $\mathbb{R}^2$, each vertex of $S \cap T$ lies on the boundaries of S or T or both (for otherwise v would be in $[\text{int}(S)] \cap [\text{int}(T)] = \text{int}(S \cap T)$). An *intersection vertex* of S and T is a vertex of $S \cap T$ which lies on the boundaries of both S and T.

Suppose π is nonseparable and let π_r and π_s be two nonseparable parts with a 2-dim intersection but neither containing the other nor being weakly separable. The outcome of CRW-sorting is then a 2-partition (π'_r, π'_s) constructed by using the following rules:

Let u and v be two arbitrary intersection vertices of conv π_r and conv π_s, let L be a line that contains these points and let L^+ and L^- be an indexation of the two open half-planes that L determines. Since π_r and π_s are not weakly separable, L satisfies the condition that at least one of π_r and π_s, say, π_r, intersects with both L^+ and L^-. Then the new partition is (L^+, L^-) with points on L joining a side which intersects π_s.

The next lemma was given by Capoyleas, Rote and Woeginger [14].

Lemma 3.3.21. *Let $d = 2$. Suppose π_r and π_s are non-separable but have a 2-dim intersection, or one is contained in the other, or the two parts are weakly separable. Then the CRW-sorting of $\pi_r \cup \pi_s$ results in the new partition π'_r and π'_s that are separable and satisfy (3.3.3).*

Proof. Lemma 3.3.16 proves that the CRW-sorting works for Case (i) and Theorem 3.1.6 proves for Case (ii) (note that (3.1.18) implies (3.3.3)). So it suffices to consider Case (iii).

For a set $U \subseteq \mathbb{R}^2$, let $U^{\pm} \equiv U \cap L^{\pm}$, $U^{\oplus} \equiv U \cap (L \cup L^+)$ and $U^{\ominus} \equiv U \cap (L \cup L^-)$. In particular, it is assumed that $\pi_s^+ \neq \emptyset$, $\pi_s^- \neq \emptyset$ and either

$\pi_r^+ \neq \emptyset$ and $\pi_r^- \neq \emptyset$ or $\pi_r \subseteq L \cup L^+$. By possibly exchanging π_r' and π_s', we shall assume that

$$\pi_r' \equiv (\pi_r \cup \pi_s) \cap (L \cup L^+) \quad \text{and} \quad \pi_s' \equiv (\pi_r \cup \pi_s) \cap L^-. \tag{3.3.4}$$

The fact that π_r' and π_s' are nonempty and separable is immediate.

Let u, v and L be chosen satisfying the conditions of a CRW-sorting and let $I(u, v)$ be the line segment that connects u and v. We next claim that $L \cap (\text{conv } \pi_s) = I(u, v)$. Trivially, $I(u, v) \subseteq L \cap (\text{conv } \pi_s)$. So, assume that w were a point in $[L \cap (\text{conv } \pi_s)] \setminus I(u, v)$. Then either u or v would be a point in bd(conv π_s) having a representation as a convex combination of two points of conv π_s, implying that all three points lie on an edge of conv π_s. As L is the line spanned by that edge, it would then follow that conv π_s must have an empty intersection with either of the two open half-spaces determined by L, a contradiction to the fact that $(\text{conv } \pi_s)^+ \supseteq \pi_s^+ \neq \emptyset$ and $(\text{conv } \pi_s)^- \supseteq \pi_s^- \neq \emptyset$. The conclusion that $L \cap (\text{conv } \pi_s) = I(u, v)$ implies that for $i = s$

$$\mathrm{peri}(\pi_i) = \mathrm{peri}[\pi_i^{\oplus} \cup \{u, v\}] + \mathrm{peri}[\pi_i^- \cup \{u, v\}] - 2\|u - v\|. \tag{3.3.5}$$

Now, if $\pi_r^+ \neq \emptyset$ and $\pi_r^- \neq \emptyset$ then the above arguments show that $L \cap (\text{conv } \pi_r) = I(u, v)$ and (3.3.5) holds for $i = r$. Alternatively, if $\pi_r \subseteq L \cup L^+$, then (3.3.5) is trite for $i = r$ as $\pi_r^- = \emptyset$ and $\pi_r^{\oplus} = \pi_r$. So, (3.3.5) holds for both $i = r$ and $i = s$.

Next, Lemma 3.3.16 implies that

$$\begin{aligned} \mathrm{peri}(\pi_r') &= \mathrm{peri}[(\pi_r \cup \pi_s)^{\oplus}] \leq \mathrm{peri}[(\pi_r \cup \pi_s)^{\oplus} \cup \{u, v\}] \\ \mathrm{peri}(\pi_s') &= \mathrm{peri}[(\pi_r \cup \pi_s)^-] < \mathrm{peri}[(\pi_r \cup \pi_s)^- \cup \{u, v\}]. \end{aligned} \tag{3.3.6}$$

Also, as a perimeter of a polytope is at least twice the length of any of its edges, we have that

$$\begin{aligned} \mathrm{peri}(P^{\oplus}) &\geq 2\|u - v\| \\ \mathrm{peri}(P^{\ominus}) &\geq 2\|u - v\|. \end{aligned} \tag{3.3.7}$$

Next observe that

$$\begin{aligned} P^{\oplus} &= [\mathrm{conv}(\pi_r^{\oplus} \cup \{u, v\})] \cap [\mathrm{conv}(\pi_s^{\oplus} \cup \{u, v\}) \\ P^{\ominus} &= [\mathrm{conv}(\pi_r^- \cup \{u, v\})] \cap [\mathrm{conv}(\pi_s^- \cup \{u, v\}) \end{aligned}$$

and therefore Corollary 3.3.18 implies that

$$\begin{aligned} &\mathrm{peri}[(\pi_r \cup \pi_s)^{\oplus} \cup \{u, v\}] + \mathrm{peri}(P^{\oplus}) \\ \leq\ &\mathrm{peri}(\pi_r^{\oplus} \cup \{u, v\}) + \mathrm{peri}(\pi_s^{\oplus} \cup \{u, v\}) \text{ and} \\ &\mathrm{peri}[(\pi_r \cup \pi_s)^- \cup \{u, v\}] + \mathrm{peri}(P^{\ominus}) \\ \leq\ &\mathrm{peri}(\pi_r^- \cup \{u, v\}) + \mathrm{peri}(\pi_s^- \cup \{u, v\}). \end{aligned} \tag{3.3.8}$$

It now follows from (3.3.6)–(3.3.7), (3.3.8) and (3.3.5), respectively, that

$$\begin{aligned}&\mathrm{peri}(\pi_{\mathrm{r}}') + \mathrm{peri}(\pi_{\mathrm{s}}') + 4\|\mathrm{u}-\mathrm{v}\| \\ &< \mathrm{peri}[(\pi_{\mathrm{r}} \cup \pi_{\mathrm{s}})^{\oplus} \cup \{\mathrm{u},\mathrm{v}\}] + \mathrm{peri}[(\pi_{\mathrm{r}} \cup \pi_{\mathrm{s}})^{-} \cup \{\mathrm{u},\mathrm{v}\}] + \mathrm{peri}(\mathrm{P}^{\oplus}) + \mathrm{peri}(\mathrm{P}^{\ominus}) \\ &\leq \mathrm{peri}[\pi_{\mathrm{r}}^{\oplus} \cup \{\mathrm{u},\mathrm{v}\}] + \mathrm{peri}[\pi_{\mathrm{s}}^{\oplus} \cup \{\mathrm{u},\mathrm{v}\}] + \mathrm{peri}[\pi_{\mathrm{r}}^{-} \cup \{\mathrm{u},\mathrm{v}\}] + \mathrm{peri}[\pi_{\mathrm{s}}^{-} \cup \{\mathrm{u},\mathrm{v}\}] \\ &= \mathrm{peri}(\pi_{\mathrm{r}}) + \mathrm{peri}(\pi_{\mathrm{s}}) + 4\|\mathrm{u}-\mathrm{v}\|,\end{aligned}$$

establishing (3.3.3). □

The following example demonstrates that the rule under which points in $(\pi_r \cup \pi_s) \cap L$ are assigned cannot be modified.

Example 3.3.1. Let ϵ be a small positive number. Consider the set of 10 vectors $W = \{\binom{-2}{0}\ \binom{-2}{\epsilon}, \binom{-2}{2}, \binom{2}{2}, \binom{2}{\epsilon}, \binom{2}{0}, \binom{-1}{1}, \binom{1}{1}, \binom{1}{-1}, \binom{-1}{-1}\}$ and the 2-partition $\pi = (\pi_1, \pi_2)$ where π_1 consists of the first 6 vectors and π_2 consists of the remaining 4; in particular, $\mathrm{peri}(\pi_1) + \mathrm{peri}(\pi_2) = 12 + 8 = 20$. With $u = \binom{-1}{0}$ and $v = \binom{1}{0}$, the construction $\pi' = (\pi_1', \pi_2')$ of Lemma 3.3.21 has $L = \{\binom{x}{y} : y = 0\}$, $L^+ = \{\binom{x}{y} : y > 0\}$ and $L^- = \{\binom{x}{y} : y < 0\}$, one part as the first 8 vectors in the above list and the other part as the remaining 2 vectors; in particular, $\mathrm{peri}(\pi_1') + \mathrm{peri}(\pi_2') = 12 + 4 = 16 < 20$. On the other hand, a partition $\pi'' = (\pi_1'', \pi_2'')$ with $\{\pi_1'', \pi_2''\} = \{W \cap (L \cup L^-), W \cap L^+\}$ has $\mathrm{peri}(\pi_1'') + \mathrm{peri}(\pi_2'') = (12 - 2\epsilon) + (6 + 2\sqrt{2}) > 20$. □

We now extend Lemma 3.3.21 to cases when at least one of π_r and π_s is of dimension < 2 (hence no 2-dim intersection). To this aim, we define the extended CRW-sorting (ECRW-sorting) by combining the CRW-sorting with the sorting defined in the following.

Observe that neither π_r nor π_s can be a single point, i.e., of dimension 0, since it would belong to either case(i) or case(ii) of Lemma 3.3.21. Without loss of generality, assume conv π_s is of dimension 1 (conv π_s is a line). Since π_r and π_s are non-separable, there are essentially three cases.

(a) conv π_r is of dimension 2 and contains exactly one endpoint of conv π_s.
(b) conv π_r is of dimension 2 and contains no endpoint of conv π_s.
(c) conv π_r is of dimension 1 and intersects conv π_s at one point.

Define L_{uv} to be the line connecting u and v, and L_q, $q \in \{u, v\}$, to be the line overlapping the edge of π_r containing q. Sort $\pi_r \cup \pi_s$ into the new partition π_r' and π_s' as follows:

(a) Set $L = L_u$ and assign points on L to π_r'.

(b) Set L to be any of L_{uv}, L_u or L_v (in the last two cases, assign points on L to π'_r).
(c) Set $L = L_u$ and assign points on L to π'_r.

Figure 3.3.4 demonstrates such an example in which a polygon is drawn as a triangle.

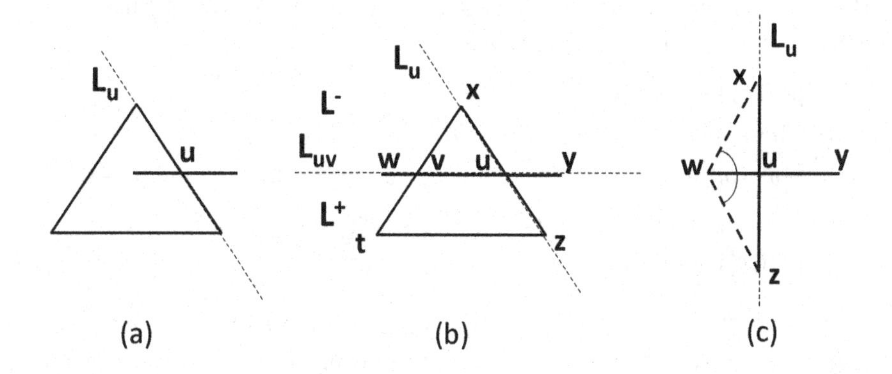

Figure 3.3.4

Theorem 3.3.22. *Let $d = 2$. Suppose π_r and π_s are non-separable parts. Then the ECRW-sorting of $\pi_r \cup \pi_s$ results in the new partition π'_r and π'_s which are nonempty, separable and satisfy (3.3.3).*

Proof. By Lemma 3.3.21, it suffices to show the three cases mentioned above.

In the case (a), $\text{peri}(\pi'_r) = \text{peri}(\pi_r)$ and $\text{peri}(\pi'_s) < \text{peri}(\pi_s)$ as conv $\pi'_r =$ conv π_r and conv $\pi'_s \subset$ conv π_s; consequently (3.3.3) holds.

Consider the case (b). Set $L = L_{uv}$. Without loss of generality, assume $\pi'_r = (\pi_r \cup \pi_s) \cap L^{\oplus}$ and $\pi'_s = (\pi_r \cup \pi_s) \cap L^-$. Then

$$\begin{aligned}&[\text{peri}(\pi_r) + \text{peri}(\pi_s)] - [\text{peri}(\pi'_r) + \text{peri}(\pi'_s)]\\&> (\|t-v\| + \|v-w\| + \|z-u\| + \|u-y\|) - (\|t-w\| + \|z-y\|)\\&> 0.\end{aligned}$$

Set $L = L_u$. It follows that conv $\pi'_r = \text{conv}(\pi_r \cup \{w\})$ and conv $\pi'_s \subset L_{yu}$.

Then

$$\begin{aligned}&[\text{peri}(\pi_{\text{r}})+\text{peri}(\pi_{\text{s}})]-[\text{peri}(\pi'_{\text{r}})+\text{peri}(\pi'_{\text{s}})]\\ &>(2\|w-v\|+\|x-v\|+\|t-v\|)-(\|x-w\|+\|w-t\|)\\ &>0.\end{aligned}$$

The case of $L = L_v$ is similar.

Consider the case (c) where conv π_r, the line segment xz, and conv π_s, the line segment wy, are of dimension 1. According to the sorting, $\pi'_r \subseteq$ conv$(\{x, w, z\})$ and $\pi'_s \subset L_{yu}$. Then

$$\begin{aligned}&\text{peri}(\pi_{\text{r}})+\text{peri}(\pi_{\text{s}})\\ &=2\|w-u\|+2\|u-y\|+2\|x-u\|+2\|u-z\|\\ &=(\|w-u\|+\|x-u\|)+(\|w-u\|+\|u-z\|)+\|x-z\|+2\|u-y\|\\ &>\|w-x\|+\|w-z\|+\|x-z\|+2\|u-y\|\\ &>\text{peri}(\pi'_{\text{r}})+\text{peri}(\pi'_{\text{s}}).\end{aligned}$$

□

In both Theorem 3.3.22 and Lemma 3.3.21, the ECRW-sorting leads to a strict reduction of the sum of perimeters over all parts; hence the sum can be used as the strictly decreasing statistics for the sortability argument. We have

Corollary 3.3.23. *For d = 2, S is (sort-specific, 2, support)-sortable with respect to the ECRW-sorting.*

Chapter 4

Clustering Problems over Multi-parameter Spaces

In this chapter, we examine multi-parameter partition problems, and identify suitable sets of conditions that assure the existence of optimal partitions that satisfy properties specified in Section 3.1. The approaches we follow are parallel to what was done in Section 5.4 and Chapter 7 of Vol. I about single-parameter problems. Since most of the results are about clustering problems, which are usually presented as to minimize some distance functions, we will also assume $F(\pi)$ to be minimized in this chapter.

The partition problems we consider may allow or prohibit empty parts. In Section 4.1, the issue of allowing or not allowing empty parts is handled along the guidelines spelt out in Section 1.2 of Vol. I. For shape problems, the situation is implicitly handled in the set of feasible shapes – empty parts are prohibited if this set is restricted to positive vectors and empty parts are allowed when the set of shapes allows 0 coordinates in the set of shapes. For size problems, results are given for single-size problems or relaxed single-size problems specifically. But sometimes there is no need to distinguish them since if the objective function is supermodular (see Vol. I, p. 126), then there always exists a nonempty 2-partition as good as a 2-partition with an empty part. A statement like "a result is for (relaxed) single-size problems" implies that the result covers for both problems. Further, to avoid discussions on trivial cases, we assume that in a p-partition problem, the number of distinct elements in the partitioned set is at least p; so that whenever a partition has an empty part, it must also have a part of size at least 2.

4.1 Geometric Properties of Optimal Partitions

Let $\{A^1, \ldots, A^n\}$ be a set of n vectors in $\mathbb{R}^d$ and let $F(\cdot)$ denote an objective function over partitions of these vectors. We will often state the results for weak properties and rely on Lemma 3.1.5 for proper extensions to corresponding strong properties, i.e., if A is in general position, then a result stands for the strong version of the property. Similarly, we rely on Lemma 1.2.1 of Vol. I and the comments that follow it to reduce the proof for a constrained-shape problem to a single-shape problem.

The next result extends Theorem 2.2.1

Theorem 4.1.1. *Suppose*

$$F(\pi) = f(|\pi_1|, \ldots, |\pi_p|, A_\pi), \tag{4.1.1}$$

where f is (edge-)quasi-concave in the $d \times p$ variables in A_π for each fixed set of values $(|\pi_1|, \ldots, |\pi_p|)$. Then every constrained-shape problem has an almost-separable optimal partition. If furthermore, f is strictly (edge-) quasi-concave, then every optimal partition for a constrained-shape problem is almost-separable.

Proof. We can view the given shape $(|\pi_1|, \ldots, |\pi_p|)$ as parameters of the cost function and write $F(\pi)$ as $f_{|\pi_1|,\ldots,|\pi_p|}(A_\pi)$. As the (edge-)quasi-concavity assumptions refer to $f_{|\pi_1|,\ldots,|\pi_p|}$, Theorem 4.1.1 follows from Theorem 2.2.1 (except that the maximum-minimum exchange induces a convex-concave exchange). □

Recall the use of "$\widehat{\ }$" introduced in (3.1.3)–(3.1.5).

Corollary 4.1.2. *Suppose*

$$F(\pi) = f(|\pi_1|, \ldots, |\pi_p|, \widehat{A}_\pi) , \tag{4.1.2}$$

where f is (edge-)quasi-concave in the $(d+1) \times p$ variables of $\widehat{A}_\pi$ for each fixed $(|\pi_1|, \ldots, |\pi_p|)$. Then every constrained-shape problem has an almost-sphere-separable optimal partition. Further, if f is strictly (edge-) quasi-convex, then every optimal partition for a constrained-shape problem is almost-sphere-separable.

Proof. The conclusion of the corollary is immediate from Theorems 4.1.1 and 3.1.3. □

Theorem 4.1.1 and Corollary 4.1.2 were first proved by Golany, Hwang and Rothblum [41] for the case where the columns of A are distinct (in which case, almost-separable reduces to separable and almost-sphere-separable reduces to sphere-separable).

If the function f in Corollary 4.1.2 is linear in the variables corresponding to $\widehat{A}_\pi$ and in the shape-variable, the results of Section 1.1 can be used; that is, bounded-shape partition problems with such objective functions can be solved by LP methods, applied in the enlarged $(d+2)$-dim space where the partitioned vectors are $\binom{\widehat{A}^i}{1}$. (We recall that the partitioned vectors were appended by 1's in Section 2.1 of a different purpose, namely, to facilitate a computational method for testing general position.)

Following Golany, Hwang and Rothblum [41], we next use Corollary 4.1.2 to address some clustering problems.

Theorem 4.1.3. *Let $w_1, \ldots, w_p$ be positive numbers. Suppose $F(\cdot)$ has either one of the following three expressions:*

(i) $F(\pi) = \sum_{j=1}^{p} w_j \sum_{i,k\in\pi_j} \|A^i - A^k\|^2,$

(ii) $F(\pi) = \sum_{j=1,|\pi_j|\neq 0}^{p} w_j \sum_{i\in\pi_j} \|A^i - \bar{A}_j\|^2$ *for* $j = 1, \ldots, p$, $\bar{A}_j = \frac{\sum_{k\in\pi_j} A^k}{|\pi_j|},$

(iii) $F(\pi) = \sum_{j=1}^{p} w_j \sum_{i\in\pi_j} \|A^i - t_j\|^2$, *where* $t_1, \ldots, t_p$ *are prescribed d-vectors.*

Then every constrained-shape problem has an almost-sphere-separable optimal partition.

Proof. By Lemma 1.2.1 of Vol. I, it suffices to consider the single-shape problem. For easy writing, the following proof assumes no empty parts (which affects (ii) only). But the adaptation to having empty parts is straightforward.

For a (nonempty) subset S of $A^1, \ldots, A^n$ and a vector t in $\mathbb{R}^d$, we have that

$$\sum_{i,k\in S} \|A^i - A^k\|^2 = 2|S| \sum_{i\in S} \|A^i\|^2 - 2\|\sum_{i\in S} A^i\|^2,$$

$$\begin{aligned} \sum_{i\in S} \|A^i - \frac{\sum_{k\in S} A^k}{|S|}\|^2 &= \sum_{i\in S} \|A^i\|^2 - 2\frac{(\sum_{i\in S} A^i)^T(\sum_{k\in S} A^k)}{|S|} \\ &\quad + \frac{\|\sum_{k\in S} A^k\|^2}{|S|} \\ &= \sum_{i\in S} \|A^i\|^2 - \frac{\|\sum_{k\in S} A^k\|^2}{|S|}, \end{aligned}$$

$$\sum_{i \in S} \|A^i - t\|^2 = \sum_{i \in S} \|A^i\|^2 - 2(\sum_{i \in S} A^i)^T t + |S| \|t\|^2.$$

Given a partition $\pi = (\pi_1, \ldots, \pi_p)$, define

$$X^j = \sum_{i \in \pi_j} A^i \text{ and } Y^j = \sum_{i \in \pi_j} \|A^i\|^2.$$

It follows that $F(\pi)$ in (i), (ii), (iii) can be written, respectively, as

(i) $2\sum_{j=1}^{p} w_j \left(|\pi_j| Y^j - \|X^j\|^2\right)$,

(ii) $\sum_{j=1}^{p} w_j \left[Y^j - \frac{\|X^j\|^2}{|\pi_j|}\right]$,

(iii) $\sum_{j=1}^{p} w_j \left[Y^j - 2(X^j)^T t_j + |\pi_j| \|t_j\|^2\right]$.

It is easily verified that these functions are quasi-concave in

$$\begin{pmatrix} X^1 \ldots X^p \\ Y^1 \ldots Y^p \end{pmatrix}$$

and the conclusions of the theorem follow from Corollary 4.1.2. □

Note that the functions (i), (ii) and (iii) are not strict quasi-convex due to the existence of the w_j term. Also, the proof of Theorem 4.1.3 expresses the third objective function as a linear function of $\widehat{A}_\pi$ and of the parts' sizes. Consequently, the comment following Corollary 4.1.2 applies and the corresponding bounded-shape partition problems can be solved by LP methods.

The next corollary was first given by Boros and Hammer [9] for single-size partition problems with distinct A^i's.

Corollary 4.1.4. *Suppose $A^1, \ldots, A^n$ lie on the boundary of a sphere. Then every constrained-shape problem with any one of the objective functions listed in Theorem 4.1.3 has an almost-separable optimal partition.*

Proof. The conclusion of the corollary is immediate from Theorem 4.1.3 and Lemma 3.1.7. □

We next consider the *uniform* versions of the objective functions that are listed in Theorem 4.1.3, where uniform means that the w_j's are constant.

Theorem 4.1.5. *Consider uniform versions of the objective functions that are listed in Theorem 4.1.3. Then:*

(i) for a constrained-shape problem with constant part-sizes and with the first function, every optimal partition is almost-separable,
(ii) for a constrained-shape problem with the second function, every optimal partition is almost-separable, and
(iii) for a constrained-shape problem with the third function, there exists an almost-separable optimal partition.

Proof. Without loss of generality assume that $w_1 = \cdots = w_p = 1$. The expressions for the objective functions obtained in the proof of Theorem 4.1.3 become:

(i) $2\left(\sum_{j=1}^{p} |\pi_j| Y^j - \sum_{j=1}^{p} \|X^j\|^2\right)$,

(ii) $\sum_{j=1}^{p} Y^j - \sum_{j=1}^{p} \|X^j\|^2/|\pi_j|$,

(iii) $\sum_{j=1}^{p} Y^j + \sum_{j=1}^{p} |\pi_j| \|t_j\|^2 - 2\sum_{j=1}^{p} (X^j)^T t_j$.

For the first function with constant part-sizes and for the second function we have that the first term is constant as it is proportional (equal in the second case) to $\sum_{i=1}^{n} \|A^i\|^2$. We further observe that the second term in either of these cases is strictly concave in $X^1 \ldots X^p$. Also, the third function is linear in $X^1 \ldots X^p$. So, the conclusions of the theorem follow from Theorem 4.1.1. □

The reason that the conclusion of (iii) is weaker than that of (i) and (ii) is due to the fact that its last term (without sign) is not necessarily increasing in X^j.

Boros and Hammer [9] considered case (i) of Theorem 4.1.5 and proved that every single-size problem has a weakly separable optimal partition. Theorem 4.1.5 strengthens their result in several ways: a stronger property is satisfied for every optimal partition, and it holds for a wider class of partition problems.

In view of Theorem 4.1.5, one would hope that the almost-sphere-separable property proved in Theorem 4.1.3 can also be upgraded to almost-separable. But this hope is dashed by Examples 7.4.2 and 7.4.3 of Vol. I for $\mathbb{R}^1$ (note that sphere-separable is reduced to noncrossing and separable to consecutive).

We next turn our attention to (relaxed) single-size problems. We update some notion introduced in Chapter 2 of Vol. I for index-partitions. For a p-partition π and $i \in \mathcal{N}$, let $j_\pi(i)$ denote the index of the part of π that contains A^i (here we distinguish between multiple copies of the same vector).

Lemma 4.1.6. *Let $t_1, \ldots, t_p$ be prescribed distinct vectors in $\mathbb{R}^d$. Then, for the relaxed single-size problem, there is a separable partition π such that*

$$\|A^i - t_{j_\pi(i)}\| \leq \|A^i - t_{j_{\pi'}(i)}\| \quad \textit{for every partition } \pi' \textit{ and } i \in \mathcal{N}. \quad (4.1.3)$$

Further, a partition π satisfies (4.1.3) if and only if each vector A^i is assigned to a part π_j for which $\|A^i - t_j\| = \min_{1 \leq u \leq p} \|A^i - t_u\|$; each such partition is weakly separable.

Proof. The fact that a partition satisfies (4.1.3) if and only if each vector A^i is assigned to a part π_j for which $\|A^i - t_j\| = \min_{1 \leq u \leq p} \|A^i - t_u\|$ is trite. Next, consider any p-partition that satisfies this condition and let $r, s \in \{1, \ldots, p\}$. Then the hyperplane H_{rs} that is perpendicular to the line connecting t_r and t_s and contains the mid-point of that line demonstrates the weak separability of π_r and π_s. Next, let π be the partition that assigns each vector A^i to the part π_j with the lowest index j among those for which $\|A^i - t_j\|$ is minimized. By the above argument, such a partition satisfies (4.1.3) and is weakly separable. Consider any pair of parts π_r and π_s of π and let H_{rs} be the hyperplane that weakly separates π_r and π_s. As all points of $(\pi_r \cup \pi_s) \cap H_{rs}$ are assigned to the same part (the one with the lower index), the 2-partition $(\pi_r \cap H_{rs}, \pi_s \cap H_{rs})$ is (trivially) separable; consequently, Lemma 3.1.4 assures that (π_r, π_s) is separable. Since r and s are arbitrary, it follows that π is separable. □

Theorem 4.1.7. *Let $t_1, \ldots, t_p$ be prescribed distinct vectors in $\mathbb{R}^d$ and consider the relaxed single-size problem, which has*

$$F(\pi) = f(\|A^1 - t_{j_\pi(1)}\|, \ldots, \|A^n - t_{j_\pi(n)}\|) \quad (4.1.4)$$

where $f : \mathbb{R}^n \to \mathbb{R}$ is nondecreasing. Then the partition π which assigns each vector A^i to the part π_j with the lowest index j among those for which $\|A^i - t_j\|$ is minimized, is both optimal and separable. Further, if f is increasing, then a partition π is optimal if and only if each vector A^i is assigned to a part π_j for which $\|A^i - t_j\| = \min_{1 \leq u \leq p} \|A^i - t_u\|$, and every optimal partition is weakly separable.

Proof. Evidently, (4.1.3) is sufficient for optimality when f is nondecreasing and it is both necessary and sufficient for optimality when f is increasing. The conclusion of the theorem follows immediately from Lemma 4.1.6. □

Corollary 4.1.8. *The conclusions of Theorem 4.1.7 hold for the relaxed single-size problem, which has*

$$F(\pi) = \sum_{j=1,\pi_j \neq \emptyset}^{p} \sum_{i \in \pi_j} h(\|A^i - t_j\|), \tag{4.1.5}$$

where $t_1, \ldots, t_p$ are prescribed distinct vectors in $\mathbb{R}^d$ and $h : \mathbb{R} \to \mathbb{R}$ is nondecreasing/increasing.

Proof. Apply Theorem 4.1.7 to the nondecreasing/increasing function $f : \mathbb{R}^n \to \mathbb{R}$ with $f(x) = \sum_{i=1}^n h(x_i)$ for each $x \in \mathbb{R}^n$. □

Theorem 4.1.7 and Corollary 4.1.8 specify a relaxed single-size separable partition that is uniformly optimal for all underlying partition problems, independently of the functions f and h that occur in (4.1.4) and (4.1.5).

The characterization of optimal partitions of Theorem 4.1.7 (and Corollary 4.1.8) allows one to construct all optimal solutions by assigning each vector A^i to any part π_j for which t_j is the closest to A^i. More specifically, associate each vector $A^i \in \mathbb{R}^d$ with the vector $D^i \equiv (\|A^i - t_1\|, \ldots, \|A^i - t_p\|)^T \in \mathbb{R}^p$. A partition is then optimal if and only if it assigns A^i to π_j with D^i_j being any minimal coordinate of D^i, and all such partitions are weakly separable. But, these partitions need not be almost-separable, let alone separable; still, the proof of Theorem 4.1.7 shows that breaking ties by using any part-ranking will produce separable partitions. We note that the amount of effort needed to compute each vector D^i is $O(dp)$, so the total amount of effort to compute all of the D^i's (and thereby solve the partition problem) is $O(dpn)$.

The next example demonstrates that the characterizing condition for optimality in Theorem 4.1.7 is not overly restrictive in the sense that it is possible that all partitions of n points satisfy it; so, the structure does not guarantee a set of partitions that is polynomial in the number of partitioned vectors. In particular, the example has optimal partitions that are not (almost-) separable, implying that the necessary condition for optimality in Theorem 4.1.7 cannot be strengthened from weak separability to (almost-) separability.

Example 4.1.1. Consider the partition problem of Theorem 4.1.7 with $p =$

2, $t_1 = \binom{1}{0}$, $t_2 = \binom{-1}{0}$ and $A^i = \binom{0}{i}$ for $i = 1, \ldots, n$. In this example all 2^n partitions satisfy the optimality condition of Theorem 4.1.7. In particular, if $n \geq 3$, the partition $\pi = (\{1, n\}, \{2, 3, \ldots, n-1\})$ is optimal; but, this partition is not separable. □

The tie-breaking scheme given in Theorem 4.1.7 reaches partitions that are both optimal and separable, but it reaches only a special subclass of such partitions. We next outline a polynomial method that generates all (almost-) separable optimal partitions for the case where h is increasing. The method will first enumerate all (almost-) separable partitions, using the (polynomial) enumeration method described in Section 2.2. By Theorem 4.1.7, π is optimal if and only if

$$[i \in \pi_j] \Leftrightarrow [\|A^i - t_j\|^2 = \nu_i \equiv \min_{1 \leq u \leq p} \|A^i - t_u\|^2]$$

(we use squares of the norms so as to keep computation restricted to arithmetic). Determining the ν_i's is easily executed with $O(npd)$ arithmetic operations and comparisons. Also, for each of the generated (almost-) separable partition π, testing if $\|A^i - t_{j_\pi(i)}\|^2 = \nu_i$ for $i = 1, \ldots, n$ is easily executed by $O(nd)$ arithmetic operations.

The sortability approach, as developed in Chapter 7 of Vol. I and Chapter 3 of this volume, uses sortability of k-partitions for a small $k < p$ to extend the existence of a property within k-partitions to its existence within p-partitions; this is accomplished through the use of a monotone statistics $s(\pi)$ to guarantee that sequential k-sorting does not cycle. In Theorem 4.1.7, we p-sort directly, thus eliminating the need of an iterative procedure and of a statistics $s(\pi)$. Since finding the corresponding statistics is crucial in the use of k-sortability approach, it is perhaps more suitable to view this p-sortability as a rival approach rather than a special case of k-sortability with $k = p$ (which one is more preferable depends on which can be shown to preserve optimality). Nevertheless, in the proof of Theorem 4.1.7, a lexicographically decreasing statistics $s(\pi) = (\sum_{j=1}^{p} \sum_{i \in \pi_j} \|A^i - t_j\|, |\pi_1|, \ldots, |\pi_p|)$ is implied in the proof which establishes (strongly, 2, support)-sortability, and thus allowing us to use the sortability approach. However, the process of merging $\binom{p}{2}$ 2-sortings is tedious as the number of 2-partitions for a given pair (π_r, π_s) is large.

The geometric figure of the partition of $\mathbb{R}^d$ into p convex subsets $S_1, \ldots, S_p$ with S_j consisting of all points in $\mathbb{R}^d$ closest to t_j (a point on a boundary is closest to all t_j whose S_j shares that boundary) is known in the literature as the Voronoi diagram. All weakly separable optimal partitions can be generated by arbitrary assignments of points in the boundaries to their neighboring S_j's. The assignment of all points on a boundary to the neighboring S_j with minimum index then generates a separable optimal partition as specified in Theorem 4.1.7. Efficient constructions of Voronoi diagrams are well-studied in theoretical computer science (see Fortune [36] for a survey). In particular, a Voronoi diagram for p given points $t_1, \ldots, t_p$ can be constructed in $O(d^{\frac{p+1}{2}})$-time. For $d = 2$, Shamos and Hoey [87] showed that the Voronoi diagram can be constructed in $O(p \log p)$-time.

We recall that a generalization of the objective function of Theorem 4.1.5 (iii) was considered in Example 1.1.1. Specifically, let $t_1, \ldots, t_p$ be given vectors, let $D(\cdot, \cdot)$ be an arbitrary (distance) function on $\mathbb{R}^d \times \mathbb{R}^d$ and consider the objective function $F(\cdot)$ with

$$F(\pi) = \sum_{j=1}^{p} \sum_{i \in \pi_j} D(A^i, t_j) \quad \text{for each partition } \pi.$$

It was observed in Example 1.1.1 that partition problems with this objective function can be converted into problems with linear objective function by substituting the partitioned vectors $A^1, \ldots, A^n$ with the vectors $D^1, \ldots, D^n$, where $D^i = (D(A^i, t_1), \ldots, D(A^i, t_p))^T \in \mathbb{R}^p$ for $i = 1, \ldots, n$. Consequently, all the methods developed in Chapter 1 for partition problems with linear objective functions apply; in particular, corresponding bounded-shape problems can be solved by linear programming. Further, Theorem 4.1.1 assures that corresponding constrained-shape problems have optimal solutions that are almost-separable (with the property referring to the vectors $D^1, \ldots, D^n$, and not to $A^1, \ldots, A^n$).

Theorem 4.1.7 showed that for p-partitions (allowing empty parts) we can generalize the objective function of Theorem 4.1.5 (iii) from the square of norm to a general nondecreasing function f. We now show that one can do the same for case (ii). For case (i) we show that even dropping the square from $\|A^i - A^k\|^2$ is impossible except for $d = 1$ (which is Theorem 7.4.1 of Vol. I). Boros and Hammer [9] gave the following example to show that Theorem 7.4.1 of Vol. I cannot be extended to $d \geq 2$.

Example 4.1.2. Consider the single-size partition problem with objective function

$$F(\pi) = \sum_{j=1}^{p} \sum_{i,k \in \pi_j} \|A^i - A^k\|$$

and with the multiset of vectors to be partitioned described in the following table:

point	*coordinates*	*multiplicity*
x^1	$(-20, 0)$	100
x^2	$(20, 0)$	100
x^3	$(-20, 12)$	1
x^4	$(0, 8)$	1
x^5	$(20, 12)$	1
x^6	$(0, 20)$	100
x^7	$(0, 40)$	100
x^8	$(0, 9)$	1
x^9	$(20, 13)$	1
x^{10}	$(60, 17)$	1

Boros and Hammer [9] stated that easy calculation shows that the unique optimal 2-partition is $\pi_1 = \{x^1, x^2, x^3, x^4, x^5\}$ and $\pi_2 = \{x^6, x^7, x^8, x^9, x^{10}\}$. But x^8 is in $\text{conv}\{x^3, x^4, x^5\}$ and x^4 in $\text{conv}\{x^8, x^9, x^{10}\}$; in particular, this (unique optimal) partition is not non-crossing, assuring that it is neither weakly separable, nor almost-separable, nor separable. □

We next develop a tool for using results about properties of optimal partitions of problems with objective functions that depend on prescribed vectors to results where the prescribed vectors are replaced by some minimizing vectors. To express a formal result we need two new definitions. A function $g : \mathbb{R}^d \to \mathbb{R}$ is *inverse-bounded* if for every $K \in \mathbb{R}$, $\{x \in \mathbb{R}^d : |g(x)| \leq K\}$ is a bounded set of $\mathbb{R}^d$. Evidently, an inverse-bounded function that is continuous is guaranteed to attain a minimum over $\mathbb{R}^d$. Also, let $P^*(\mathbb{R}^d)$ be the set of finite subsets of $\mathbb{R}^d$, that is, $P^*(\mathbb{R}^d) = \{\Omega \subset \mathbb{R}^d : |\Omega| \text{ is finite}\}$.

Theorem 4.1.9. *Let Q be a partition property, $f : \mathbb{R}^p \to \mathbb{R}$ be nondecreasing and $g : P^*(\mathbb{R}^d) \times \mathbb{R}^d \to \mathbb{R}$ with the property that for every $\Omega \in P^*(\mathbb{R}^d)$, $g(\Omega, \cdot)$ is inverse-bounded and continuous and $g(\emptyset, \cdot) \leq g(\Omega, \cdot)$ for any Ω. Let Γ be a set of p-integer vectors with coordinate-sum n. Assume that for any $t_1, \ldots, t_p \in \mathbb{R}^d$, some optimal partition of the constrained-shape problem corresponding to Γ and having objective function $F^{t_1,\ldots,t_p}(\cdot)$ given by*

$$F^{t_1,\ldots,t_p}(\pi) = f[g(\pi_1, t_1), \ldots, g(\pi_p, t_p)] \text{ for each partition } \pi \qquad (4.1.6)$$

satisfies Q. Then some optimal partition for the corresponding constrained-shape problem with objective function $F(\cdot)$ given by

$$F(\pi) = f\left[\min_{x_1 \in \mathbb{R}^d} g(\pi_1, x_1), \ldots, \min_{x_p \in \mathbb{R}^d} g(\pi_p, x_p)\right] \quad \text{for each partition } \pi \tag{4.1.7}$$

satisfies Q. Further, the above holds with "every" replacing "some".

Proof. We first verify the "some"-variant of the theorem. Consider any optimal partition π^* of the constrained-shape problem with objective function $F(\cdot)$ and for $j = 1, \ldots, p$, let t_j^* be a minimizer of $g(\pi_j^*, x)$ over $x \in \mathbb{R}^d$. The assumptions of the theorem assure that there exists a partition $\pi^\#$ that satisfies property Q and is optimal for the constrained-shape problem with objective function $F^{t_1^*,\ldots,t_p^*}(\cdot)$. As $f(\cdot)$ is nondecreasing, we conclude that

$$\begin{aligned} F(\pi^\#) &= f\left[\min_{x_1 \in \mathbb{R}^d} g(\pi_1^\#, x_1), \ldots, \min_{x_p \in \mathbb{R}^d} g(\pi_p^\#, x_p)\right] \\ &\leq f\left[g(\pi_1^\#, t_1^*), \ldots, g(\pi_p^\#, t_p^*)\right] = F^{t_1^*,\ldots,t_p^*}(\pi^\#) \\ &\leq F^{t_1^*,\ldots,t_p^*}(\pi^*) = F(\pi^*) \leq F(\pi^\#), \end{aligned}$$

implying that the partition $\pi^\#$ which satisfies Q is also optimal for the constrained-shape problem with objective function $F(\cdot)$.

We next consider the "every"-variant of the theorem. So, assume that for any $t_1, \ldots, t_p \in \mathbb{R}^d$, every optimal partition of the constrained-shape problem corresponding to Γ and having objective function $F^{t_1,\ldots,t_p}(\cdot)$ given by (4.1.6) satisfies Q. Consider any optimal partition π^* of the constrained-shape problem with objective function $F(\cdot)$ and for $j = 1, \ldots, p$, let t_j^* be a minimizer of $g(\pi_j^*, x)$ over $x \in \mathbb{R}^d$. It then follows that for every partition π,

$$F^{t_1^*,\ldots,t_p^*}(\pi^*) = F(\pi^*) \leq F(\pi) \leq F^{t_1^*,\ldots,t_p^*}(\pi).$$

So, the partition π^* is also optimal for the constrained-shape problem with objective function $F^{t_1^*,\ldots,t_p^*}(\cdot)$ and it therefore must satisfy Q. □

The proof of Theorem 4.1.9 can be applied to a partition that is not optimal. Specifically, if π is any partition and $t_1, \ldots, t_p$ are minimizers of $g(\pi_1, \cdot), \ldots, g(\pi_p, \cdot)$, respectively, then any partition $\pi^\#$ that satisfies $F^{t_1,\ldots,t_p}(\pi^\#) \leq F^{t_1,\ldots,t_p}(\pi)$ satisfies $F(\pi^\#) \leq F(\pi)$. In particular, any sorting method that is applicable to partition problems with objective function $F^{t_1,\ldots,t_p}(\cdot)$ for fixed t_j's which does not rely on a statistics depending on the t_j's, applies to the partition problem with objective function $F(\cdot)$. This is the case for p-sortings for which no statistics is needed.

Recall that if $h : \mathbb{R} \to \mathbb{R}$, an *h-centroid* of a finite set $\Omega \subseteq \mathbb{R}^d$ is a minimizer of $\sum_{A^i \in \Omega} h(\|A^i - x\|)$ over $x \in \mathbb{R}^d$. When h is continuous, bounded from below and inverse-bounded, each finite set Ω has an h-centroid. If π_j is an empty part, then the sum of the $h(\cdot)$ function is defined to be $h(0)$.

Theorem 4.1.10. *Consider the single-size problem which has*

$$F(\pi) = \sum_{j=1}^{p} \sum_{i \in \pi_j} h(\|A^i - c_j\|), \tag{4.1.8}$$

where $h : \mathbb{R} \to \mathbb{R}$ is nondecreasing, continuous and inverse-bounded and for $j = 1, \ldots, p$, c_j is the h-centroid of π_j. Then there exists a separable optimal partition. Further, if h is increasing, then every optimal partition is weakly separable.

Proof. The assumptions about h assure that it is bounded from below by $h(0)$; as h is assumed to be continuous and inverse-bounded, it follows that each $\Omega \in P^*(\mathbb{R}^d)$ has an h-centroid.

We first consider the case where h is nondecreasing. Corollary 4.1.8 guarantees that for each $t_1, \ldots, t_p \in \mathbb{R}^d$, some optimal solution of the single-size problem with objective function

$$F^{t_1,\ldots,t_p}(\pi) = \sum_{j=1}^{p} \sum_{i \in \pi_j} h(\|A^i - t_j)\|)$$

is separable. Consider the function $g : P^*(\mathbb{R}^d) \times \mathbb{R}^d \to \mathbb{R}$ with $g(\Omega, x) = \sum_{y \in \Omega} h(\|y - x\|)$ for $\Omega \in P^*(\mathbb{R}^d)$ and $x \in \mathbb{R}^d$. The assumptions about the function $h(\cdot)$ assure that for every $\Omega \in P^*(\mathbb{R}^d)$, the function $g(\Omega, \cdot)$ is nondecreasing, continuous and inverse-bounded. As (4.1.8) can be rewritten as

$$F(\pi) = \sum_{j=1}^{p} \left[\min_{x_j \in \mathbb{R}^d} \sum_{i \in \pi_j} h(\|A^i - x_j\|) \right] = \sum_{j=1}^{p} \left[\min_{x_j \in \mathbb{R}^d} g(\pi_j, x_j) \right],$$

it follows from Theorem 4.1.9 with f as the summation function and the aforementioned function g that some optimal solution of the underlying partition problem (with objective function $F(\cdot)$) is separable. Finally, if h is increasing, then the above arguments apply with "weakly separable" replacing "separable" and "every" replacing "some". □

Corollary 4.1.11. *Consider the "relaxed version" of Theorem 4.1.10. Then there exists a separable optimal partition without empty parts. Further, if h is increasing, then every optimal partition has no empty parts.*

Proof. Let π be a partition with an empty part π_u. By the tacit assumption that $\mathcal{N}$ contains at least p distinct points, there exists another part π_w which contains at least two distinct points. Let v be a vertex of $\text{conv}\pi_w$ with $v \neq c_w$ and let V be the set of all copies of v. Set $\pi'_u = V$, $\pi'_w = \pi_w \setminus V$ and $\pi'_j = \pi_j$ for $j \notin \{u, w\}$. Then

$$F(\pi) - F(\pi') = |V| h(\|v - c_w\|) \geq 0$$

where strict inequality holds if h is increasing. Use the same argument to eliminate all empty parts. Note that if π is separable then π' is also separable. Further, the existence of a separable optimal partition is assured by Theorem 4.1.10. So, the conclusions of the corollary now follow immediately. □

Theorem 4.1.10 asserts the existence of a separable optimal partition that is specific to the function h which occurs in (4.1.8). This type of result is weaker than the uniform optimality of a specific separable partition asserted in Theorem 4.1.7.

The paragraph following Theorem 4.1.9 shows that the assignment of each vector to its closest t_j, used in Theorem 4.1.7 and Corollary 4.1.8, can be used to replace a partition π that is not separable with one that is and has improved F-value where $F(\cdot)$ is given by (4.1.8).

We next consider partition problems with objective functions that depend on the least radii of spheres that, respectively, include the parts. We start by showing that given a partition whose parts are included in prescribed spheres, there is a separable partition with the same property. The proof of this result follows an approach of Capoyleas, Rote and Woeginger [14] who proved the case $d = 2$ (and claimed its extension to general d).

Lemma 4.1.12. *Let $t_1, \ldots, t_p$ be prescribed distinct vectors in $\mathbb{R}^d$. Consider the relaxed single-size problem. Then for each partition π, there exists a separable partition π' such that*

$$\max_{i \in \pi'_j} \|A^i - t_j\| \leq \max_{i \in \pi_j} \|A^i - t_j\| \quad \textit{for } j = 1, \ldots, p. \tag{4.1.9}$$

Further, if $r_j \equiv \max_{i \in \pi_j} \|A^i - t_j\|$ for $j = 1, \ldots, p$ (that is, r_j is the smallest radius of a sphere that is centered at t_j and contains π_j), then π' can be *selected by the following rule:* $i \in \pi'_j$ *if* j *is the lowest index of those for which* $\|A^i - t_j\|^2 - r_j^2 = \min_u[\|A^i - t_u\|^2 - r_u^2]$ *(of course, one may use any a-priori permutation of the part-indices).*

Proof. Let π be a given partition. For $j = 1, \ldots, p$, let $r_j = 0$ if π_j is empty and let $r_j = \max_{i \in \pi_j} \|A^i - t_j\|$ otherwise. Let π' be the partition which assigns each vector A^i to the part with the lowest index j of those for which $\|A^i - t_j\|^2 - r_j^2 = \min_u[\|A^i - t_u\|^2 - r_u^2]$. Consider part π'_j of π' and $i \in \pi'_j$. Assume that A^i is assigned to part π_w under the partition π. Then $\|A^i - t_w\| \leq r_w$ and $\|A^i - t_j\|^2 - r_j^2 \leq \|A^i - t_w\|^2 - r_w^2 \leq 0$, assuring that $\|A^i - t_j\| \leq r_j$. So, $\max_{i \in \pi'_j} \|A^i - t_j\| \leq r_j = \max_{i \in \pi_j} \|A^i - t_j\|$, verifying (4.1.9).

We next show that π' is separable. Consider two nonempty parts of π', say π'_j and π'_w with $j < w$, and let $i \in \pi'_j$ and $u \in \pi'_w$. Then

$$\|A^i - t_j\|^2 - r_j^2 \leq \|A^i - t_w\|^2 - r_w^2$$

and

$$\|A^u - t_w\|^2 - r_w^2 < \|A^u - t_j\|^2 - r_j^2.$$

Summing the above two inequalities, using the standard formula $\|x - y\|^2 = \|x\|^2 - 2x^T y + \|y\|^2$ for vectors $x, y \in \mathbb{R}^d$ and canceling identical terms, we conclude that

$$-2(t_j)^T A^i - 2(t_w)^T A^u < -2(t_w)^T A^i - 2(t_j)^T A^u.$$

So, with $C \equiv t_j - t_w$, we have that $C^T A^i > C^T A^u$. This proves the separability of π'_j and π'_w and completes the proof. □

Capoyleas et al. [14] did not mention a tie-breaking rule; as such, the "power diagram" they used to repartition the vectors is weak rather than strict separability (and we recall from Section 3.2 that the number of weakly separable partitions is not necessarily polynomial in the number of partitioned vectors).

The construction in Lemma 4.1.12 is specific to the underlying partition π (unlike the uniform construction of Lemma 4.1.6). The following example shows that a partition π' that achieves (4.1.9) uniformly for all partitions π needs not exist.

Example 4.1.3. Let $n = 3, d = 1, p = 2, A^1 = -1, A^2 = 0, A^3 = 1, t_1 = -1, t_2 = 1$. Then $(\max_{i \in \pi_1} \|A^i - t_1\|, \max_{i \in \pi_2} \|A^i - t_2\|)$ can take values $(1,0)$,

$(0, 1)$, $(2, 2)$, $(1, 2)$ and $(2, 1)$ (the first three pairs are attained by separable partitions and last two by non-separable partitions as well). Also, if empty parts are allowed there are two additional pairs (attained by separable partitions) – $(2, 0)$ and $(0, 2)$. Whether or not empty parts are allowed, there is no partition π' satisfies (4.1.9) uniformly for all partitions π. □

The separating hyperplane between parts π'_j and π'_w of the partition π' constructed in Lemma 4.1.12 is given by

$$\begin{aligned} H &\equiv \{x \in \mathbb{R}^d : \|x - t_j\|^2 - r_j^2 = \|x - t_w\|^2 - r_w^2\} \\ &= \{x \in \mathbb{R}^d : 2(t_w - t_j)^T x = \|t_w\|^2 - \|t_j\|^2 + r_j^2 - r_w^2\}. \end{aligned} \tag{4.1.10}$$

This hyperplane is called the *power-hyperplane* of the underlying spheres $\{x \in \mathbb{R}^d : \|x - t_s\| \leq r_s\}$ and $\{x \in \mathbb{R}^d : \|x - t_w\| \leq r_w\}$. Further, the construction of π' from π in the proof of Lemma 4.1.12 is a sorting operation of all p-parts; called *power-hyperplane sorting*. The use of p-sorting has the advantage that it achieves the goal in one step and therefore does not require a monotone statistics that shows iterative sortings will not cycle.

A natural variant of the use of power-hyperplane p-sorting to prove the conclusions of Lemma 4.1.12 is to iteratively apply power-hyperplane 2-sorting (of pairs of parts). For this approach to work, that is to avoid the possibility of cycling, it is necessary to identify a statistics over partitions that is reduced at each iteration. Unfortunately, we are not aware of such a statistics. Still, power-hyperplane 2-sortings can be used to prove a weaker result. Specifically, consider $t_1, \ldots, t_p \in \mathbb{R}^d$ and $r_1, \ldots, r_p \in \mathbb{R}$. Lemma 4.1.12 implies that power-hyperplane p-sorting will convert a partition π whose parts are included, respectively, in the spheres $\{x \in \mathbb{R}^d : \|x - t_j\| \leq r_j\}$, $j = 1, \ldots, p$, into a separable partition having the same property. Here, one can use iterative power-hyperplane 2-sorting and assure that the process does not cycle because of strict lexicographic reduction of the statistics

$$s(\pi) \equiv \left(\sum_{j=1}^{p} \sum_{i \in \pi_j} [\|A^i - t_j\|^2 - r_j^2], -|\pi_1|, \ldots, -|\pi_p| \right)$$

in each step. In particular, this proves that S is (sort-specific, 2, support)-sortable with "sort-specific" referring to the power-hyperplane method with fixed centers and fixed radii.

Theorem 4.1.13. *Let $t_1, \ldots, t_p$ be prescribed distinct vectors in $\mathbb{R}^d$ and consider the relaxed single-size problem, which has*

$$F(\pi) = f[\max_{i \in \pi_1} \|A^i - t_1\|, \ldots, \max_{i \in \pi_p} \|A^i - t_p\|], \tag{4.1.11}$$

where $f : \mathbb{R}^p \to \mathbb{R}$ is nondecreasing. Let π be a given partition. Then the partition π' which satisfies the conclusions of Lemma 4.1.12 is separable and satisfies $F(\pi') \leq F(\pi)$. In particular, the underlying single-size problem has a separable optimal partition.

Proof. Let π be a given partition. As f is nondecreasing, the partition π' that satisfies the conclusions of Lemma 4.1.12 is separable and satisfies

$$\begin{aligned} F(\pi) &= f[\max_{i\in\pi_1} \|A^i - t_1\|, \ldots, \max_{i\in\pi_p} \|A^i - t_p\|] \\ &\geq f[\max_{i\in\pi'_1} \|A^i - t_1\|, \ldots, \max_{i\in\pi'_p} \|A^i - t_p\|] = F(\pi'). \end{aligned}$$

Applying the above string to an optimal partition π, yields a partition π' which is both separable and optimal. □

Theorem 4.1.13 provides a construction (appearing explicitly in the statement of Lemma 4.1.12) of a separable partition that improves on the objective function of a given partition π. This result can be cast as a p-sortability result, same as the results of Theorem 4.1.7 and Corollary 4.1.8. We note that the construction is independent of the f appearing in (4.1.11), but unlike the construction of Theorem 4.1.7 (and Corollary 4.1.8), it is specific to the underlying partition π. In particular, Theorem 4.1.13 asserts the existence of a separable optimal partition that is specific to f and the next example shows that a (separable) partition that is optimal under all functions f needs not exist.

Example 4.1.4. Consider Example 4.1.3 with the parametric objective function over partitions given by $F(\pi) = B \max_{i\in\pi_1} \|A^i - t_1\|^2 + \max_{i\in\pi_2} \|A^i - t_2\|^2$ where B is a positive number. The optimal objective value is then $\min\{B, 0\}$ and $\pi^1 = (\{1,2\},\{3\})$ is the only optimal partition if $0 < B < 1$, while $\pi^2 = (\{1\},\{2,3\})$ is the only optimal partition if $B > 1$. The optimal partitions when $B = 1$ are π^1 and π^2. Of course, in either case, all of the optimal partitions are separable (and have no empty parts). □

For a bounded set $\Omega \subset \mathbb{R}^d$, let $r(\Omega)$ denote the minimum radius of a sphere that includes Ω; in particular, when Ω is finite,

$$r(\Omega) = \inf_{x\in\mathbb{R}^d} [\sup_{y\in\Omega} \|x - y\|]$$

and the "inf" and "sup" are attained (and can therefore be replaced by "min" and "max"). For a p-partition π, let $r(\pi) = (r(\pi_1), \ldots, r(\pi_p))$. The next result considers partition problems with objective function that is determined by these partition-characteristics; Capoyleas et al. considered the case $d = 2$.

Theorem 4.1.14. *Consider the relaxed single-size problem which has $F(\pi) = f[r(\pi)]$ where $f : \mathbb{R}^p \to \mathbb{R}$ nondecreasing. Then there exists a separable optimal partition.*

Proof. Consider the function $g : P^*(\mathbb{R}^d) \times \mathbb{R}^d \to \mathbb{R}$ having $g(\Omega, x) = \max_{y \in \Omega} \|y - x\|$ for $\Omega \in P^*(\mathbb{R}^d)$ and $x \in \mathbb{R}^d$. It is easily verified that for every $\Omega \in P^*(\mathbb{R}^d)$, the function $g(\Omega, \cdot)$ is nondecreasing, continuous and inverse-bounded. Now, Theorem 4.1.13 guarantees that for each $t_1, \dots, t_p \in \mathbb{R}^d$, the relaxed single-size problem which allows for empty parts and has objective function

$$F^{t_1,\dots,t_p}(\pi) = f[g(\pi_1, t_1), \dots, g(\pi_p, t_p)]$$

has a separable optimal solution. For each partition π

$$F(\pi) = f[\min_{x_1 \in \mathbb{R}^d} g(\pi_1, x_1), \dots, \min_{x_p \in \mathbb{R}^d} g(\pi_p, x_p)];$$

hence, the conclusions of the theorem follow from Theorem 4.1.9. □

Given functions f and F as those appearing in the statement of Theorem 4.1.7, the paragraph following Theorem 4.1.9 shows that power-hyperplane sorting, used in Theorem 4.1.7 and Corollary 4.1.8, can be used to replace a partition π that is not separable with one that is and has improved F-value.

We next generalize a result of Pfersky, Rudolf and Woeginger [80]; their result is then deduced as a corollary. The generalization is stated in our regular context of minimization problems.

Theorem 4.1.15. *Consider the single-size problem which has*

$$F(\pi) = \sum_{\substack{u,v=1 \\ u \neq v}}^{p} \sum_{i \in \pi_u} \sum_{k \in \pi_v} h(A^i, A^k), \tag{4.1.12}$$

where $h(\cdot, \cdot)$ is a metric. Then there exists a nearly monopolistic optimal partition. If h is strict, then every optimal partition is nearly monopolistic. Finally, for the relaxed version, every monopolistic partition is optimal.

Proof. Since $N_e M_p$ is (strongly, 2, support)-sortable, it suffices to prove that for any two parts, say, π_1 and π_2, for which (π_1, π_2) is not nearly monopolistic, there exists a nearly monopolistic partition of $\pi_1 \cup \pi_2$ which preserves optimality.

Without loss of generality, assume that $\pi = (\pi_1, \pi_2)$ with $|\pi_1| = n_1$, $|\pi_2| = n_2$, $2 \le n_1 \le n_2$, $\pi_1 = \{1, \ldots, n_1\}$ and $\pi_2 = \{n_1 + 1, \ldots, n\}$. For $i = 1, \ldots, n_1$, consider the partition $\pi^i = (\pi_1^i, \pi_2^i)$ with $\pi_1^i = \{i\}$ and $\pi_2^i = \{1, \ldots, n\} \setminus \{i\}$ and let

$$S_i = \sum_{\substack{k \in \pi_1 \\ A^k \neq A^i}} h(A^i, A^k) \qquad \text{and} \qquad T_i = \sum_{\substack{k \in \pi_1 \\ A^k \neq A^i}} \sum_{j \in \pi_2} h(A^k, A^j)\,.$$

Then $F(\pi) - F(\pi^i) = T_i - S_i$. Note that

$$\begin{aligned}
\sum_{i=1}^{n_1} S_i &= \sum_{i=1}^{n_1} \sum_{\substack{k=1 \\ k \neq i}}^{n_1} h(A^i, A^k) \\
&\le \sum_{i=1}^{n_1} \sum_{\substack{k=1 \\ k \neq i}}^{n_1} \left[h\left(A^i, A^{k+n_1}\right) + h\left(A^{k+n_1}, A^k\right) \right] \\
&= \sum_{i=1}^{n_1} \sum_{\substack{k=1 \\ k \neq i}}^{n_1} h\left(A^i, A^{k+n_1}\right) + (n_1 - 1) \sum_{k=1}^{n_1} h\left(A^k, A^{k+n_1}\right) \\
&\le (n_1 - 1) \sum_{i=1}^{n_1} \sum_{\substack{k=1 \\ k \neq i}}^{n_1} h\left(A^i, A^{k+n_1}\right) + (n_1 - 1) \sum_{i=1}^{n_1} h\left(A^i, A^{i+n_1}\right) \\
&= (n_1 - 1) \sum_{i=1}^{n_1} \sum_{k=1}^{n_1} h\left(A^i, A^{k+n_1}\right) \\
&= \sum_{k=1}^{n_1} \left[\sum_{\substack{i=1 \\ i \neq k}}^{n_1} \sum_{j=n_1+1}^{2n_1} h(A^i, A^j) \right] \le \sum_{k=1}^{n_1} T_k = \sum_{i=1}^{n_1} T_i.
\end{aligned}$$

As $\sum_{i=1}^{n_1} S_i \le \sum_{i=1}^{n_1} T_i$, it follows that $S_i \le T_i$ for some $i \in \{1, \ldots, n_1\}$. The corresponding nearly monopolistic partition π^i preserves optimality. The strict version follows immediately and the relaxed version is trivial. □

We next derive the result by Pfersky, Rudolf and Woeginger [80]. It is stated in the original framework of a maximization problem (of course, multiplying the objective function by -1 converts a maximization problem into an equivalent minimization problem).

Corollary 4.1.16. *Consider the single-size problem in which*

$$F(\pi) = \sum_{j=1}^{p} \sum_{i,k \in \pi_j} \|A^i - A^k\|^2 \tag{4.1.13}$$

is to be maximized. Then the conclusions are the same as Theorem 4.1.15.

Proof.

$$\sum_{j=1}^{p} \sum_{i,k\in\pi_j} \|A^i - A^k\|^2 = \sum_{\substack{i,k=1\\ i\neq k}}^{n} \|A^i - A^k\|^2 - \sum_{\substack{u,v=1\\ u\neq v}}^{p} \sum_{i\in\pi_u} \sum_{i\in\pi_v} \|A^i - A^k\|^2.$$

On the right-hand side, the first term is a constant and the second term is a special case of (4.1.12) with $h(A^i, A^k) = -\|A^i - A^k\|^2$. Note that the negative sign represents the switch to a maximization problem in Corollary 4.1.16. □

When empty parts are allowed in Corollary 4.1.16, then trivially, the monopolistic partition is an optimal partition.

We next present a totally different kind of argument introduced by Boros and Hwang [10] to prove properties of optimal partitions for single-shape problems. So, assume that a particular shape and a function F over partitions are given. Let π be a shape-partition and let π_r and π_s be two parts of π. Let π' be a partition obtained from π by interchanging two points $i \in \pi_r$ and $k \in \pi_s$, and express $F(\pi') - F(\pi)$ as $\nabla_F(A^i, A^k)$, in particular, $\nabla_F : \pi_r \times \pi_s \to \mathbb{R}$. Further, let us consider a continuous extension of ∇_F mapping $\mathbb{R}^d \times \mathbb{R}^d$ satisfying the following condition:

$$\nabla_F(z, z) = 0 \text{ for all } z \in \mathbb{R}^d.$$

When F is given in an algebraic format, the extended format of ∇_F can be determined naturally (but note that, in general, such a continuous extension always exists). For vectors $\bar{x}, \bar{y} \in \mathbb{R}^d$, consider the restrictions $\nabla_F(\bar{x}, \cdot)$ and $\nabla_F(\cdot, \bar{y})$ of ∇_F.

Theorem 4.1.17. *Consider the constrained-shape problem with $F(\cdot)$ satisfying the following condition: for every partition π and two of its parts, say π_r and π_s, either $\nabla_F(A^i, \cdot)$ for any $i \in \pi_r$ or $\nabla_F(\cdot, A^k)$ for any $k \in \pi_s$ is strictly quasi-concave. Then every optimal partition is weakly convex separable.*

Proof. By Lemma 1.2.1 of Vol. I, it suffices to consider single-shape problems. Let π^* be an optimal partition and let π_r^* and π_s^* be two of its parts. The case where either part is a singleton is trite and excluded. We consider only the case where the corresponding functions $\nabla_F(\cdot, A^k)$ for any $k \in \pi_s^*$ are strictly quasi-concave. We observe that the function

$g(\cdot) \equiv \min_{k \in \pi_s^*} \nabla_F(\cdot, A^k)$, as the minimum of finitely many strictly quasi-concave functions, is continuous and strictly quasi-concave on $\mathbb{R}^d$. Also, let $C \equiv \{x \in \mathbb{R}^d : g(x) \geq 0\}$. We next make the following observations.

(a) As $g(\cdot)$ is quasi-concave, C is convex.
(b) The optimality of π^* assures that $g(\cdot)$ is nonnegative on π_r^*, that is, $C \supseteq \pi_r^*$.
(c) As π_r^* has at least two points and $g(\cdot)$ is strictly quasi-concave, there is a point in $x \in \text{conv}\pi_r^*$ with $g(x) > 0$ (in fact, $g(x) > 0$ for each point of C which is not an extreme point of C).
(d) As $g(\cdot)$ is strictly quasi-concave and continuous and $\{x \in \mathbb{R}^d : g(x) > 0\} \neq \emptyset$, we have that $\dim C = d$ and

$$\{x \in \mathbb{R}^d : g(x) > 0\} = \text{int } \{x \in \mathbb{R}^d : g(x) \geq 0\} = \text{int } C.$$

(e) As $g(A^k) = \nabla_F(A^k, A^k) = 0$ for each $k \in \pi_s^*$, we have that $\pi_s^* \cap (\text{int } C) = \emptyset$.

This completes the proof that π^* is weakly convex separable. □

The arguments used to establish convex separable optimal partitions are also applicable for nonpenetrating optimal partitions.

Theorem 4.1.18. *Consider the constrained-shape problem with $F(\cdot)$ satisfying the following condition: for every partition π and two of its parts, say π_r and π_s, $\nabla_F(A^i, \cdot)$ for any $i \in \pi_r$ and $\nabla_F(\cdot, A^k)$ for any $k \in \pi_s$ is strictly quasi-concave. Then every optimal partition is weakly nonpenetrating.*

Proof. By Lemma 1.2.1 of Vol. I, it suffices to consider single-shape problems. Now, given an optimal partition and two of its parts, say π_r and π_s, the proof of Theorem 4.1.17 shows the existence of two convex sets C_r and C_s such that $C_r \supseteq \pi_r$, $(\text{int } C_r) \cap \pi_s = \emptyset$, $C_s \supseteq \pi_s$, $(\text{int } C_s) \cap \pi_r = \emptyset$, that is, π_r and π_s are weakly nonpenetrating. □

Boros and Hwang [10] applied Theorems 4.1.17 and 4.1.18 to the weighted sum-of-square (Theorem 4.1.3(ii)) and the sum-of-square (Theorem 4.1.5(ii)) objective functions, respectively. These results improve the earlier results of Boros and Hammer [9] but were themselves improved by Hwang, Onn and Rothblum [53] (see Theorems 4.1.3 and 4.1.5).

By Lemma 1.2.1 of Vol. I, the weak properties established in Theorems 4.1.17 and 4.1.18 for shape partition problems hold also for constrained-shape problems.

4.2 Geometric Properties of Optimal Partitions for $d = 2$

The current section focuses on single-size problems with $d = 2$. We consider problems with objective functions that depend on parts' perimeters and parts' diameters and identify conditions that guarantee the existence of separable optimal partitions; consequently, the results of Section 3.2 imply that the corresponding partition problems can be solved in polynomial time in the number of partitioned vectors.

Perimeter is introduced in Section 3.3. We introduce diameter here. The *diameter* of a set $\Omega \subseteq \mathbb{R}^2$, denoted $\text{diam}(\Omega)$, is defined as

$$\sup_{x,y \in \Omega} \|x - y\|$$

(of course, if Ω is compact, "sup" can be replaced by "max"). The perimeter and diameter are insensitive to the presence of more than a single copy of a vector and they decrease weakly when a point is removed. So, when considering size problems (that allow moving points between parts), it suffices to restrict attention to partitions that assign multiple copies of vectors to the same part and to consider such multiple copies as a single element. Further, our goal is to establish separability of optimal partitions and separability requires that multiple vectors are assigned to the same part. For these reasons we adopt throughout the current section the assumption that the columns of the matrix $A \in \mathbb{R}^{2 \times n}$ are distinct.

Monotonicity of $\text{peri}(\cdot)$ is established in Lemma 3.3.16. We now establish the much simpler monotonicity of diameter.

Lemma 4.2.1. *Let S and T be two polytopes in $\mathbb{R}^2$ where $\emptyset \neq S \subseteq T$. Then* $\text{diam(S)} \leq \text{diam(T)}$.

Proof. Since any two points defining diam(S) are in T, Lemma 4.2.1 follows immediately. □

While many results on perimeter are given in Section 3.3, the discussion is focused on sortability-related material and does not touch on other issues. In this section we show that most constraints imposed on these results cannot be relaxed. Example 4.2.1 shows that even though S and T have a 2-dim intersection, strict inequality in (3.3.1) does not hold.

Example 4.2.1. Let $S = \{1, 1', 3, 3'\}$ and $T = \{2, 2', 4, 4'\}$ be polygons in $\mathbb{R}^2$ as shown in Figure 4.2.1. Then it is easily seen that S and T have a 2-dim intersection but $\text{peri(S} \wedge \text{T)} + \text{peri(S} \vee \text{T)} = \text{peri(S)} + \text{peri(T)}$. □

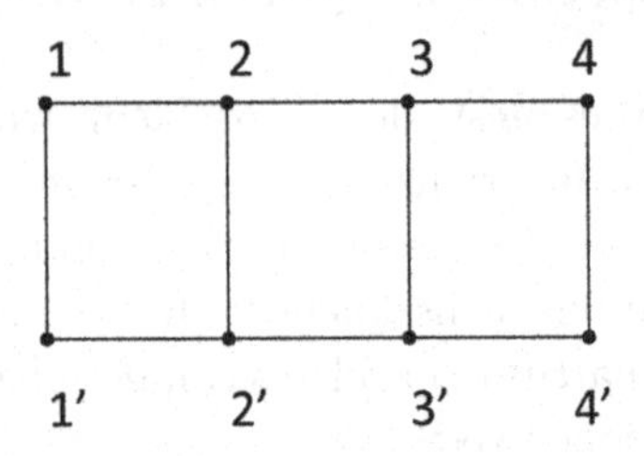

Figure 4.2.1 2-dim intersection not guaranteeing strict inequality in (3.3.1)

Note that (3.3.1) essentially coincides with the inverse of the inequalities in (7.3.2) of Vol. I (that define submodular set functions), except that the domain of peri$(\cdot)$ are polytopes rather than subsets of a given finite set and "$\vee$" is not the union operator "$\cup$" (but, "$\wedge$" is the intersection operator "$\cap$").

The next example demonstrates that the assumption in Lemma 3.3.17 that the underlying convex sets are not disjoint is necessary.

Example 4.2.2. Let S be the line connecting $\binom{-2}{-1}$ and $\binom{-2}{1}$ and let T be the line connecting $\binom{2}{-1}$ and $\binom{2}{1}$. Then $S \vee T$ is the rectangular with vertices $\binom{-2}{-1},\binom{-2}{1},\binom{2}{-1},\binom{2}{1}$, $S \wedge T = \emptyset$ and

$$\text{peri}(\text{S} \wedge \text{T}) + \text{peri}(\text{S} \vee \text{T}) = 0 + 12 > 4 + 4 = \text{peri}(\text{S}) + \text{peri}(\text{T}).$$

□

The next example demonstrates that Lemma 3.3.17 is false when the definition of perimeter of a line segment is taken as its length and not as twice its length. Example 4.2.3 (with $\pi = (S,T)$) demonstrates that the conclusion of Lemma 3.3.19 is false when the perimeter of a line segment is defined to be its length.

Example 4.2.3. Let S be the line connecting $\binom{-4}{0}$ and $\binom{4}{0}$ and let T be the line connecting $\binom{0}{-3}$ and $\binom{0}{3}$. Then $S \vee T$ is the quadrangle with consecutive vertices $\binom{-4}{0}$, $\binom{0}{3}$, $\binom{4}{0}$, $\binom{0}{-3}$, $S \wedge T$ is the origin and

$$\text{peri}(\text{S} \wedge \text{T}) + \text{peri}(\text{S} \vee \text{T}) = 0 + 20 > 8 + 6 = \text{length}(\text{S}) + \text{length}(\text{T}).$$

□

Define the *surface* of a polytope $S \subset \mathbb{R}^d$ of dimension $d > 2$, denoted surf(S), to be the sum of the volumes of its facets. The following example demonstrates that Lemma 3.3.17 does not extend to $d > 2$, even in the case that the intersection of S and T is d-dim.

Example 4.2.4. Let S be the "stick" with 8 vertices $\{(\pm 100, \pm\epsilon, \pm\epsilon)\}$ and T the "disc" with 8 vertices $\{(\pm\epsilon, \pm 1, \pm 1)\}$, where $\pm z$ means that two alternatives $+z$ and $-z$ are taken. Ignoring ϵ-terms, we have that $\text{surf(S)} = 0$, $\text{surf(T)} = 8$, $\text{surf(S} \vee \text{T)} = 400$ and $\text{surf(S} \cap \text{T)} = 0$; so, $\text{surf(S)} + \text{surf(T)} = 8 < 400 = \text{surf(S} \vee \text{T)} + \text{surf(S} \cap \text{T)}$. □

The next two results demonstrate the power of Lemma 3.3.19.

Lemma 4.2.2. *Suppose π is a partition that is not separable. Then there exists a separable partition π' having no empty parts and satisfying*

$$\sum_{j=1}^{p} \text{peri}(\pi_j) > \sum_{j=1}^{p} \text{peri}(\pi'_j). \tag{4.2.1}$$

Proof. By Lemma 3.3.19, each JS-sorting of two parts that either are not separable or have one part empty not only strictly decreases the sum of perimeters (over the p parts), but also guarantees that the sorted 2-partition contains no empty part. Lemma 4.2.2 follows by iterated applications of JS-sorting and the finiteness of the set of p-partitions. □

Theorem 4.2.3. *Consider the (relaxed) single-size problem where $\sum_{j=1}^{p} \text{peri}(\pi_j)$ is to be minimized. Then every optimal partition is separable and has no empty parts.*

Proof. If π were an optimal partition, then Lemma 4.2.2 would imply that there is a partition π' that satisfies (4.2.1), contradicting the optimality of π. □

The next example demonstrates that Lemma 4.2.2 and Theorem 4.2.3 are false when the definition of the perimeter of a line segment is taken as its length. In fact, in this example no separable partition minimizes the modified version of partition-perimeter.

Example 4.2.5. Consider the following 6 points in the plane $s = \binom{-1}{0}, t = \binom{-.5}{0}, u = \binom{.5}{0}, v = \binom{1}{0}$, $x = \binom{0}{\epsilon}$ and $y = \binom{0}{-\epsilon}$ with ϵ positive and small; in particular, s, t, u, v lie on the horizontal line connecting $\binom{-1}{0}$ and $\binom{1}{0}$, while x and y form a "small" vertical line segment around $\binom{0}{0}$. The 2-partition π

with $\pi_1 = \{s, t, u, v\}$ and $\pi_2 = \{x, y\}$ is not separable. Now suppose that the perimeter of lines is their length. Ignoring ϵ terms, $\text{peri}(\pi_1) + \text{peri}(\pi_2) = 2$. Next we will compute $\text{peri}(\pi'_1) + \text{peri}(\pi'_2)$ for all separable 2-partitions π'. Due to symmetry of the two sides of the line segment, we need only consider the following cases:

case (i): $\pi'_1 = \emptyset$. Then $\text{peri}(\pi'_1) + \text{peri}(\pi'_2) = 4$.
case (ii): $\pi'_1 = \{s\}$. Then $\text{peri}(\pi'_1) + \text{peri}(\pi'_2) = 3$.
case (iii): $\pi'_1 = \{s, t\}$. Then $\text{peri}(\pi'_1) + \text{peri}(\pi'_2) = 5/2$.
case (iv): $\pi'_1 = \{s, x\}$. Then $\text{peri}(\pi'_1) + \text{peri}(\pi'_2) = 5/2$.
case (v): $\pi'_1 = \{s, t, x\}$. Then $\text{peri}(\pi'_1) + \text{peri}(\pi'_2) = 4$.
case (vi): $\pi'_1 = \{x\}$. Then $peri(\pi'_1) + peri(\pi'_2) = 4$.

Thus, no separable partition π' satisfies (4.2.1) and no separable partition minimizes the modified peri(·); in particular, (4.2.1) is not satisfied for either of the four partitions that are obtained from π by JS-sorting (using the vertices s, x, y, v of $\text{conv}(\pi_1 \cup \pi_2)$). □

Observe that Theorem 4.2.3 does not extend to partitions over points in $\mathbb{R}^d$ with $d > 2$ where surf(·) replaces peri(·). Consider Example 4.2.4 with the partitioned vectors as the 16 vertices of S and T, described therein. It is easily verified that no separable partition has lower total surface area than (S, T).

We next use Lemma 3.3.19 to establish the optimality of separable partitions for a broader class of partition problems than the sum-partition problem considered in Theorem 4.2.3.

Theorem 4.2.4. *Consider a (relaxed) single-size problem with* $F(\pi) = f[\text{peri}(\pi_1), ..., \text{peri}(\pi_p)]$ *to be minimized where* $f(\cdot)$ *is Schur concave and nondecreasing. Then there exists a separable optimal partition whose parts are all nonempty. Further, if* $f(\cdot)$ *is strictly increasing, (in the sense that* $f(x) < f(y)$ *whenever* $x \leq y$ *and* $x \neq y$*), then every optimal partition has those properties.*

Proof. Let π be an optimal partition that minimizes $\sum_{j=1}^{p} \text{peri}(\pi_j)$ among all optimal partitions. To see that π satisfies the asserted conditions, assume that it does not and let π_r and π_s be parts of π which are either not separable or have $\pi_r = \emptyset$ and $|\pi_s| \geq 2$. Let π'_r and π'_s be the outcome obtained from the application of JS-sorting to (π_r, π_s) with π'_r consisting of a vertex of $\pi_r \cup \pi_s$ and let π' be the resulting p-partition. Let x and x' be the vectors in $\mathbb{R}^p$ with $x_j = \text{peri}(\pi_j)$ and $x'_j = \text{peri}(\pi'_j)$ for

$j = 1, \dots, p$. In particular,

$$\min\{x'_r, x'_s\} = x'_r = 0 \le \min\{x_r, x_s\}$$

and (from Lemma 3.3.19) $x'_s < x_r + x_s - x'_r$, implying that (x_r, x_s) strictly majorizes $(x'_r, x_r + x_s - x'_r)$. Since f is Schur concave and nondecreasing, it follows that $F(\pi) = f(x) \ge f(x_1, \cdots, x_{r-1}, x'_r, \cdots, x_{s-1}, x_r + x_s - x'_r, \cdots, x_p) \ge f(x') = F(\pi')$, implying that π' is another optimal partition. Next, as $x'_r + x'_s < x_r + x_s$, π' satisfies (4.2.1), yielding a contradiction to the selection of π as a minimizer of $\sum_{j=1}^{p} \text{peri}(\pi_j)$ among all optimal partitions. This contradiction proves the separability of π.

For the final conclusion of the theorem assume that π is an optimal partition which is not separable or has an empty part and another that has at least two points. Let π', x and x' be as in the above paragraph. It then follows that when f is Schur concave and strictly increasing $F(\pi') = f(x') < f(x) = F(\pi)$, a contradiction to the optimality of π. □

Observe that Theorem 4.2.3 is an instance of Theorem 4.2.4 with f being the summation function which is Schur concave and strictly increasing.

We next consider problems whose objective function depends on the parts' diameters. We start by stating a geometric lemma that will be used in our analysis. Recall the definition of intersection vertices in Section 3.3.

Lemma 4.2.5. *Suppose S and T are compact convex sets in $\mathbb{R}^2$ and $d_{ST} \equiv \sup\{\|s - t\| : s \in S \text{ and } t \in T\}$. Then there exist $s^* \in S$ and $t^* \in T$ that satisfy $\|s^* - t^*\| = d_{ST}$; further, such s^* and t^* are vertices of S and T, respectively, and $C \equiv t^* - s^*$ satisfies $S \setminus \{s^*\} \subseteq \{x \in \mathbb{R}^d : C^T x < C^T s^*\}$ and $T \setminus \{t^*\} \subseteq \{x \in \mathbb{R}^d : C^T t^* < C^T x\}$.*

Proof. Compactness arguments show that there exist $s^* \in S$ and $t^* \in T$ that satisfy $\|s^* - t^*\| = d_{ST}$. Now, if $s^* \in S$ and $t^* \in T$ are such points, then S is included in a circle of radius $d_{ST} = \|t^* - s^*\|$ centered at t^*. The tangent to this circle at s^* is the line $\{x \in \mathbb{R}^d : (t^* - s^*)^T x = (t^* - s^*)^T s^*\}$ and no point x within the circle other than s^* satisfies $(t^* - s^*)^T x \ge (t^* - s^*)^T s^*$, in particular, this is the case for points in S. So, with $C = t^* - s^*$,

$$S \setminus \{s^*\} \subseteq \{x \in \mathbb{R}^d : C^T x < C^T s^*\}.$$

As $C^T x < C^T s^*$ for each $x \in S \setminus \{s^*\}$ we have that s^* is the unique maximizer of a linear function over S, assuring that s^* is a vertex of S. The conclusions about T and t^* follow from symmetric arguments and the observation that $s^* - t^* = -(t^* - s^*)$. □

Corollary 4.2.6. *If* Ω *is a finite set, then* $\mathrm{diam}(\mathrm{conv}\,\Omega) = \mathrm{diam}(\Omega)$.

Proof. As Ω is contained in convΩ, $\mathrm{diam}(\mathrm{conv}\Omega) \geq \mathrm{diam}(\Omega)$. Next, by Lemma 4.2.5, diam(convΩ) is realized by vertices of Ω; as such vertices are in Ω, the reverse inequality follows. □

A *quadrangle* with distinct consecutive vertices a, b, c, d will be denoted by $\square abcd$. If no internal angle exceeds 180°, then a quadrangle is convex. Note that our definition of a quadrangle also includes the degenerate case where one of the internal angles is 180° (the shape is reduced to a triangle). Standard geometry shows that if $\square abcd$ is a regular or degenerate quadrangle, then

$$\|a - c\| + \|b - d\| > \|a - b\| + \|c - d\|. \tag{4.2.2}$$

The following result is key to the analysis of diameter problems. Unfortunately, the proof is complicated and lengthy.

Lemma 4.2.7. *Let* S *and* T *be two polytopes in* $\mathbb{R}^2$ *which have 2-dim intersection and satisfy* $\mathrm{diam}(S \cup T) > \max\{\mathrm{diam}(S), \mathrm{diam}(T)\}$. *Then there exist two intersection vertices of* S *and* T, *say* u *and* v, *an indexation of the two open half-planes defined by the line* L *that contains* u *and* v, *say* L^+ *and* L^-, *and nonempty sets* S' *and* T' *such that*

$$\{S', T'\} = \{(S \cup T) \cap L^+, (S \cup T) \cap L^-\} \tag{4.2.3}$$

and

$$\mathrm{diam}(S') \leq \mathrm{diam}(S), \mathrm{diam}(T') \leq \mathrm{diam}(T). \tag{4.2.4}$$

Proof. To prove the Lemma, we may assume that $S \cap T \neq \emptyset$, $S \not\subseteq T$ and $T \not\subseteq S$ as otherwise the conclusion is trivial. Then there exist two interlacing sequences of polygons $S(1), T(1), S(2) \ldots, S(k), T(k)$ such that

$$\bigcup_{i=1}^{k} S(i) = (S \setminus \mathrm{int(T)}), \ \bigcup_{\mathrm{i=1}}^{\mathrm{k}} \mathrm{T(i)} = (\mathrm{T} \setminus \mathrm{int(S)})$$

and for $i = 1, \ldots, k$, $S(i) \subseteq S$, $S(i) \cap (S \setminus \mathrm{int(T)}) \neq \emptyset$, $T(i) \subseteq T$ and $T(i) \cap (T \setminus \mathrm{int(S)}) \neq \emptyset$. We refer to these polygons as S- and T-half-moons. Sometimes a half-moon consists of just a piece of $\mathrm{bd(S \cap T)}$ which happens when bd(S) and bd(T) overlap. Such a half-moon is called "flat". The question is how to classify a flat half-moon as S or T – The rule is if both of its neighbors are of the same type, then it should be classified as the other type; if the two neighbors are of different type, then merge the flat half-moon into its preceding half-moon. By so doing, we preserve the interlacing feature of the cyclic sequence of half-moons. The intersection of two adjacent half-moons contains a single point which is an intersection vertex of S and T. For $i = 1, \ldots, k$, let u_{2i-1} and u_{2i} be the points where $S(i)$ touches $T(i-1)$ and $T(i)$, respectively (notice that $T(0) = T(k)$); see Figure 4.2.2 for a configuration of half-moons and u-points ($T(3)$ is flat).

If distinct points a, b, c, d are, respectively, selected from 4 distinct cyclicly-ordered half-moons, then these points form a quadrangle (degeneracy allowed) $\Box abcd$ with a, b, c, d as consecutive vertices. We prove that $\Box abcd$ is convex so that (4.2.2) can be used.

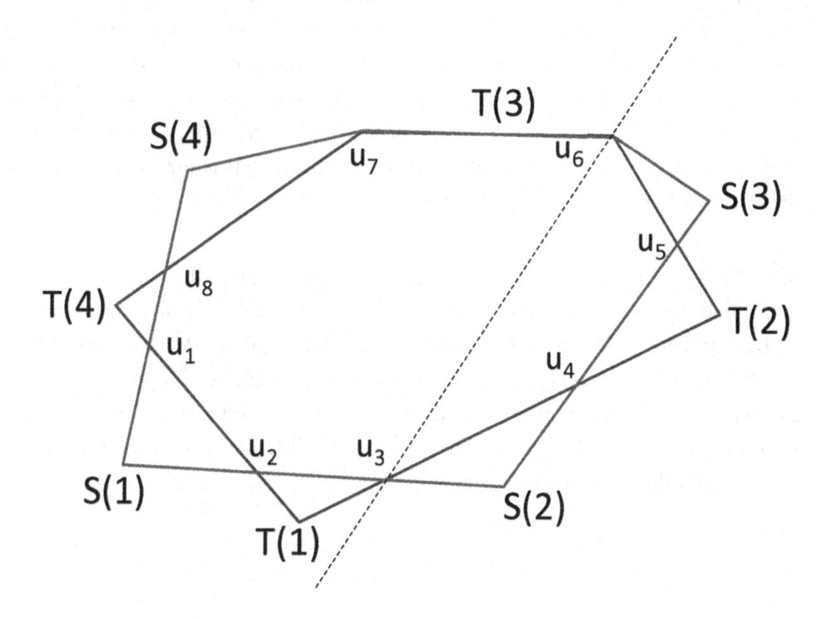

Figure 4.2.2 Configuration of half-moons and u-points

Claim 0: $\Box abcd$ is convex.

Each half-moon rests on a piece of bd(S∩T), called the supporting path of that half-moon. Note that the supporting paths of two consecutive half-moons are continuous and the union of all supporting paths is bd(S ∩ T). Since both S and T are convex, so is $S \cap T$. Consequently, if a supporting path consists of several edges, then the angles forming by those edges must all be less than 180° (if = 180°, then the two edges are considered the same one). Extend the two edges (could be the same one) at the two ends of a supporting path indefinitely, then this extended supporting path separate $\mathbb{R}^2$ into two open spaces. In particular, it weakly separates the half-moon owning that supporting path with all other half-moons ("weakly" since half-moons can have points on the path).

Suppose that $\Box abcd$ is not convex, say, $\sphericalangle abc > 180°$. Then the extended supporting path of b cannot separate b from either a or c, a contradiction. This completes the proof of Claim 0.

Let $d_{S \cup T} \equiv \text{diam}(\text{S} \cup \text{T})$, $d_S \equiv \text{diam}(\text{S})$ and $d_T \equiv \text{diam}(\text{T})$. Without loss of generality we will assume that $d_S \geq d_T$. Call a pair $(S(i), T(j))$ of half-moons a *bad pair* if $\text{diam}(\text{S(i)} \cup \text{T(j)}) > \text{d}_\text{S}$, in this case we also say that $S(i)$ and $T(j)$ are *bad half-moons* and that they are *bad partners to each other*. As (by assumption) $d_{S \cup T} > d_S$, a bad pair exists. In the following, we use common superscripts to index bad partners, e.g., we refer to a bad pair (S^i, T^i). We emphasize that a bad half-moon can have more than one bad partner. Two bad pairs (S^i, T^i) and (S^j, T^j) are called *nonoverlapping* if $S^i \neq S^j$ and $T^i \neq T^j$. They are said to be *crossing* if their cyclic sequence is either S^i, S^j, T^i, T^j or S^i, T^j, T^i, S^j.

Claim 1: Two nonoverlapping bad pairs must be crossing.

Suppose to the contrary that (S^i, T^i) and (S^j, T^j) are nonoverlapping bad pairs that are not crossing, and let s^i, t^i, s^j, t^j be elements of S^i, T^i, S^j, T^j, respectively, such that $\|s^i - t^i\| > d_S$ and $\|s^j - t^j\| > d_S$; evidently, none of these points is in $S \cap T$. The cyclic order of these bad half-moons is then either (S^i, T^i, S^j, T^j) or (S^i, T^i, T^j, S^j); see Figure 4.2.3. We next consider each of these two cases.

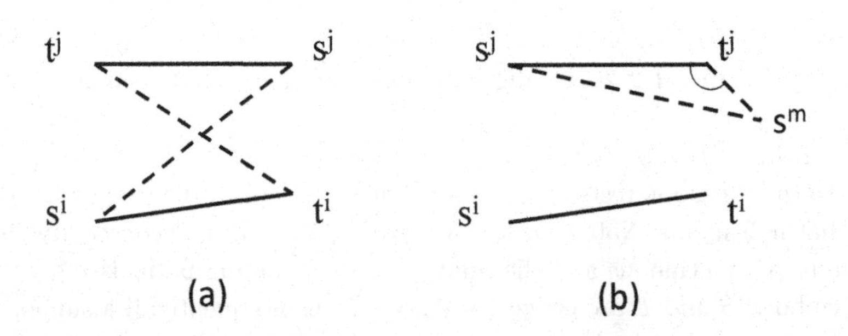

Figure 4.2.3 Configuration of noncrossing bad pairs

In the configuration illustrated in Figure 4.2.3(a), we have

$$d_S + d_T \geq \|s^i - s^j\| + \|t^i - t^j\| > \|s^i - t^i\| + \|s^j - t^j\| > 2d_S\,,$$

contradicting the assumption $d_S \geq d_T$.

In the configuration illustrated in Figure 4.2.3(b), the quadrangle $\square\, s^i t^i t^j s^j$ has an angle $\geq 90°$. Assume that $\sphericalangle\, s^i t^i t^j \geq 90°$. Then there exists a half-moon S^m between T^i and T^j. Let s^m be a point in S^m which is not a u-point. It then follows that s^m is on the other side of line containing t^i and t^j as versus s^i and $\sphericalangle\, s^i t^i s^m > \sphericalangle\, s^i t^i t^j \geq 90°$, implying that

$$d_S \geq \|s^i - s^m\| > \|s^i - t^i\| > d_S\,,$$

a contradiction. Next assume that $\sphericalangle\, s^j s^i t^i \geq 90°$. Then it is possible to identify a point t^m of a half-moon T^m that lies between S^j and S^i and observe that $\sphericalangle\, t^m s^i t^i > \sphericalangle\, s^j s^i t^i \geq 90°$, implying that

$$d_S \geq d_T \geq \|t^i - t^m\| > \|s^i - t^i\| > d_S\,,$$

a contradiction. Similar arguments work, respectively, if $\sphericalangle\, t^i t^j s^j \geq 90°$ or $\sphericalangle\, t^j s^j s^i \geq 90°$. This completes the proof of Claim 1.

The cyclic sequence of bad half-moons consists of alternative runs of bad S-half-moons and bad T-half-moons, to which we refer as S- and *T-runs*.

An S-run and a T-run are called a pair if one consists of all partners of the other's bad half-moons. Claim 1 implies that the runs form an interlacing sequence of S- and T-runs. In particular, it follows that the number of runs is even, say $2k$. Now, consider any S-run. Claim 1 assures that each of the remaining $k-1$ pairs of runs must be represented between this S-run and the T-run of its partners, and between the run of its partners and the given S-run. Since runs are interlacing, there is an even number of runs between any S-run and its partner T-runs, implying that $k-1$ is even and therefor k is odd, i.e., the number of pairs of runs is odd.

Claim 2: If two half-moons are in a common run, then their partners are also in a common run.

Suppose that S^i and S^j are in the same S-run, $S^i = S^j$ allowed. We next argue that the bad partners of S^i and S^j, say T^i and T^j, are also in the same T-run. Indeed, suppose to the contrary that T^i and T^j are in different runs. Then $T^i \neq T^j$ and there exists a bad half-moon S^k such that the cyclic order of half-moons has either (T^i, S^k, T^j) or (T^j, S^k, T^i); without loss of generality, assume the former. Let T^k be a bad partner of S^k. Depending on whether T^k is between the S-run containing S^i, S^j and S^k or between S^k and that S-run, we have that (S^k, T^k) is a nonoverlapping, noncrossing pair with either (S^j, T^j) or (S^i, T^i), contradicting Claim 1. A symmetric argument shows that if T^i and T^j are in the same T-run, then their bad partners are in the same S-run. This completes the proof of Claim 2.

Let S^i be a last bad half-moon in an S-run, and T^i be the last bad partner of S^i. Also, let T^j be the first bad half-moon after S^i (it has to be of type T) and let S^j be its first bad partner. Claim 1 implies that S^j is between T^i and S^i, inclusive, thus S^i, T^j, T^i, S^j are cyclicly ordered, with $T^i = T^j$ and/or $S^i = S^j$ possible. Next, let u be the u-point that "starts" T^j and let v be the u-point that "ends" T^i (if $T^j = T^i$, then the u-point of T^j is also the u-point of T^i). Let L be the line that contains u and v. It then follows that S^i and T^i are included in opposite closed half-planes that L defines. Let L^+ and L^- be an indexation of these open half-planes such that $S^i \subseteq L^\oplus \equiv L \cup L^+$ and $T^i \subseteq L^\ominus \equiv L \cup L^-$. The partition induced by L is $(L^+, L^\ominus)$, every half-moon that is between T^j and T^i, exclusive, is included in L^- while T^j (like T^i) is included in $L^\ominus$. Also, every half-moon that is between T^i and T^j, exclusive, is included in L^+. Let

$$S' \equiv (S \cup T) \cap L^+ \quad \text{and} \quad T' \equiv (S \cup T) \cap L^\ominus. \qquad (4.2.5)$$

Note that $S' \supseteq S^i \cap L^+ \neq \emptyset$ and $T' \supseteq T^i \cap L^- \neq \emptyset$. So, S' and T' are nonempty.

Claim 3: $d_{S'} \equiv \text{diam}(S') \leq d_S$.

For two half-moons U and V, let $I(U, V)$ denote the open interval in the cyclic sequence from U to V (replace a "(" by a "[" to show that U or V is included). At times, we give more detail by replacing U (or V) with a point in U (or V).

We prove that no two points in $(S \cup T) \cap L^+$ can have a distance $> d_s$. Clearly, two such points must be in two half-moons of different types. Let $s^k \in S^k$ and $t^k \in T^k$ be two such points. Then (S^k, T^k) is a bad pair. Note that both S^k and T^k can neither lie in $I(S^i, u)$ by the definition of T^j, nor in $I[u, v]$ since that interval is in $L^\ominus$; so they must lie in $I(v, S^i]$. In fact, we will relax the restriction to $I[v, S^i]$.

There are two possible cyclic orderings of the four bad half-moons: either S^i, T^i, S^k, T^k or S^i, T^i, T^k, S^k. For the former case, examine the $\Box s^i t^i s^k t^k$ where $s^i \in S^i$ and $t^i \in T^i$ have a distance $> d_s$. By Claim 0, the $\Box s^i t^i s^k t^k$ is convex. Hence Claim 1 is violated. For the second case, note that T^k and T^i cannot be in the same run by the definition of T^i. Hence there exists a bad pair (S^m, T^m) with S^m lying in $I(T^i, T^k)$ and with T^m not in the same run of T^i. Again, by Claim 0, the $\Box s^i t^i t^k s^k$ is convex.

But (S^m, T^m) does not cross (S^i, T^i), contradicting Claim 1 again. This completes the proof of Claim 3.

We will next prove that $d_{T'} \leq d_T$. If $T^j = T^i$, then $T' = T^j$ and the inequality is trite. So, assume that $T^j \neq T^i$. Let s^i, t^i, s^j, t^j be points in S^i, T^i, S^j, T^j, respectively, with $\|s^i - t^i\| > d_S$ and $\|s^j - t^j\| > d_S$ ($S^j = S^i$ and $s^j = s^i$ allowed, but $T^j \neq T^i$ assuring $t^j \neq t^i$); evidently, neither of these points is in $S \cap T$.

Claim 4: Assume $s^k \in S^k$ and $t^m \in T^m$ are in $I[u, v]$ and T^j and T^i are not in the same run. Then $\|s^k - t^m\| \leq d_r$.

Consider the two possible orders of S^k and T^m:

(i) If S^k follows T^m, then the cyclic order of the 4 half-moons is (T^j, S^k, T^m, S^j). Consequently, Claim 1 is violated in that convex quadrangle.

(ii). If T^m follows S^k, then the cyclic order of the 4 half-moons is (S^i, T^m, S^k, T^i). Again, Claim 1 is violated in that convex quadrangle. This completes the proof of Claim 4.

Claim 5: Assume s and s' lie in S-half-moons in $I(T^j, T^i)$ (neither half-moon is necessarily bad, also note that $T^j \neq T^i$ or s and s' would not exist). Then $\|s - s'\| \leq d_T$.

We will show that s and s' lie between the two lines that are orthogonal to L and contain t^j and t^i, respectively. We consider only the point s since the case of s' is similar. Assume that L is horizontal and L^- lies above L. If s were to the left of the vertical line containing t^j, then the $\sphericalangle\, s^j t^j s$ would be larger than 90°, implying that $d_S \geq \|s - s^j\| > \|t^j - s^j\| > d_S$, a contradiction. Similarly, if s were to the right of the vertical line containing t^i, then the $\sphericalangle\, s t^i s^i$ would be larger than 90°, implying that $d_S \geq \|s - s^i\| > \|t^i - s^i\| > d_S$, a contradiction. Finally, as T^j, the half-moon containing s, T^i and S^i are cyclically ordered, t^j,s,t^i and s^i are consecutive vertices of a quadrangle; as s^i lies (strictly) below the line connecting t^j and t^i and s lies (weakly) above that line.

Let U be the region restricted by the two circles of radius d_T with centers at t^i and t^j, by the two lines that are orthogonal to L and contain t^j and t^i and by the closed half-plane determined by the line containing t^i and t^j and having empty intersection with S', see Figure 4.2.4. By Preparata and Shamos [81], the diameter of a convex set U is the maximum distance between two vertices of U that are on parallel lines, each intersecting U at only one point. It can then be easily seen in Figure 4.2.4 that one of the two lines achieving maximum must contain either t^i or t^j. As U is contained in the circle of radius d_T with center at either one of these two points, it follows that diam(U) $\leq d_T$. Claim 4 and the above paragraph show that every S-half-moon that lies strictly between T^j and T^i is included in U and therefore $\|s - s'\| \leq$ diam(U) $\leq d_T$, completing the proof of Claim 5. □

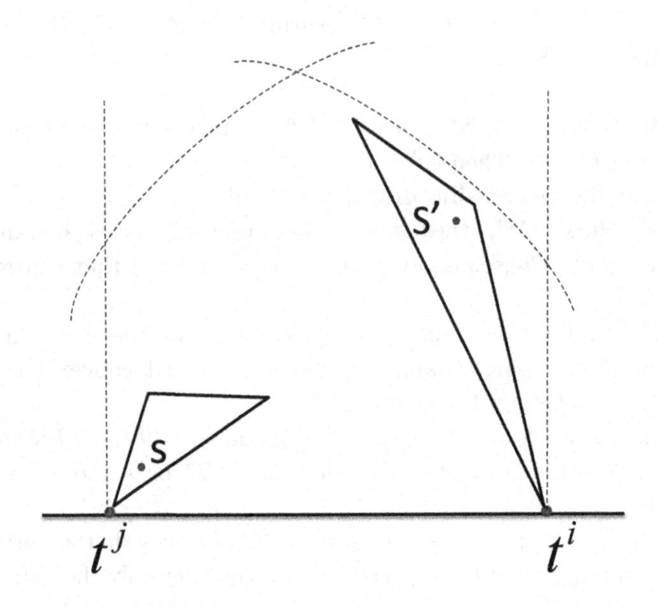

Figure 4.2.4 Configuration of the set U.

The proof of Lemma 4.2.7 essentially follows Capoyleas, Rote and Woeginger [14]. We add in Claim 0 to solidify the use of (4.2.2).

Theorem 4.2.8. *Suppose S and T in $\mathbb{R}^2$ are nonseparable. Then there exists a separable partition (S', T') of $S \cup T$, where S' and T' are nonempty, such that* diam(S') $\leq$ diam(S), diam(T') $\leq$ diam(T).

Proof. Lemma 4.2.7 proved the special case of Theorem 4.2.8 when $S \cap T$ is 2-dim and at least one bad pair exists. This proof will deal with the uncovered cases.

(i) $S \cap T$ is 2-dim, but no bad pair exists.
Assume $d_s \geq d_t$. If $T \not\subseteq S$, then take any T-half-moon T^i and set L to be the line connecting the two intersection vertices of T^i and set $T' = T^i$. Then trivially, diam(T') $\leq$ diam(T). On the other hand, $S' = (S \cup T) \cap L^+$.

But L^+ has no bad pair, implying $\mathrm{diam(S')} \leq \mathrm{diam(S)}$.
If $T \subseteq S$, then set $T' = \{s\}$ where s is a vertex of S and set $S' = (S \cup T) \setminus T'$. Obviously, $\mathrm{diam(T')} \leq \mathrm{diam(T)}$ and $\mathrm{diam(S')} \leq \mathrm{diam(S)}$.

The only case left is that $\dim(\mathrm{S} \cap \mathrm{T}) \leq 1$ with at least a bad pair. This is actually the following two cases.
(ii) S and T are weakly separable with $\dim(\mathrm{S}) = \dim(\mathrm{T}) = 2$.
Use the CRW(= the sorting given in Theorem 3.1.6)-sorting. Theorem 4.2.8 follows from Theorem 3.1.6) and Lemma 4.2.1.
(iii) One of S and T, say, T, is of dimension 1 (note that if T is of dimension 0, then either no bad pair exists or S and T are separable).
There are three possible cases (see Figure 4.2.5 in which a polygon is drawn as a triangle).

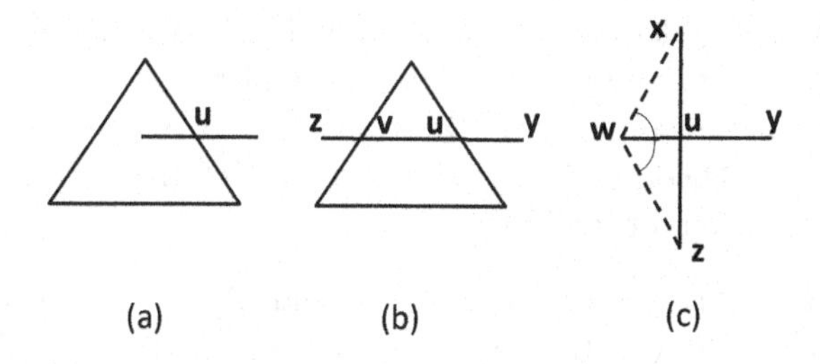

Figure 4.2.5

Figure 4.2.5(a). Set $T' = T \setminus S$ and $S' = S$. (4.2.4) is obviously satisfied.
Figure 4.2.5(b). The arguments used in Claims 3,4,5 do not discriminate against the case that a T-half-moon is degenerated into a line.
Figure 4.2.5(c). In this case S, the line segment xz, and T, the line segment wy, are of dimension 1. Consider $\Box wxyz$. Then one of the angles $\geq 90°$, say, the angle at w. Set S' to be triangle wxz (xz being the longest line) and T' the intersection of T and the line segment uy. Thus, $\mathrm{diam(S')} \leq \mathrm{diam(S)}$ and $\mathrm{diam(T')} = \mathrm{diam(\{u, y\})} < \mathrm{diam(T)}$. □

We are now ready for the main result about partition problems whose objective depends on parts' diameters.

Theorem 4.2.9. *Consider a (relaxed) single-size problem with* $F(\pi) = f(\mathrm{diam}(\pi_1), ..., \mathrm{diam}(\pi_p))$ *is to be minimized where* $f(\cdot)$ *is nondecreas-*

ing. Then there exists a separable optimal partition whose parts are all nonempty.

Proof. Consider the case where $p = 2$ and let $\pi = (\pi_1, \pi_2)$ be an optimal partition that is nonseparable. Let $\pi' = (\pi'_1, \pi'_2)$ with $\pi'_1 \cup \pi'_2 = \pi_1 \cup \pi_2$ be the resulting partition obtained by the sorting in Theorem 4.2.8. Then π'_1 and π'_2 are separable, $\text{diam}(\pi'_1) \leq \text{diam}(\pi_1)$ and $\text{diam}(\pi'_2) \leq \text{diam}(\pi_2)$; as f is nondecreasing, it follows that $F(\pi') \leq F(\pi)$. As Corollary 3.3.23 shows that S is (ECRW-sort-specific, 2, support)-sortable, the conclusion extends to $p > 2$. Finally, if a separable optimal partition has empty parts (hence it must have a part of size at least 2), the latter can be split into the two parts by JS (=CRW)-sorting without increasing diameters. □

Capoyleas, Rote and Woeginger [14] first gave Corollary 3.3.23 and Theorem 4.2.9 for the version that all parts are of dimension 2. Since a JS-sorting will bring in a singleton part (dimension 0), and a CRW-sorting can bring in a doubleton part (dimension 1), Theorem 4.2.8 becomes a necessary part for these two results to be complete.

The following example, given by Capoyleas et al., demonstrates that the extension of Theorem 4.2.8 and Theorem 4.2.9 from $d = 2$ to $d > 2$ is impossible, not even for $d = 3$ and $p = 2$.

Example 4.2.6. Let $n = 6$, $d = 3$ and $A^1 = (-1, 0, 0)^T$, $A^2 = (1, 0, 0)^T$, $A^3 = (0, \sqrt{3}, 0)^T$, $A^4 = (-1, 0, -\epsilon)^T$, $A^5 = (1, 0, -\epsilon)^T$ and $A^6 = (0, \epsilon, \sqrt{2} + \epsilon)^T$ where $\epsilon < 1/100$. Consider the single-size problem over 2-partitions with $F(\pi) = \max\{\text{diam}(\pi_1), \text{diam}(\pi_2)\}$ to be minimized. Then $\pi^* = (\{1, 2, 3\}, \{4, 5, 6\})$ is a unique optimal partition (with $F(\pi^*) = \text{diam}(\pi^*_1) = \text{diam}(\pi^*_2) = 2$). But, π^* is not separable. □

Polynomial algorithms have been established for some related geometrical problems. Asano, Bhattacharya, Keil and Yao [4] solved the minimization problem of maximum diameter between two parts. Monma and Suri [72] solved the minimization problem of maximum radius for two parts; Drezner [28] extended it to general p parts. Hershberger and Suri [46] considered the problem whether there exists a 2-partition satisfying $W(\pi_1) \leq b_1$, $W(\pi_2) \leq b_2$, where b_1 and b_2 are specified bounds given in advance, for various W functions.

In general, the open partition versions are not solvable efficiently. Supowit [90] proved that the maximum diameter problem is NP-complete. Megiddo and Supowit [69] proved that the maximum radius problem is NP-complete. Megiddo and Tamir [70] proved that the maximum area and the sum of area problems are both NP-complete. Brucker (see Garey and Johnson [39, p. 281]) proved that the maximum within-part distance problem is NP-complete.

For shape problems, only the uniform shape has been treated. Pfersky, Rudolf and Woeginger [80] proved that the sum of perimeter problems for constant part-size 3 is NP-complete, but a polynomial algorithm exists if the points lie on a convex polygon. Further, if the points lie on a circle, then a polynomial algorithm exists for all constant part-size k.

We finally consider some 2-dim partition problems which can be reduced to 1-dim problems. For this purpose we need another 1-dim partition property. A 1-dim partition π is called *cyclic separable* (called $c(\mathcal{N})$-separable in the proof of Theorem 6.2.3 of Vol. I), if the separability requirement is satisfied for all pair of parts with the possible exception of pairs involving one special part, say π_w, and this part satisfies the following requirements: (i) π_w is the union of two separable sets, (ii) each of these two sets forms a separable pair with each of the other parts of π, and (iii) $\mathrm{conv}\pi_w$ contains all of the elements that are to be partitioned. For example, the partition $(\{1,2,8\},\{3,4,5\},\{6,7\})$ is a cyclic separable 3-partition of $\{1,2,3,4,5,6,7,8\}$. If the partitioned elements $\theta^1,\ldots,\theta^n$ are distinct and are renumbered so that $\theta^1 < \cdots < \theta^n$, cyclic separability of a partition π means that the index sets $\{i : \theta^i \in \pi_j\}$ are consecutive with the possible exception of one part which is the union of two consecutive sets, one containing 1 and the other containing n. When referring to partitions of indices, the adjective corresponding to "cyclic separable" is "cyclic consecutive".

The *polar representation* of a 2-dim vector $\binom{a}{b}$ is the pair (r,φ) where $r = \sqrt{(a)^2+(b)^2}$ and $0 \le \varphi < 2\pi$ satisfies $a = r\cos(\varphi)$ and $b = r\sin(\varphi)$; in this case we write $\binom{a}{b} = re^{\imath\varphi}$ with $\imath$ standing for (the imaginary number) $\sqrt{-1}$. Of course, the polar representation of a nonzero vector is unique.

We next provide a characterization of cone-separability of 2-dim partitions in terms of cyclic separability. For that purpose, we shall find it useful to use the polar representation of the partitioned vectors. Consider a 2-dim partition problem where the partitioned vectors are $A^i = \binom{a_i}{b_i}$ for $i = 1,\ldots,n$. To avoid degenerate situations we shall assume that the partitioned vectors are all nonzero. Using the polar representation of these vectors, say $A^1 = r_1e^{\imath\varphi_1},\ldots,A^n = r_ne^{\imath\varphi_n}$ to renumber them so that

$$\varphi^1 \le \cdots \le \varphi^n. \tag{4.2.6}$$

Under such enumeration we have that the vectors are split into two (possibly empty) groups $A^1,\ldots,A^s$ and $A^{s+1},\ldots,A^n$ with $s \in \{0,1,\ldots,n\}$,

$b_1, \ldots, b_s$ nonnegative, $b_{s+1}, \ldots, b_n$ negative, and

$$\frac{a_1}{b_1} \geq \frac{a_2}{b_2} \geq \cdots \geq \frac{a_s}{b_s}$$
$$\text{and} \tag{4.2.7}$$
$$\frac{a_{s+1}}{b_{s+1}} \geq \frac{a_{s+2}}{b_{s+2}} \geq \cdots \geq \frac{a_n}{b_n}$$

(here $\frac{a_u}{b_u}$ is taken as ∞ if $a_u > 0 = b_u$ and as $-\infty$ if $a_u < 0 = b_u$).

Lemma 4.2.10. *Assume that the partitioned vectors $A^1 = r_1 e^{\imath\varphi_1}, \ldots, A^n = r_n e^{\imath\varphi_n}$ are renumbered so that (4.2.6) is satisfied. A partition $\pi = (\pi_1, \ldots, \pi_p)$ is then cone-separable if and only if the partition $(\{\varphi_i : i \in \pi_1\}, \ldots, \{\varphi_i : i \in \pi_p\})$ is cyclic separable (as a 1-dim partition of $\{\varphi^1, \ldots, \varphi^n\}$). Further, if $A^i = \binom{a_i}{b_i}$ for $i = 1, \ldots, n$ and either all a_i's or all b_i's are one-sided (that is, nonnegative or nonpositive), then "cyclic separable" can be replaced with "separable" (and if $\varphi_1 \leq \cdots \leq \varphi_n$ then "separable" means "consecutive" in the sense of single-parameter partition problems studied in Chapters 2–7 of Vol. I).*

Proof. The general conclusion of the lemma is immediate from the characterization of polyhedral cones in $\mathbb{R}^2$ as sets with the (polar-coordinate-) representation $\{re^{\varphi} : a \leq \varphi \leq b\}$ for $0 \leq a \leq b < 2\pi$, or the union of two such sets– one with $a = 0$ and the other with $b = 2\pi$. The specialized conclusions of the lemma (for the cases where the a_i's/b_i's are one sided) follow from the observations that for each $i = 1, \ldots, n$: (i) $b_i \geq 0 \Leftrightarrow 0 \leq \varphi_i \leq \pi$, (ii) $b_i \leq 0 \Leftrightarrow \pi \leq \varphi_i \leq 2\pi$, (iii) $a_i \leq 0 \Leftrightarrow \frac{\pi}{2} \leq \varphi_i \leq \frac{3\pi}{2}$, and (iv) $a_i \geq 0 \Leftrightarrow [0 \leq \varphi_i \leq \frac{\pi}{2}$ or $\frac{3\pi}{2} \leq \varphi_i \leq 2\pi]$. □

Theorem 4.2.11. *Assume that the vectors $A^1 = r_1 e^{\imath\varphi_1}, \ldots, A^n = r_n e^{\imath\varphi_n}$ to be partitioned are renumbered so that (4.2.6) is satisfied. Consider a single-size partition problem with $F(\pi) = f(A_\pi)$ to be minimized, where $f(\cdot)$ is concave (jointly in its 2p variables). Then there exists an optimal partition π for which $(\{\varphi_i : i \in \pi_1\}, \ldots, \{\varphi_i : i \in \pi_p\})$ is cyclic separable (as a 1-dim partition of $\{\varphi^1, \ldots, \varphi^n\}$). Further, if $A^i = \binom{a_i}{b_i}$ for $i = 1, \ldots, n$ and either all a_i's or all b_i's are one-sided (that is, nonnegative or nonpositive), then "cyclic separable" can be replaced with "separable" (and if $\varphi_1 \leq \cdots \leq \varphi_n$ then "separable" means "consecutive" in the sense of single-parameter partition problems studied in Chapters 2–7 of Vol. I).*

Proof. Theorem 2.4.1 proves the existence of an optimal partition that is cone-separable. Lemma 4.2.10 shows that the latter property is equivalent to corresponding cyclic separability condition in the general case, and to separability requirement in the restricted (one-sided) case. □

Chakravarty, Orlin and Rothblum [19] proved that the conclusion of Theorem 4.2.11 for problems which satisfy the "one-sided" condition. For general problems (without that condition) they proved a weaker conclusion than the one stated in Theorems 2.4.1 and 4.2.11. Specifically, their result refers to index-partitions and establishes the existence of optimal partitions that are "order-nonpenetrating", (see Section 6.1 of Vol. I).

Chapter 5

Sum-Multipartition Problems over Single-Parameter Spaces

The previous chapters extended the partition model studied in Chapters 2–7 of Vol. I, from single-parameter to multi-parameter problems. In this chapter, we consider a generalization of single-parameter problems in a different direction. Specifically, the partitioned set is composed of a number of sets representing distinct types of elements; the set $\mathcal{N}$ to be partitioned is divided a priori into subsets $\mathcal{N}_u$ for $u = 1, \ldots, t$, with the index u referring to the distinct *types* of elements of $\mathcal{N}$. In such situations, each partition represents a group of partitions, one for each type (except that here our default position is allowing empty parts), and we refer to partitions of $\mathcal{N}$ as *multipartitions*. A sum multipartition problem is the counterpart of a sum partition problem when the single partition becomes a multipartition. Its formal definition will be given in Section 5.1 when suitable notation is available.

Our analysis of multipartition problems parallels the approach developed in Vol. I for partition problems. However, we differ from Vol. I in considering partitions on the element-sets, not the index-sets, to be consistent with the spirit of this volume. By doing so, we demonstrate that, indeed, the index-partition is just a special case of the element-partition (vector-partition for $d = 1$) when all elements are distinct.

While there are many results developed for partitions which can be easily extended to multipartitions, we do not report them exhaustively in this chapter to avoid spending too many pages on repeating familiar arguments in slightly different environments. We rely on the readers to recognize these extensions if they are useful to them. Our goal is to present the basic theory on multipartitions and only report those results which are most likely to be useful.

5.1 Multipartitions

For $d = 1$, the notion of consecutiveness plays a crucial role in the analysis of index-partitions in Vol. I. As the index-set has distinct elements, consecutiveness is actually defined for sets (with distinct elements). In Vol. II, we no longer assume that the partition-set consists of distinct elements; so we need to extend the definition of consecutiveness to such sets. Given a set $\mathcal{N}$, a subset $S \in \mathcal{N}$ is *consecutive* if no element in $\mathcal{N} \setminus S$ strictly penetrates S, i.e., no such element lies in int$S$. Note that a consecutive index-partition is always a consecutive element-partition but the converse is not true. For example, for $\theta^1 = 1, \theta^2 = \theta^3 = 2$, the index-partition $\pi_1 = \{1, 3\}, \pi_2 = \{2\}$ is not consecutive but the corresponding element partition $\pi_1 = \{\theta^1, \theta^3\}, \pi_2 = \{\theta^2\}$ is.

Consider a set $\Theta = \{\theta^1, \cdots, \theta^n\}$. Order the θ^i's into a sequence satisfying

$$\theta^i \leq \theta^j \text{ if } \theta^i \text{ precedes } \theta^j. \tag{5.1.1}$$

While Θ can generate many such ordered sequences (due to the existence of identical elements) satisfying (5.1.1), they are equivalent in the sense that only identical elements can reverse their orders in two different sequences. So we can select an arbitrary sequence of Θ satisfying (5.1.1) as a representative. Once the selection of a sequence is done, we relabel the θ^i's according to the order in the selected sequence, i.e., the first member is labeled θ^1 and the last θ^n. The labels of θ^i's in this sequence define an index-set $\{1, \cdots, n\}$ with index i corresponds to θ^i, thus a partition can be described either based on the Θ-set or the index-set (after an ordered sequence of Θ is selected). In particular, we will write $\theta^i \in \pi_j$ or $i \in \pi_j$ interchangeably, solely depending on which usage is more suitable in that text.

Assume that nonempty disjoint sets $\mathcal{N}_1, \ldots, \mathcal{N}_t$ are given with $\mathcal{N} \equiv \cup_{u=1}^t \mathcal{N}_u$ and $|\mathcal{N}_u| = n_u$. A *multipartition* π of size p of $\mathcal{N}$ is a partition of $\mathcal{N}$ where for each $j = 1, \ldots, p$, part π_j of π is composed of t (possibly empty) subparts $\pi_{uj} \equiv \mathcal{N}_u \cap \pi_j$ for $u = 1, \ldots, t$; in particular, for each u, $(\pi_{u1}, \ldots, \pi_{up})$ is a partition of $\mathcal{N}_u$ (hence the reference to "multi" partitions). The *multishape* of multipartition π is the vector $(|\pi_{11}|, \ldots, |\pi_{1p}|, |\pi_{21}|, \ldots, |\pi_{2p}|, \ldots, |\pi_{t1}|, \ldots, |\pi_{tp}|)$. If, for all $u = 1, \cdots, t$ and $j = 1, \cdots, p$, $|\pi_{uj}|$ is a prescribed number, then the multishape is a single-multishape; if each $|\pi_{uj}|$ is bounded, then the multishape is a bounded-multishape. Note that a bounded-multishape family Γ is simply the product of Γ^u's, say $\Gamma = \Gamma^1 \times \cdots \times \Gamma^t$, where Γ^u is a family of bounded-shapes of type u for $u = 1, \cdots, t$. Finally, if the set of multishapes is arbitrary, then the multishape is a constrained-multishape.

Assume throughout this chapter that $\mathcal{N} = \{1, \dots, n\}$ is given and each $i \in \mathcal{N}$ is associated with a real number θ^i. For a multipartition π, as for partitions, let $\theta_\pi \equiv (\sum_{i \in \pi_1} \theta^i, \dots, \sum_{i \in \pi_p} \theta^i)$.

Throughout this chapter, we find it useful to enumerate, in order, the θ^i's for i in each set $\mathcal{N}_u$ as $\theta^{u1}, \dots, \theta^{u|\mathcal{N}_u|}$ (with θ^{ui} as the i-th smallest element in $\mathcal{N}_u$); in particular, we have

$$\theta^{u1} \leq \cdots \leq \theta^{u|\mathcal{N}_u|}, \; u = 1, \dots, t \; . \tag{5.1.2}$$

For convenience, for each u, we typically identify $\mathcal{N}_u$ with $\{1, \dots, |\mathcal{N}_u|\}$; for each multipartition π, the π_{uj}'s are then identified with the corresponding subsets of $\{1, \cdots, |\mathcal{N}_u|\}$ and $\pi^u \equiv (\pi_{u1}, \dots, \pi_{up})$ is viewed as a partition of $\{\theta^{u1}, \cdots, \theta^{u|\mathcal{N}_u|}\}$ (as well as a partition of $\mathcal{N}_u$). In particular, with this notation, we have that

$$\begin{aligned} \theta_\pi &= \left(\sum_{u=1}^{t} \sum_{i \in \pi_{u1}} \theta^{ui}, \sum_{u=1}^{t} \sum_{i \in \pi_{u2}} \theta^{ui}, \dots, \sum_{u=1}^{t} \sum_{i \in \pi_{up}} \theta^{ui} \right) \\ &= \sum_{u=1}^{t} \left(\sum_{i \in \pi_{u1}} \theta^{ui}, \sum_{i \in \pi_{u2}} \theta^{ui}, \dots, \sum_{i \in \pi_{up}} \theta^{ui} \right) = \sum_{u=1}^{t} \theta_{\pi^u} \; . \end{aligned} \tag{5.1.3}$$

When the objective function can be written as $F(\pi) = f(\theta_\pi)$, then the multipartition problem is called a *sum multipartition problem.*

Given a set Π of multipartitions, the *multipartition polytope* corresponding to Π, denoted P^Π, is the convex hull of the θ_π's with π ranging over Π.

Given a set Π of multipartitions and subset I of $\{1, \dots, p\}$, let

$$\theta_*^\Pi(I) \equiv \min \left\{ \sum_{j \in I} (\theta_\pi)_j : \pi \in \Pi \right\} \; , \tag{5.1.4}$$

in particular, $\theta_*^\Pi(\emptyset) = 0$ and (as $\cup_{j=1}^p \pi_j = \mathcal{N}$ for each multipartition π)

$$\theta_*^\Pi(\{1, \dots, p\}) = \sum_{u=1}^{t} \sum_{i \in n_u} \theta^{ui}. \tag{5.1.5}$$

For a permutation $\sigma = (\{j_1\}, \{j_2\}, \cdots, \{j_p\})$ of $\{1, \cdots, p\}$, let

$$[\theta_*^\Pi]_\sigma \equiv \left(\theta_*^\Pi(\bigcup_{k=1}^{1} j_k), \theta_*^\Pi(\bigcup_{k=1}^{2} j_k) - \theta_*^\Pi(\bigcup_{k=1}^{1} j_k), \cdots, \theta_*^\Pi(\bigcup_{k=1}^{p} j_k) - \theta_*^\Pi(\bigcup_{k=1}^{p-1} j_k) \right) . \tag{5.1.6}$$

Also, C^{Π} and H^{Π} are defined as the polytopes $C^{\theta_*^{\Pi}}$ and $H^{\theta_*^{\Pi}}$; in particular, $H^{\theta_*^{\Pi}} \equiv \text{conv}\{[\theta_*^{\Pi}]_{\sigma} : \sigma$ ranging over all permutations of $\{1, \cdots, p\}\}$ and C^{Π} is the solution set of the linear inequality system

$$\sum_{j \in I} x_j \geq \theta_*^{\Pi}(I) \text{ for each nonempty subset } I \text{ of } \{1, \ldots, p\} \tag{5.1.7}$$

$$\sum_{j=1}^{p} x_j = \theta_*^{\Pi}(\{1, \ldots, p\}) \, . \tag{5.1.8}$$

A multipartition π is called *consecutive* (C) if for every $u = 1, \cdots, t$, the partition of $\mathcal{N}_u$ is consecutive (as an element partition); it is called *monotone* (M) if there is an index j_1 such that for each $u = 1, \ldots, t$, π_{uj_1} consists of the smallest $|\pi_{uj_1}|$ θ^i's in $\mathcal{N}_u$, there exists an index j_2 such that for each $u = 1, \ldots, t$, π_{uj_2} consists of the subsequent smallest $|\pi_{uj_2}|$ θ^i's in $\mathcal{N}_u$, and so on. Finally, there is an index j_p such that for each $u = 1, \ldots, t$, π_{uj_p} consists of the largest $|\pi_{uj_p}|$ θ^i's in $\mathcal{N}_u$ (note that this definition allows empty parts). In particular, if we use the index-partition representation, we have that for $u = 1, \ldots, t$ and $s = 1, \ldots, p$, $I_{\pi_{uj_s}} = \{\sum_{r=1}^{s-1} |\pi_{uj_r}| + 1, \ldots, \sum_{r=1}^{s} |\pi_{uj_r}|\}$. We observe that monotone multipartitions are determined by their multishapes and by a permutation of $\{1, \ldots, p\}$ (with possibly more than a single permutation corresponding to a given multipartition). Given a set of multipartitions Π, the convex hull of $\{\theta_\pi : \pi \in \Pi$ is a monotone partition$\}$ is denoted M^{Π}.

The next lemma records some trivial relationships between three of the polytopes corresponding to a set of multipartitions.

Lemma 5.1.1. *For every set of multipartitions* Π, $M^{\Pi} \subseteq P^{\Pi} \subseteq C^{\Pi}$.

Proof. The inclusion $M^{\Pi} \subseteq P^{\Pi}$ is trite. Next, from (5.1.4), for every multipartition $\pi \in \Pi$ and subset I of $\{1, \ldots, p\}$ we have that $\theta_*^{\Pi}(I) \leq \sum_{j \in I} (\theta_\pi)_j$. Also, from (5.1.5), we have that $\sum_{i=1}^{p} (\theta_\pi)_j = \sum_{i \in \mathcal{N}} \theta^i = \theta_*^{\Pi}(\{1, \ldots, p\})$. So, $\theta_\pi \in C^{\Pi} = C^{\theta_*^{\Pi}}$ for each $\pi \in \Pi$ and therefore the convex hulls of the θ_π's, namely P^{Π}, is contained in C^{Π}. □

Let Γ be a set of vectors $(n_{11}, \ldots, n_{1p}, n_{21}, \ldots, n_{2p}, \ldots, n_{t1}, \ldots, n_{tp})$ whose coordinates are nonnegative integers. We then denote by Π^{Γ} the set of multipartitions with multishape in Γ. Further, we use superscript "Γ" for the superscript "Π^{Γ}", e.g., we write $P^{\Gamma}, \theta_*^{\Gamma}, H^{\Gamma}$ and C^{Γ}; in particular, we refer to P^{Γ} as a *constrained-multishape multipartition polytope.* If Γ consists of a single vector $(n_{11}, \ldots, n_{1p}, n_{21}, \ldots, n_{2p}, \ldots, n_{t1}, \ldots, n_{tp})$ we use superscript "$(n_{11}, \ldots, n_{tp})$" for superscript "Γ", e.g., we write $\Pi^{(n_{11}, \ldots, n_{tp})}$,

$H^{(n_{11},\ldots,n_{tp})}$ and $\theta_*^{(n_{11},\ldots,n_{tp})}$; in particular, we refer to $P^{(n_{11},\ldots,n_{tp})}$ as a *single-multishape multipartition polytope.* When a set Γ is determined by lower bound $L = (L_{11},\ldots,L_{1p}, L_{21},\ldots,L_{2p},\ldots,L_{t1},\ldots,L_{tp})$ and upper bound $U = (U_{11},\ldots,U_{1p}, U_{21},\ldots,U_{2p},\ldots,U_{t1},\ldots,U_{tp})$ (which are vectors consisting of nonnegative integers) we use the superscript "(L,U)" for superscript "Γ", e.g., we write $\Pi^{(L,U)}$, $H^{(L,U)}$ and $\theta_*^{(L,U)}$; in particular, we refer to $P^{(L,U)}$ as a *bounded-multishape multipartition polytope.*

5.2 Single-Multishape Multipartition Polytopes

In the current section we consider multipartitions of prescribed multishape and follow Hwang and Rothblum [58] for the theoretical development. So, assume throughout the section that positive integers $p, n, n_1, \ldots, n_p$ and nonnegative integers $\{n_{uj} : u = 1,\ldots,t \text{ and } j = 1,\ldots,p\}$ are given, the prescribed multishape is then $(n_{11},\ldots,n_{tp})$. Our goal is to study the multipartition polytope $P^{(n_{11},\ldots,n_{tp})} \equiv \text{conv}\{\theta_\pi : \pi \in \Pi^{(n_{11},\ldots,n_{tp})}\}$. Assume throughout this section that θ^i's are enumerated by lists $\theta^{u1},\ldots,\theta^{u|\mathcal{N}_u|}$ for $u = 1,\ldots,t$, so that (5.1.2) is satisfied.

For a subset I of $\{1,\ldots,p\}$ and $u = 1,\ldots,t$, let $n_u(I) \equiv \sum_{j\in I} n_{uj}$. From (5.1.4) and (5.1.2) we obtain an explicit expression for $\theta_*^{(n_{11},\ldots,n_{tp})}(I)$ by

$$\theta_*^{(n_{11},\ldots,n_{tp})}(I) = \sum_{u=1}^{t}\sum_{j=1}^{n_u(I)} \theta^{uj} = \sum_{u=1}^{t}[\theta_u^{(n_{u1}\ldots,n_{up})}]_*(I) , \tag{5.2.1}$$

where for $u = 1,\ldots,t$,

$$[\theta_u^{(n_{u1}\ldots,n_{up})}]_*(I) = \sum_{j=1}^{n_u(I)} \theta^{uj} \tag{5.2.2}$$

is the function $\theta_*^{\Pi^u}$ with Π^u as the set of partitions of $\mathcal{N}_u$ with shape $(n_{u1},\ldots,n_{up})$ (cf. (5.1.6) in Vol. I). Using results that follow (5.1.7) of Vol. I, the 2^p values of each function $[\theta_u^{(n_{u1}\ldots,n_{up})}]_*$ can be computed in $2^p + |\mathcal{N}_u|$ arithmetic operations, hence their union can be computed with at most $t2^p + n$ arithmetic operations. The values of $\theta_*^{(n_{11},\ldots,n_{tp})}$ are then available from (5.1.7) by additional $t2^p$ arithmetic operations. Thus, the total computational effort needed for computing the values of $\theta_*^{(n_{11},\ldots,n_{tp})}$ is bounded $O(t2^p + n)$.

We say that the prescribed multishape $(n_{11},\ldots,n_{tp})$ is *regular* if for every $j,\ k \in \{1,\ldots,p\}$, there exists a $u \in \{1,\ldots,t\}$ with $n_{uj} > 0$ and $n_{uk} > 0$.[1] We say that $(n_{11},\ldots,n_{tp})$ is *universal* if for some $u = 1,\ldots,t$, $n_{uj} > 0$ for each $j = 1,\ldots,p$; such a type u is then called a *universal type*. Obviously, the prescribed multishape $(n_{11},\ldots,n_{tp})$ is universal implies that it is regular.

Theorem 5.2.1. *The function $\theta_*^{(n_{11},\ldots,n_{tp})}$ is supermodular. Further, if either*

(i) there exists a universal type u for the prescribed multishape such that all the θ^{ui}'s are distinct, or
(ii) the prescribed multishape is regular and for every $u = 1,\ldots,t$, the θ^{ui}'s are distinct,

then $\theta_*^{(n_{11},\ldots,n_{tp})}$ is strictly supermodular.

Proof. For brevity, let $\Pi \equiv \Pi^{(n_{11},\ldots,n_{tp})}$. We first argue that each of the functions $[\theta_u^{(n_{u1}\ldots,n_{up})}]_*$ is supermodular. These functions are associated with the single-shape partition problem of $\mathcal{N}_u$, with nonnegative $n_{u1},\ldots,n_{up}$. With $\theta^i \to \theta^{ui}$, $n(\cdot) \to n_u(\cdot)$ and $\theta_*^{(n_1,\ldots,n_p)} \to [\theta_u^{(n_{u1}\ldots,n_{up})}]_*$, we conclude from Lemma 5.1.1 of Vol. I that each $[\theta_u^{(n_{u1}\ldots,n_{up})}]_*$ is supermodular. Further, (5.2.1) shows that $\theta_*^{(n_{11},\ldots,n_{tp})}$ is the sum of supermodular functions, implying that it is supermodular.

Next assume that some $u = 1,\ldots,t$ is a universal type and for that u the θ^{ui}'s are distinct. Then the second part of Theorem 5.1.3 of Vol. I assures that $[\theta_u^{(n_{u1}\ldots,n_{up})}]_*$ is strictly supermodular. As the sum of a strictly supermodular function and supermodular functions is trivially strictly supermodular, we conclude that $\theta_*^{(n_{11},\ldots,n_{tp})}$ is strictly supermodular.

Finally, assume that our prescribed multishape is regular and for every $u = 1,\ldots,t$, the θ^{ui}'s are distinct. Now, assume that I and J are subsets of $\{1,\ldots,p\}$ with

$$\theta_*^{\Pi}(I \cup J) + \theta_*^{\Pi}(I \cap J) = \theta_*^{\Pi}(I) + \theta_*^{\Pi}(J).$$

As θ_*^{Π} is the sum of the supermodular functions $[\theta_u^{(n_{u1}\ldots,n_{up})}]_*$, with the sum taken over $u \in \{1,\ldots,t\}$, it follows that for each u,

$$[\theta_u^{(n_{u1}\ldots,n_{up})}]_*(I \cup J) + [\theta_u^{(n_{u1}\ldots,n_{up})}]_*(I \cap J) \\ = [\theta_u^{(n_{u1}\ldots,n_{up})}]_*(I) + [\theta_u^{(n_{u1}\ldots,n_{up})}]_*(J)\,.$$

[1]Regularity of multishapes should not be confused with regularity of partition properties introduced in Section 6.4 of Vol. I.

Now, if I and J are not comparable, then $I \setminus J \neq \emptyset$ and $J \setminus I \neq \emptyset$ and the regularity of $(n_{11}, \ldots, n_{tp})$ assures the existence of an index $u \in \{1, \ldots, t\}$ with $n_u(I \setminus J) > 0$ and $n_u(J \setminus I) > 0$; as $n_u(I) - n_u(I \cap J) = n_u(I \setminus J)$, it then follows from the arguments of Lemma 5.1.1 of Vol. I that θ^{ui} must be constant for $n_u(I \cap J) < i \leq n_u(I \cup J)$. As

$$n_u(I \cup J) - n_u(I \cap J) = n_u(I \setminus J) + n_u(J \setminus I) \geq 2,$$

we obtained a contradiction to the assertion that the θ^{ui}'s are distinct in i. □

The next example demonstrates that strict supermodularity of $\theta_*^{(n_{11}, \ldots, n_{tp})}$ does not extend to situations where the prescribed multishape is not regular, even if for each u, all the θ^{ui}'s are distinct.

Example 5.2.1. Let $n = 4$, $p = 3$, $t = 2$, $n_{11} = n_{12} = 1$, $n_{13} = n_{21} = 0$ and $n_{22} = n_{23} = 1$. This multishape is not regular as there exists no u with $n_{u1} > 0$ and $n_{u3} > 0$. Now, for $I = \{1, 2\}$ and $J = \{2, 3\}$, we have that for both $u = 1$ and $u = 2$, $n_u(I \cup J) - n_u(I \cap J) = n_u(\{1, 2, 3\}) - n_u(\{2\}) = 2 - 1 = 1$. Thus, by Lemma 5.1.1 of Vol. I, $\theta_*^{(n_{11}, \ldots, n_{23})}$ is not strictly supermodular. □

As the multishape of the partitions we consider is prescribed, monotone multipartitions are determined by the permutations of $\{1, \ldots, p\}$. With $\sigma = (\{j_1\}, \ldots, \{j_p\})$ as a permutation of $\{1, \ldots, p\}$, the multipartition I_{π_σ} corresponding to σ has $(I_{\pi_\sigma})_{uj_s} = \{\sum_{r=1}^{s-1} n_{uj_r} + 1, \ldots, \sum_{r=1}^{s} n_{uj_r}\} = \{n_u(\{j_1, \ldots, j_{s-1}\}) + 1, \cdots, n_u(\{j_1, \ldots, j_s\})\}$ for $u = 1, \ldots, t$ and $s = 1, \ldots, p$; of course, π_σ is in Π, that is, it has the prescribed multishape.

Theorem 5.2.2. *The correspondence $\sigma \to \pi_\sigma$ maps the set of permutations of $\{1, \ldots, p\}$ onto the set of monotone multipartitions in $\Pi^{(n_{11}, \ldots, n_{tp})}$ (having our prescribed multishape), and for every permutation σ of $\{1, \ldots, p\}$, $\theta_{\pi_\sigma} = [\theta_*^{(n_{11}, \ldots, n_{tp})}]_\sigma$. Further, if the prescribed multishape is regular, then the correspondence $\sigma \to \pi_\sigma$ of permutations of $\{1, \ldots, p\}$ onto $\Pi^{(n_{11}, \ldots n_{tp})}$ is one-to-one.*

Proof. Again, let $\Pi \equiv \Pi^{(n_{11}, \ldots n_{tp})}$. The paragraph preceding our theorem explains that the correspondence $\sigma \to \pi_\sigma$ maps the set of permutations of $\{1, \ldots, p\}$ onto the set of monotone multipartitions in $\Pi^{(n_{11}, \ldots, n_{tp})}$. Let $\sigma = (\{j_1\}, \ldots, \{j_p\})$ be a permutation and $s = 1, \ldots, p$. From (5.2.1) and the definition of π_σ we have that

$$\begin{aligned}\left[(\theta_*^\Pi)_\sigma\right]_{j_s} &= \left[(\theta_*^\Pi)_\sigma\right](\{j_1,\ldots,j_s\}) - \left[(\theta_*^\Pi)_\sigma\right](\{j_1,\ldots,j_{s-1}\}) \\ &= \sum_{u=1}^{t} \sum_{j=n_u(\{j_1,\ldots,j_{s-1}\})+1}^{n_u(\{j_1,\ldots,j_s\})} \theta^{uj} \\ &= \sum_{u=1}^{t} \sum_{j\in(\pi_\sigma)_{uj_s}} \theta^{uj} = (\theta_{\pi_\sigma})_{j_s} \,.\end{aligned}$$

As $\{j_1,\ldots,j_p\} = \{1,\ldots,p\}$ we conclude that $(\theta_*^\Pi)_\sigma = \theta_{\pi_\sigma}$.

Finally, assume that the prescribed multishape is regular and that $\pi \in \Pi$ is monotone. Let j and k be a pair of indices in $\{1,\ldots,p\}$. Then there is an index $u \in \{1,\ldots,t\}$ with $n_{uj} > 0$ and $n_{uk} > 0$. It follows that the part among π_j and π_k that gets the lower indices of $\mathcal{N}_u$ must precede the other one under any permutation σ with $\pi_\sigma = \pi$, that is, if for such σ, we have that $\sigma_{r_j} = \{j\}$, $\sigma_{r_k} = \{k\}$ and π_{uj} contains smaller indices than those in π_{uk}, then $r_j < r_k$. Since j and k were selected arbitrarily, we have that the permutation σ with $\pi_\sigma = \pi$ is uniquely determined from π. □

Recall that the polytopes $H^{(n_{11},\ldots,n_{tp})}$ and $C^{(n_{11},\ldots,n_{tp})}$ are defined as $H^{\theta_*^{(n_{11},\ldots,n_{tp})}}$ and $C^{\theta_*^{(n_{11},\ldots,n_{tp})}}$, respectively.

Corollary 5.2.3. $M^{(n_{11},\ldots,n_{tp})} = H^{(n_{11},\ldots,n_{tp})}$.

Proof. Theorem 5.2.2 assures that $\{\theta_\pi : \pi$ a monotone multipartition with multishape $(n_{11},\ldots,n_{tp})\} = \{[\theta_*^{(n_{11},\ldots,n_{tp})}]_\sigma : \sigma$ a permutation of $\{1,\ldots,p\}\}$; consequently, the convex hulls of these sets coincide, namely, $M^{(n_{11},\ldots,n_{tp})} = H^{\theta_*^{(n_{11},\ldots,n_{tp})}} = H^{(n_{11},\ldots,n_{tp})}$. □

Theorem 5.2.4. $P^{(n_{11},\ldots,n_{tp})} = C^{(n_{11},\ldots,n_{tp})} = H^{(n_{11},\ldots,n_{tp})} = M^{(n_{11},\ldots,n_{tp})}$.

Proof. Again, let $\Pi \equiv \Pi^{(n_{11},\ldots n_{tp})}$. By Lemma 5.1.1 and Corollary 5.2.3, $H^\Pi = M^\Pi \subseteq P^\Pi \subseteq C^\Pi$. By Theorem 4.1.1 of Vol. I and the supermodularity of θ_*^Π(established in Theorem 5.2.1), we have that $C^\Pi = C^{\theta_*^\Pi} = H^{\theta_*^\Pi} = H^\Pi$. So, we conclude that $P^\Pi = C^\Pi = H^\Pi = M^\Pi$. □

The next corollary of Theorem 5.2.4 asserts that in nondegenerate situations, single-multishape multipartition polytopes are normally and combinatorically equivalent to the standard permutahedron. Recall that "combinatorial equivalence" is defined on p. 78 and "normal equivalence" on p. 82 in Vol. I.

Corollary 5.2.5. *Suppose either*

(i) there exists a universal type u for the prescribed multishape such that all the θ^{ui}'s distinct, or
(ii) the prescribed multishape is regular and for every $u = 1, \ldots, t$, the θ^{ui}'s are distinct.

Then $P^{(n_{11},\ldots,n_{tp})}$ is both normally and combinatorically equivalent to the standard permutahedron.

Proof. Again, let $\Pi \equiv \Pi^{(n_{11},\ldots n_{tp})}$. By Theorem 5.2.4 it suffices to prove the conclusion of the corollary with H^{Π} replacing P^{Π}. Now, by Theorem 5.2.1, the assumptions of the corollary imply that $\theta_*^{(n_{11},\ldots,n_{tp})}$ is strictly supermodular, and therefore Corollary 4.3.4 of Vol. I implies that H^{Π} satisfies that asserted conclusion. □

In order to derive further results about single-multishape multipartition polytopes under the assumptions of Corollary 5.2.5, we need the following auxiliary result.

Lemma 5.2.6. *Let π and π' be two distinct monotone multipartitions in $\Pi^{(n_{11},\ldots,n_{tp})}$ such that π and π' coincide on all but exactly two parts, say the j-th and k-th part, and $\pi_{uj} \cup \pi_{uk} = \pi'_{uj} \cup \pi'_{uk}$ is a consecutive set of points for each $u = 1, \ldots, t$. Then there exist permutations σ and σ' which coincide on all but exactly two parts which are indexed by consecutive integers under both σ and σ', and $\pi = \pi_\sigma$ and $\pi' = \pi_{\sigma'}$.*

Proof. Suppose the hypothesis of the lemma is satisfied. Let σ and σ' be two permutations of $\{1, \ldots, p\}$ with $\pi = \pi_\sigma$ and $\pi' = \pi_{\sigma'}$. For each $q = 1, \ldots, p$, let i_q and i'_q be the indices of part q under σ and σ', that is, i_q and i'_q are the unique indices with $\sigma_{i_q} = \sigma'_{i'_q} = \{q\}$. As π and π' are distinct and coincide on all parts but part j and part k, we have that the relative order of i_j and i_k is the inverse of the relative order of i'_j and i'_k; without loss of generality assume that $i_j < i_k$ and $i'_j > i'_k$.

For $q = 1, \ldots, p$, let $\text{supp}(q) \equiv \{u = 1, \ldots, t : n_{uq} > 0\}$. Consider the undirected graph $G = (V, E)$ with vertex set $\{r : i_j \le i_r \le i_k\}$ and edge set

$$E = \{\{r, s\} : r, s \in V,\ r \ne s,\ \{r, s\} \ne \{j, k\} \text{ and } \text{supp}(r) \cap \text{supp}(s) \ne \emptyset\}.$$

Note that for $\{r, s\} \in E$, the relative orders of r and s are invariant under σ and σ' (for otherwise, $\pi_r \neq \pi'_r$ or $\pi_s \neq \pi'_s$), that is, either $i_r < i_s$ and $i'_r < i'_s$, or $i_r > i_s$ and $i'_r > i'_s$.

An (undirected) path in G between $r \in V$ and $s \in V$ is a sequence of vertices $r_0 = r, r_1, \ldots, r_q = s$ in V with $\{r_i, r_{i+1}\} \in E$ for $i = 0, 1, \ldots, q-1$. When such a path exists, we have that either $i_r < i_s$ and $i'_r < i'_s$, or alternatively, $i_r > i_s$ and $i'_r > i'_s$. As $i_j < i_k$ and $i'_j > i'_k$ we have that there exists no path in G between j and k. Let A be the set of vertices v in V for which there is a path in G between j and v. It then follows that $j \in A$, $k \notin A$ and for every $r \in A$ and $s \in V \setminus A$, supp$(r) \cap$ supp$(s) = \emptyset$. Without loss of generality assume that j precedes k in the path. It then follows that σ can be modified by moving all the elements of $A \setminus \{j\}$ that lie between j and k, in order, to the locations immediately following k without changing the corresponding associated partition. Similarly, σ can also be modified by moving all the elements of $(V \setminus A) \setminus \{j\}$ that lie between j and k, in order, to the locations immediately preceding j without changing the corresponding associated partition. These changes in σ can be done jointly resulting in a partition σ with $\pi = \pi_\sigma$ where the elements indexing j and k under σ are consecutive integers. By reversing the order of these integers we get a permutation σ' where $\pi_{\sigma'}$ coincides with π on all parts but part j and part k and $\pi_{uj} \cup \pi_{uk} = (\pi_{\sigma'})_{uj} \cup (\pi_{\sigma'})_{uk}$ for each $u = 1, \ldots, t$. It follows that $\pi' = \pi_{\sigma'}$, completing the proof. □

The next example demonstrates that for a multipartition π, the requirements that for every pair of parts, say the j-th part and the k-th part, $\pi_{uj} \cup \pi_{uk}$ is a consecutive set of integers for each $u = 1, \ldots, t$ does not imply the existence of a permutation σ where j and k are indexed by consecutive integers under σ.

Example 5.2.2. Let $p = 3$, $t = 2$, $n = 4$, $n_{11} = n_{12} = n_{22} = n_{23} = 1$, $n_{13} = n_{21} = 0$, $\mathcal{N}_1 = \{1, 2\}$ and $\mathcal{N}_2 = \{3, 4\}$. Consider the monotone multipartition π associated with the identity permutation, that is, $\pi_{11} = \{1\}$, $\pi_{12} = \{2\}$, $\pi_{22} = \{3\}$ and $\pi_{23} = \{4\}$. Evidently, the identity is the only permutation associated with π. As $\pi_{11} \cup \pi_{13}$ and $\pi_{21} \cup \pi_{23}$ consist of singletons, we have that these two unions consist of consecutive elements; but, part 1 and part 3 are not indexed by consecutive integers under σ. □

We next characterize vertices and edges of single-multishape multipartition polytopes. For subset I of $\{1, \ldots, p\}$, we use the notation F_I as defined in Vol. I (4.1.4) with respect to $\lambda \equiv \theta_*^{(n_{11},\dots,n_{tp})}$, that is,

$$F_I = \{x \in P^{(n_{11},\dots,n_{tp})} : \sum_{j \in I} x_j = \theta_*^{(n_{11},\dots,n_{tp})}(I)\} \subseteq \mathbb{R}^p.$$

Theorem 5.2.7.

(a) *For $v \in \mathbb{R}^p$ the following are equivalent:*

(i) *v is a vertex of $P^{(n_{11},\dots,n_{tp})}$,*

(ii) *there is a permutation σ of $\{1,\dots,p\}$ with $v = [\theta_*^{(n_{11},\dots,n_{tp})}]_\sigma$,*

(iii) *there is a chain of length $p-1$, say $I_1,\dots,I_{p-1}$, with $\{v\} = \cap_{t=1}^{p-1} F_{I_t}$, and*

(iv) *there is a monotone multipartition π with multishape $(n_{11},\dots,n_{tp})$ such that $v = \theta_\pi$.*

Further, if the above equivalent conditions hold, then the monotone multipartition π_σ satisfies $v = \theta_{\pi_\sigma}$ and for every multipartition $\pi' \in \Pi^{(n_{11},\dots,n_{tp})}$ with $\theta_{\pi'} = v$ and every $\alpha \in \{\theta^{u1},\dots,\theta^{un_u}\}$, we have that $|\{i \in \pi'_{uj} : \theta^{ui} = \alpha\}| = |\{i \in (\pi_\sigma)_{uj} : \theta^{ui} = \alpha\}|$ for $u = 1,\dots,t$ and $j = 1,\dots,p$.

(b) *For distinct vertices v and v' of P^Π the following are equivalent:*

(i) *$\text{conv}\{v, v'\}$ is an edge of $P^{(n_{11},\dots,n_{tp})}$,*

(ii) *there exist permutations σ and σ' such that $\{v, v'\} = \{[\theta_*^{(n_{11},\dots,n_{tp})}]_\sigma, [\theta_*^{(n_{11},\dots,n_{tp})}]_{\sigma'}\}$ and σ and σ' coincide on all but exactly two parts which are indexed by consecutive integers under both σ and σ',*

(iii) *there exist two chains of length $p-1$, say $I_1,\dots,I_{p-1}$ and $I'_1,\dots,I'_{p-1}$ such that $\{v, v'\} = \{\cap_{t=1}^{p-1} F_{I_t}, \cap_{t=1}^{p-1} F_{I'_t}\}$ and $I_1,\dots,I_{p-1}$ and $I'_1,\dots,I'_{p-1}$ have a common subchain of length $p-2$, and*

(iv) *there exist monotone multipartitions π and π' with multishape $(n_{11},\dots,n_{tp})$ such that $\{v, v'\} = \{\theta_\pi, \theta_{\pi'}\}$, π and π' coincide on all but exactly two parts, say the j-th and k-th parts, and $\pi_{uj} \cup \pi_{uk} = \pi'_{uj} \cup \pi'_{uk}$ is a consecutive set of integers for each $u = 1,\dots,t$.*

Further, if the above equivalent conditions hold and j and k are as in (iv), then $v - v'$ is a scalar multiple of $(e^j - e^k)$.

Proof. Again, let $\Pi \equiv \Pi^{(n_{11},\dots,n_{tp})}$.

(a) The equivalence of (i)–(iii) follows from Theorem 4.1.5 of Vol. I, the supermodularity of θ_*^Π (Theorem 5.2.1) and the equality $P^\Pi = C^{\theta_*^\Pi} = H^{\theta_*^\Pi}$

(Theorem 5.2.4). Further, the equivalence of (ii) and (iv) is established in Theorem 5.2.2. Now assume that $\sigma = (\{j_1\}, \ldots, \{j_p\})$ is a permutation of $\{1, \ldots, p\}$ with $v = (\theta_*^\Pi)_\sigma$. From Theorem 5.2.2, we then have that $(\theta_*^\Pi)_\sigma = \theta_{\pi_\sigma}$, implying that $v = \theta_{\pi_\sigma}$. It remains to establish the "uniqueness" of the representation of v. So, let π' be a multipartition in Π with $\theta_{\pi'} = v = \theta_{\pi_\sigma}$ and let $\alpha \in \{\theta^{u1}, \ldots, \theta^{un_u}\}$. As $| \cup_{r=1}^s \pi'_{uj_r}| = n_u(\{j_1, \ldots, j_s\})$, (5.1.2) implies that for $s = 1, \ldots, p$ and $u = 1, \ldots, t$, we have that (with $\pi_\sigma^u \equiv (\pi_\sigma)^u$)

$$\sum_{i=1}^{s} (\theta_{\pi'^u})_{j_r} = \sum_{i \in \cup_{r=1}^s \pi'_{uj_r}} \theta^{ui} \geq \sum_{i=1}^{n_u(\{j_1,\ldots,j_s\})} \theta^{ui} = \sum_{r=1}^{s} (\theta_{\pi_\sigma^u})_{j_r}; \qquad (5.2.3)$$

consequently, from (5.1.3), for $s = 1, \ldots, p$,

$$\sum_{i=1}^{s} (\theta_{\pi'})_{j_r} = \sum_{u=1}^{t} \sum_{r=1}^{s} (\theta_{\pi'^u})_{j_r} \geq \sum_{u=1}^{t} \sum_{r=1}^{s} (\theta_{\pi_\sigma^u})_{j_r} = \sum_{r=1}^{s} (\theta_{\pi_\sigma})_{j_r}. \qquad (5.2.4)$$

As $\theta_{\pi'} = v = \theta_{\pi_\sigma}$, we conclude that the inequalities of (5.2.3) and (5.2.4) must hold as equalities; this conclusion together with (5.1.2) implies that for each $\alpha \in \{\theta^{u1}, \ldots, \theta^{un_u}\}$, $u = 1, \ldots, t$ and $s = 1, \ldots, p$,

$$\begin{aligned} |\{i \in \cup_{r=1}^s \pi'_{uj_r} : \theta^{ui} = \alpha\}| &= |\{1 \leq i \leq n_u(\{j_1, \ldots, j_s\}) : \theta^{ui} = \alpha\}| \\ &= \left|\{i \in \cup_{r=1}^s (\pi_\sigma)_{uj_r} : \theta^{ui} = \alpha\}\right|. \end{aligned}$$

It follows that $|\{i \in \pi'_{uj_r} : \theta^{ui} = \alpha\}| = |\{i \in (\pi_\sigma)_{uj_r} : \theta^{ui} = \alpha\}|$ for $u = 1, \ldots, t$ and $r = 1, \ldots, p$; that is (as $\{j_1, \ldots, j_p\} = \{1, \ldots, p\}$),

$$|\{i \in \pi'_{uj} : \theta^{ui} = \alpha\}| = |\{i \in (\pi_\sigma)_{uj} : \theta^{ui} = \alpha\}|$$

for $u = 1, \ldots, t$ and $j = 1, \ldots, p$.

(b) Here again, the equivalence of (i)–(iii) follows from Theorem 4.1.5 of Vol. I, the supermodularity of θ_*^Π (Theorem 5.2.1) and the equality $P^\Pi = C^{\theta_*^\Pi} = H^{\theta_*^\Pi}$ (Theorem 5.2.4).

Next assume that (ii) holds with σ and σ' as permutations which satisfy $\{v, v'\} = \{(\theta_*^\Pi)_\sigma, (\theta_*^\Pi)_{\sigma'}\}$ and coincide on all but exactly two parts which are indexed by consecutive integers under both σ and σ'. Let j and k be the indices of the parts of σ and σ' that do not coincide. Let π and π' be the monotone multipartitions with shape $(n_{11}, \ldots, n_{tp})$ corresponding to σ and σ', respectively. As the indices of the parts containing j and k under both σ and σ' are consecutive, we have that $\pi_{uj} \cup \pi_{uk} = \pi'_{uj} \cup \pi'_{uk}$ is a consecutive set of integers for each $u = 1, \ldots, t$.

Finally assume that (iv) holds and π and π' are monotone multipartitions such that π and π' coincide on all but exactly two parts, say part j and part k, and $\pi_{uj} \cup \pi_{uk} = \pi'_{uj} \cup \pi'_{uk}$ is a consecutive set of integers for each $u = 1, \ldots, t$. By Lemma 5.2.6, there exist permutations σ and σ' which coincide on all but exactly two parts which are indexed by consecutive integers under both σ and σ' and $\pi = \pi_\sigma$ and $\pi' = \pi_{\sigma'}$. As $\theta_\pi = (\theta_*^\Pi)_\sigma$ and $\theta_{\pi'} = (\theta_*^\Pi)_{\sigma'}$ (Theorem 5.2.2), we have that $\{(\theta_*^\Pi)_\sigma, (\theta_*^\Pi)_{\sigma'}\} = \{\theta_\pi, \theta_{\pi'}\} = \{v, v'\}$, establishing (ii). Further, by Theorem 4.1.5 of Vol. I, in the above situation $v - v'$ is a scalar multiple of $(e^j - e^k)$. □

Theorem 5.2.7 implies that the set of vertices of $P^{(n_{11},\ldots,n_{tp})}$ equals the set $\{\theta_\pi : \pi$ ranging over the set of monotone multipartitions in $\Pi^{(n_{11},\ldots,n_{tp})}\}$ and the directions of the edges of $P^{(n_{11},\ldots,n_{tp})}$ are differences of unit vectors.

We next show that if θ^{ui}'s are distinct for each $u = 1, \ldots, t$, then monotone multipartitions are in a one-to-one correspondence with the set of vertices of $P^{(n_{11},\ldots,n_{tp})}$.

Corollary 5.2.8. *Suppose the multishape satisfies that for all $u = 1, \ldots, t$, θ^{ui}'s are distinct. Then for every vertex v of $P^{(n_{11},\ldots,n_{tp})}$, there exists a unique multipartition $\pi \in \Pi^{(n_{11},\ldots,n_{tp})}$ with $\theta_\pi = v$; in particular, π is monotone.*

Proof. The proof follows immediately from Theorem 5.2.7(a). □

Recall the one-to-one correspondence, introduced in Section 4.1 of Vol. I, of chains of length $p-1$ onto permutations of $\{1, \ldots, p\}$. In view of Theorem 5.2.2, this correspondence translates into a map of chains of length $p-1$ onto monotone multipartitions (with the prescribed multishape). Specifically, chain $I_1, \ldots, I_{p-1}$ having $I_s \setminus I_{s-1} = \{j_s\}$ for $s = 1, \ldots, p$, corresponds to the multipartition π with $(\pi)_{uj_s} = (n_u(I_{s-1}) + 1, \ldots, n_u(I_s))$ for $u = 1, \ldots, t$ and $s = 1, \ldots, p$. When the prescribed multishape is regular, the correspondence of permutations of $\{1, \ldots, p\}$ onto monotone multipartitions is one-to-one; consequently, in such cases, the correspondence of chains of length $p - 1$ to multipartitions is one-to-one. We say that monotone multipartition π is *consistent with chain* $I_1, \ldots, I_k$ if $I_1, \ldots, I_k$ is a subchain of a chain of length $p - 1$ that corresponds to π.

In view of Theorems 5.2.1 and 5.2.4, the results of Sections 4.1 and 4.2 of Vol. I about the faces of permutation polytopes corresponding to supermodular functions apply to the polytope $P^{(n_{11},\ldots,n_{tp})} = C^{(n_{11},\ldots,n_{tp})} = H^{(n_{11},\ldots,n_{tp})}$ and the results of Section 4.3 of Vol. I apply when either assumption (i) or (ii) of Theorem 5.2.1 are satisfied (cf., Section 5.1 of Vol. I where single-shape partition polytopes are considered). In particular, we obtain the face characterization of permutation polytopes obtained in Theorem 4.1.2 of Vol. I, and the maximal representation of faces obtained in Theorem 4.1.3 of Vol. I. In applying these results to single-multishape multipartition polytopes, the reference to "λ_σ with σ as a permutation of $\{1,\ldots,p\}$ which is consistent with chain $I_1,\ldots,I_k$" translates to "θ_π with π as a monotone multipartition which is consistent with chain $I_1,\ldots,I_k$". (Here, the results we stated explicitly translated results about vertices and edges, cf., Theorem 5.1.9 and Corollary 5.1.10 in Vol. I where single-shape partition polytopes are considered.)

5.3 Constrained-Multishape Multipartition Polytopes

Hwang and Rothblum [58] extended the results in Section 5.2 to the bounded-multishape polytopes. In the section, we further extend them to the constrained-shape multishape polytopes. Our development resembles some, but not all, of the results we obtained for constrained-shape partition polytopes.

Throughout this section we assume that positive integers n, p, t and nonnegative integers $n_1,\ldots,n_t$ are given with $\sum_{u=1}^{t} n_u = n$. Also, it is assumed that the θ^i's are enumerated by lists $\theta^{u1},\ldots,\theta^{u|\mathcal{N}_u|}$ for $u = 1,\ldots,t$ so that (5.1.2) is satisfied.

Our first result asserts that the vertices of all constrained-multishape multipartition polytopes have representations by vectors associated with monotone multipartitions.

Theorem 5.3.1. *Let Γ be a set of nonnegative integer tp-vectors $(n_{11},\ldots,n_{tp})$ satisfying $\sum_{j=1}^{p} n_{uj} = n_u$ for $u = 1,\ldots,t$ and let $\Pi \equiv \Pi^\Gamma$. Then each vertex v of P^Π has a representation as θ_π with π as a monotone multipartition in Π; in particular, $M^\Pi = P^\Pi$.*

Proof. Let v be a vertex of $P \equiv P^\Pi$. Then there exists a p-vector c such that the linear function f_c mapping $x \in \mathbb{R}^p$ into $c^T x$ attains a unique maximum over P at v. Let π^* be a multipartition in Π which maximizes $c^T\theta_\pi$ over π in Π. It follows that θ_{π^*} maximizes $c^T x$ over $x \in P$ (Proposition 3.1.1(h) of Vol. I), implying that $c^T\theta_{\pi^*} = c^T v$. Let $(n_{11},\ldots,n_{tp})$ be the multishape of π^*. Now, f_c attains a maximum over $P^{(n_{11},\ldots,n_{tp})}$ at a vertex (Proposition 3.1.1(i) of Vol. I) and such a vertex has a representation as $\theta_{\pi'}$ for some monotone multipartition π' with multishape $(n_{11},\ldots,n_{tp})$ (Theorem 5.2.7). Then $\theta_{\pi'} \in P$ and $c^T\theta_{\pi'} = c^T\theta_{\pi^*} = c^T v$; as v is the unique maximizer of f_c over P^Π, we conclude that $\theta_{\pi'} = v$.

It follows that the vertices of P^Π are all in M^Π. As P^Π is the convex hull of its vertices (Proposition 3.1.1(a) of Vol. I), we have that $P^\Pi \subseteq M^\Pi$. As the inverse inclusion follows without any restrictions on Π, we have that $M^\Pi = P^\Pi$. □

Example 5.2.1 in Vol. I demonstrates that even for partitions, that is the case with $t = 1$, the inverse of Theorem 5.3.1 needs not hold and the vectors corresponding to monotone partitions need not be vertices of the corresponding multipartition polytope.

We say that a set of multipartitions Π is *complete* if for every permutation σ of $\{1,\ldots,p\}$, there exists a multipartition π in Π with $\theta_\pi = (\theta_*^\Pi)_\sigma$. Observe that for a permutation $\sigma = (\{j_1\},\ldots,\{j_p\})$ of $\{1,\ldots,p\}$ and $s = 1,\ldots,p$,

$$\sum_{k=1}^{s} \left[(\theta_*^\Pi)_\sigma\right]_{j_k} = (\theta_*^\Pi)(\{j_1,\ldots,j_s\}) = \min_{\tau\in\Pi} \sum_{k=1}^{s} (\theta_\tau)_{j_k} ;$$

hence, $(\theta_*^\Pi)_\sigma = \theta_\pi$ for a partition π if and only if

$$\sum_{k=1}^{s} (\theta_\pi)_{j_k} = \min_{\tau\in\Pi} \sum_{k=1}^{s} (\theta_\tau)_{j_k} \quad \text{for each } s = 1,\ldots,p .$$

Note that completeness depends both on the set of partitions and on the θ^{ui}'s.

Example 5.3.1 demonstrates that sets of multipartitions with a single-multishape are always complete.

Example 5.3.1. Let $n_{11},\ldots,n_{tp}$ be nonnegative integers with $\sum_{j=1}^{p} n_{uj} = n_u$ for $u = 1,\ldots,t$. Theorem 5.2.2 assures that the set $\Pi^{(n_1,\ldots,n_p)}$ of all partitions with multishape $(n_{11},\ldots,n_{tp})$ is complete, regardless of the θ^i's. □

Next, we introduce an auxiliary result which parallels Lemma 5.2.3 in Vol. I.

Lemma 5.3.2. *Let Π be a complete set of multipartitions. Then for every chain $I_1,\ldots,I_m$ of subsets of $\{1,\ldots,p\}$, there exists a multipartition π in Π satisfying $(\theta_*^\Pi)(I_s) = \sum_{j\in I_s} (\theta_\pi)_j$.*

Proof. As every chain is a subchain of a chain of length $p-1$, attention can be restricted to chains of length $p-1$. Let $I_1,\ldots,I_{p-1}$ be a chain of subsets of $\{1,\ldots,p\}$ and let σ be the corresponding permutation. If

$\sigma = (\{j_1\}, \ldots, \{j_p\})$, then for $s = 1, \ldots, p$, $I_s = \{j_1, \ldots, j_s\}$. Let π be the multipartition corresponding to σ through the consistency of Π. Then for each $s = 1, \ldots, p$,

$$\sum_{j \in I_s} (\theta_\pi)_j = \sum_{k=1}^{s} (\theta_\pi)_{j_k} = \sum_{k=1}^{s} \left[(\theta_*^\Pi)_\sigma\right]_{j_k} = (\theta_*^\Pi)(\{j_1, \ldots, j_s\}) = (\theta_*^\Pi)(I_s) .$$

□

The next result, which parallels Theorem 5.2.4 in Vol. I, demonstrates the usefulness of completeness.

Theorem 5.3.3. *Let Π be a complete set of multipartitions. Then:*

(a) θ_*^Π is supermodular.

(b) $M^\Pi \subseteq P^\Pi = H^\Pi = C^\Pi$, and if Π is a constrained-multishape set of multipartitions, then $M^\Pi = P^\Pi$.

(c) The direction of each edge of P^Π is the difference of two unit vectors.

(d) v is a vertex of P^Π (hence a vertex of M^Π and $v = \theta_\pi$ for some monotone multipartition π) if and only if $v = (\theta_*^\Pi)_\sigma$ for some permutation σ of $\{1, \ldots, p\}$.

(e) If each multishape in Π satisfies either condition (i) or (ii) in Theorem 5.2.1, then θ_*^Π is strictly supermodular and for each vertex v there is a unique permutation σ of $\{1, \ldots, p\}$ satisfying $v = (\theta_*^\Pi)_\sigma$; in particular, P^Π has exactly $p!$ vertices.

Proof.

(a) Let I and J be subsets of $\{1, \ldots, p\}$. Applying Lemma 5.3.2 to the 2-chain $I \cap J$, $I \cup J$, we conclude the existence of a multipartition π in Π with $(\theta_*^\Pi)(S) = \sum_{j \in S} (\theta_\pi)_j$ for $S = I \cap J$ and $S = I \cup J$. Let $(n_{11}, \ldots, n_{tp})$ be the multishape of π. Now, for every $S \subseteq \{1, \ldots, p\}$,

$$(\theta_*^\Pi)(S) \leq \theta_*^{(n_{11}, \ldots, n_{tp})}(S) \leq \sum_{j \in S} (\theta_\pi)_j,$$

hence the assumption about π implies that

$$(\theta_*^\Pi)(I \cap J) = \theta_*^{(n_{11}, \ldots, n_{tp})}(I \cap J) = \sum_{j \in I \cap J} (\theta_\pi)_j$$

and

$$(\theta_*^\Pi)(I \cup J) = \theta_*^{(n_{11}, \ldots, n_{tp})}(I \cup J) = \sum_{j \in I \cup J} (\theta_\pi)_j.$$

It now follows from the supermodularity of $\theta_*^{(n_{11},\ldots,n_{tp})}$ (established in Theorem 5.2.1) that

$$\begin{aligned}(\theta_*^\Pi)(I \cup J) + (\theta_*^\Pi)(I \cap J) &= \theta_*^{(n_{11},\ldots,n_{tp})}(I \cup J) + \theta_*^{(n_{11},\ldots,n_{tp})}(I \cap J) \\ &\geq \theta_*^{(n_{11},\ldots,n_{tp})}(I) + \theta_*^{(n_{11},\ldots,n_{tp})}(J) \\ &\geq (\theta_*^\Pi)(I) + (\theta_*^\Pi)(J)\,, \qquad (5.3.1)\end{aligned}$$

proving that θ_*^Π is supermodular.

(b) The inclusion relation $M^\Pi \subseteq P^\Pi \subseteq C^\Pi$ can be proved by using an argument similar to the one used in the proof of Lemma 5.1.1. As part (a) assures that θ_*^Π is supermodular, Theorem 4.1.1 of Vol. I implies that $H^\Pi = C^\Pi$ and that the vertices of this polytope are precisely the $(\theta_*^\Pi)_\sigma$'s with σ ranging over the permutations of $\{1,\ldots,p\}$. By the completeness of Π, each such vertex has a representation as θ_π for some π in Π. It follows that $H^\Pi = \text{conv}\{(\theta_*^\Pi)_\sigma : \sigma \text{ a permutation}\} \subseteq \text{conv}\{(\theta_*^\Pi)_\sigma : \sigma \text{ a permutation}\} \subseteq \text{conv}\{\theta_\pi : \pi \in \Pi\} = P^\Pi$. We conclude that $M^\Pi \subseteq P^\Pi \subseteq C^\Pi = H^\Pi \subseteq P^\Pi$, proving that $M^\Pi \subseteq P^\Pi = H^\Pi = C^\Pi$. The equality $M^\Pi = P^\Pi$ when Π is a constrained-multishape set of partitions follows from Theorem 5.3.1.

(c) The fact that the direction of each edge of P^Π is the difference of two unit vectors follows from Theorem 4.1.5 of Vol. I and the established supermodularity of θ_*^Π.

(d) If $v = (\theta_*^\Pi)_\sigma$ for some permutation σ, then trivially, v is a vertex of H^Π, hence a vertex of P^Π by (b). On the other hand, suppose v is a vertex. By (a), θ_*^Π is supermodular. Then $v = (\theta_*^\Pi)_\sigma$ by Theorem 4.1.5 of Vol. I.

(e) By Theorem 5.2.1, for each multishape $(n_{11},\ldots,n_{tp})$, $\theta_*^{(n_{11},\ldots,n_{tp})}$ is strictly supermodular. Let $(n_{11},\ldots,n_{tp})$ denote the multishape of π in (a). Then the first inequality in (5.3.1) is strict. Hence $\theta_*^{(n_{11},\ldots,n_{tp})}$ is strictly supermodular; (e) now follows from Theorem 4.3.3 of Vol. I. □

Note that for the single-multishape case, Corollary 5.2.8 gives a sufficient condition for unique representation of a vertex of the multipartition polytope. The following example shows that the same condition is not sufficient for the constrained-multishape case. Whether it is sufficient for the bounded-multishape case remains an open problem.

Example 5.3.2. Consider the case $p = u = 2$ and $\Theta = \begin{pmatrix}\theta^{11}\ \theta^{12}\ \theta^{13} & \\ \theta^{21}\ \theta^{22}\ \theta^{23}\ \theta^{24}\end{pmatrix} = \begin{pmatrix}1\ 7\ 8 & \\ 2\ 3\ 4\ 9\end{pmatrix}$. Clearly, θ^{ui}'s are all distinct. Let $\Gamma = \{(2,1,1,3),(1,2,3,1)\}$ and $\Pi = \Pi^\Gamma$. Then $v = (10,24)$ is a vertex of P^Γ and $\sigma = (j_1, j_2) = (1,2)$ is the unique permutation such that $(10,24) = (\theta_*^\Pi)_\sigma$.

But $v = (10, 24)$ has two monotone multipartitions $(\{1,7,2\},\{8,3,4,9\})$ and $(\{1,2,3,4\},\{7,8,9\})$ corresponding to the two different multishapes $(2,1,1,3)$ and $(1,2,3,1)$, respectively. □

In the single partition problems, unique representation of a vertex is assured in Theorem 5.2.4(f)(ii) of Vol. I by adding the extra conditions that Γ is complete and θ^i's are one-sided. Note that θ^i's are one-sided in the above example, and Γ is easily checked to be complete since $p = 2$ (see Vol. I Example 5.2.7). These two conditions are indeed satisfied in the above example, so there must be a more fundamental difference between the partition and the multipartition problems.

In the single partition problems, i.e., $t = 1$, the equality $(\theta_*^\Pi)(I_s) = \sum_{i=1}^{n_+^\Gamma(I_s)} \theta^i$ holds for every $I_s = \{j_1, j_2, \cdots, j_s\}$ which implies that all partitions π with $\theta_\pi = (\theta_*^\Pi)_\sigma$ have the same shape if θ^i's are one-sided. Then the equal shape conclusion, together with the condition θ^i's are all distinct, leads to the unique representation of a vertex. Nevertheless, in the multipartition problems, we cannot make the equal shape conclusion because the equality $(\theta_*^\Pi)(I_s) = \sum_{i=1}^{n_+^\Gamma(I_s)} \theta^i$ needs not hold. Actually, we have only $(\theta_*^\Pi)(I_s) = \min_{(n_{11},\cdots,n_{tp})\in\Gamma} \left\{ \sum_{u=1}^{t} \sum_{j=1}^{n_u(I_s)} \theta^{uj} \right\}$ and the minimum can be achieved by two distinct multishapes even when θ^{ui}'s are one-sided and distinct for each $u = 1, \cdots, t$.

In Section 5.2 of Vol. I we showed that shape-completeness is sufficient for completeness when the θ^i's are one-sided. We next develop a parallel condition for multipartitions. Recall from Section 5.2 of Vol. I that for a set Γ of p-shapes for partitions of $\{1, \ldots, n\}$, we defined for each subset I of $\{1, \ldots, p\}$

$$n_-^\Gamma(I) = \max_{(n_1,\ldots,n_p)\in\Gamma} \sum_{j\in I} n_j \tag{5.3.2}$$

and

$$n_+^\Gamma(I) = \min_{(n_1,\ldots,n_p)\in\Gamma} \sum_{j\in I} n_j, \tag{5.3.3}$$

and for each permutation $\sigma = (\{j_1\}, \{j_2\}, \cdots, \{j_p\})$ of $\{1, \cdots, p\}$,

$$(n_-^\Gamma)_\sigma = \left(n_-^\Gamma(\bigcup_{k=1}^{1} j_k), n_-^\Gamma(\bigcup_{k=1}^{2} j_k) - n_-^\Gamma(\bigcup_{k=1}^{1} j_k), \cdots, n_-^\Gamma(\bigcup_{k=1}^{p} j_k) - n_-^\Gamma(\bigcup_{k=1}^{p-1} j_k) \right) \tag{5.3.4}$$

and

$$(n_+^\Gamma)_\sigma = \left(n_+^\Gamma(\bigcup_{k=1}^{1} j_k), n_+^\Gamma(\bigcup_{k=1}^{2} j_k) - n_+^\Gamma(\bigcup_{k=1}^{1} j_k), \cdots, n_+^\Gamma(\bigcup_{k=1}^{p} j_k) - n_+^\Gamma(\bigcup_{k=1}^{p-1} j_k) \right). \tag{5.3.5}$$

We then defined Γ to be *shape-complete* if for any permutation σ of $\{1, \cdots, p\}$, $(n_-^\Gamma)_\sigma$ and $(n_+^\Gamma)_\sigma$ are vectors in Γ. It was noted that with $\Pi = \Pi^\Gamma$, if $\theta \leq 0$ then

$$\theta_*^\Pi(I) = \sum_{u=1}^{t} \sum_{i=1}^{n_-^\Gamma(I)} \theta^{ui}, \tag{5.3.6}$$

and if $\theta \geq 0$, then

$$\theta_*^\Pi(I) = \sum_{u=1}^{t} \sum_{i=1}^{n_+^\Gamma(I)} \theta^{ui}. \tag{5.3.7}$$

For a multipartition π, we let $\pi^1, \ldots, \pi^t$ be the partitions of $\mathcal{N}_1, \ldots, \mathcal{N}_t$, respectively. We say that a set Γ of nonnegative integer tp-vectors $(n_{11}, \ldots, n_{tp})$ satisfying $\sum_{j=1}^{p} n_{uj} = n_u$ for $u = 1, \ldots, t$ is in *product form* if there exist sets $\Gamma^1, \ldots, \Gamma^t$ of nonnegative integer p-vectors such that $\Gamma = \Gamma^1 \times \cdots \times \Gamma^t$. In such cases, let Π^{Γ^u}, for $u = 1, \ldots, t$, denote the set of p-partitions of $\mathcal{N}_u$ with shape in Γ^u; we then have that $\Pi^\Gamma = \Pi^{\Gamma^1} \times \cdots \times \Pi^{\Gamma^t}$, that is, a multipartition π is in Π^Γ if and only if $\pi^u \in \Pi^{\Gamma^u}$ for $u = 1, \ldots, t$. Further, for each subset I of $\{1, \ldots, p\}$,

$$\begin{aligned}
\theta_*^{\Pi^\Gamma}(I) &= \min\left\{ \sum_{j \in I} (\theta_\pi)_j : \pi \in \Pi \right\} = \min\left\{ \sum_{j \in I} \sum_{u=1}^{t} (\theta_{\pi^u})_j : \pi^u \in \Pi^{\Gamma^u} \right\} \\
&= \sum_{u=1}^{t} \min\left\{ \sum_{j \in I} (\theta_{\pi^u})_j : \pi^u \in \Pi^{\Gamma^u} \right\} = \sum_{u=1}^{t} \theta_*^{\Pi^{\Gamma^u}}(I),
\end{aligned} \tag{5.3.8}$$

that is, $\theta_*^{\Pi^\Gamma}$ is the sum of the $\theta_*^{\Pi^{\Gamma^u}}$'s; in particular, if all θ^i's are either nonpositive or nonnegative, (5.3.6) or (5.3.7) provides one with an efficient method for computing of $\theta_*^{\Pi^\Gamma}(I)$.

Lemma 5.3.4. *Let $\Gamma = \Gamma^1 \times \cdots \times \Gamma^t$ be a set of nonnegative integer tp-vectors $(n_{11}, \ldots, n_{tp})$ satisfying $\sum_{j=1}^{p} n_{uj} = n_u$ for $u = 1, \ldots, t$. Let σ be* *a permutation of* $\{1,\ldots,p\}$ *and let* π *be a multipartition in* Π^Γ. *Then* π *satisfies* $(\theta_*^\Pi)_\sigma = \theta_\pi$ *if and only if for each* $u = 1,\ldots,t$, $(\theta_*^{\Pi^{\Gamma^u}})_\sigma = \theta_{\pi^u}$.

Proof. Let $\Pi \equiv \Pi^\Gamma$ and for $u = 1,\ldots,t$, let $\Pi^u \equiv \Pi^{\Gamma^u}$. Also, let $\sigma = (\{j_1\},\ldots,\{j_p\})$. From (5.3.8), $(\theta_*^\Pi)_\sigma = \sum_{u=1}^t (\theta_*^{\Pi^u})_\sigma$. Also, from (5.1.3), for every multipartition π, $\theta_\pi = \sum_{u=1}^t \theta_{\pi^u}$. Now, for each $s = 1,\ldots,p$, $I_s \equiv \{j_1,\ldots,j_s\}$ satisfies $(\theta_*^\Pi)(I_s) \le \sum_{j\in I_s}(\theta_\pi)_j$ and $(\theta_*^{\Pi^u})(I_s) \le \sum_{j\in I_s}(\theta_{\pi^u})_j$. It follows that for each $s = 1,\ldots,p$, $\sum_{u=1}^t (\theta_*^{\Pi^u})(I_s) = \sum_{u=1}^t \sum_{j\in I_s}(\theta_{\pi^u})_j$ if and only if for all u, $(\theta_*^{\Pi^u})(I_s) = \sum_{j\in I_s}(\theta_\pi)_j$. We conclude that the following are equivalent:

(i) $(\theta_*^\Pi)_\sigma = \theta_\pi$,
(ii) $[(\sum_{u=1}^t \theta_*^{\Pi^u})_\sigma]_{j_r} = (\sum_{u=1}^t \theta_{\pi^u})_{j_r}$ for each $r = 1,\ldots,p$,
(iii) $[(\sum_{u=1}^t \theta_*^{\Pi^u})](I_s) = \sum_{u=1}^t \sum_{j\in I_s}(\theta_{\pi^u})_j$ for each $s = 1,\ldots,p$,
(iv) $\sum_{u=1}^t (\theta_*^{\Pi^u})(I_s) = \sum_{u=1}^t \sum_{j\in I_s}(\theta_{\pi^u})_j$ for each $s = 1,\ldots,p$,
(v) $(\theta_*^{\Pi^u})(I_s) = \sum_{j\in I_s}(\theta_{\pi^u})_j$ for each $s = 1,\ldots,p$ and for each $u = 1,\ldots,t$,
(vi) $[(\theta_*^{\Pi^u})_\sigma]_{j_r} = (\theta_{\pi^u})_{j_r}$ for each $r = 1,\ldots,p$ and for each $u = 1,\ldots,t$, and
(vii) $(\theta_*^{\Pi^u})_\sigma = \theta_{\pi^u}$, for each $u = 1,\ldots,t$,

where the equivalence of (v) and (vi) follows from the fact that for $r = 1,\ldots,p$, $\{j_r\} = I_r \setminus I_{r-1}$. The established equivalence of (i) and (vii) proves the lemma. □

Corollary 5.3.5. *Let* $\Gamma = \Gamma^1 \times \cdots \times \Gamma^t$ *be a set of nonnegative integer tp-vectors* $(n_{11},\ldots,n_{tp})$ *satisfying* $\sum_{j=1}^p n_{uj} = n_u$ *for* $u = 1,\ldots,t$. *Then,* Π^Γ *is complete if and only if for every* $u = 1,\ldots,t$, Π^{Γ^u} *is complete (where for each* $i \in \mathcal{N}^u$, θ^i *is the element associated with* i*).*

Proof. The result is immediate from Lemma 5.3.4 and the observation $\pi = (\pi^1,\ldots,\pi^t)$ is in Π^Γ if and only if for each $u = 1,\ldots,t$, $\pi^u \in \Pi^{\Gamma^u}$. □

We say that Γ is *multishape-complete*, if it is in product form and in its representation $\Gamma = \Gamma^1 \times \cdots \times \Gamma^t$, each Γ^u is shape-complete (as a set of shapes of partitions). In this case, multishape-completeness means that for every permutation σ of $\{1,\ldots,p\}$, the vectors $((n_-^{\Gamma^1})_\sigma,\ldots,(n_-^{\Gamma^p})_\sigma)$ and $((n_+^{\Gamma^1})_\sigma,\ldots,(n_+^{\Gamma^p})_\sigma)$ are in Γ.

We next show that when the θ^i's are one-sided, multishape-completeness is a sufficient condition for completeness. The result parallels Theorem 5.2.5

in Vol. I.

Theorem 5.3.6. *Let Γ be a set of nonnegative integer tp-vectors $(n_{11}, \ldots, n_{tp})$ satisfying $\sum_{j=1}^{p} n_{uj} = n_u$ for $u = 1, \ldots, t$ and let $\Pi \equiv \Pi^{\Gamma}$. Suppose further that Γ is multishape-complete and either $\theta \leq 0$ or $\theta \geq 0$. Then, Π is complete and the conclusions of Theorem 5.3.3 apply.*

Proof. By the multishape-completeness of Γ, it has a representation $\Gamma = \Gamma^1 \times \cdots \times \Gamma^t$ where each Γ^u is shape-complete. We conclude from Theorem 5.2.5 of Vol. I that for each $u = 1, \ldots, t$, $\Pi^u \equiv \Pi^{\Gamma_u}$ is complete. Corollary 5.3.5 now implies that $\Pi = \Pi^{\Gamma}$ is complete.

Next assume that σ is a permutation. We first consider the case where $\theta \leq 0$. For $u = 1, \ldots, t$, the proof of Theorem 5.2.5 in Vol. I assures that the consecutive partition $\pi^u(\sigma)$ determined by the shape $(n_-^{\Gamma^u})_\sigma$ and the permutation σ has $(\theta_*^{\Pi^u})_\sigma = \theta_{\pi^u(\sigma)}$. As the same order, namely σ, determines each of the consecutive partitions $\pi^u(\sigma)$ for $u \in \{1, \ldots, t\}$, we have that the partition $\pi(\sigma) \equiv (\pi^1(\sigma), \ldots, \pi^t(\sigma))$ is monotone; further, $\text{shape}[\pi(\sigma)] = [(n_-^{\Gamma^1})_\sigma, \ldots, (n_-^{\Gamma^t})_\sigma] \in \Gamma^1 \times \cdots \times \Gamma^t = \Gamma$, so $\pi(\sigma) \in \Pi$. Also, by Lemma 5.3.4, $(\theta_*^{\Pi})_\sigma = \theta_\pi$. The case where $\theta \geq 0$ follows from the same arguments with $n_+^{\Gamma^u}$ replacing $n_-^{\Gamma^u}$. □

Observe that Theorem 5.3.6 can be split into a result about the case $\theta \leq 0$ and another about the case $\theta \geq 0$, each requiring only part of the two conditions about multishape-completeness, see the remark following Theorem 5.2.5 in Vol. I.

Corollary 5.3.7. *Let $L = (L_{11}, \ldots, L_{tp})$ and $U = (U_{11}, \ldots, U_{tp})$ be nonnegative integer tp-vectors satisfying $\sum_{j=1}^{p} L_{uj} \leq n_u \leq \sum_{j=1}^{p} U_{uj}$ for $u = 1, \ldots, t$ and let $\Pi \equiv \Pi^{(L,U)}$ be the set of corresponding bounded-multishape multipartitions. If either $\theta \leq 0$ or $\theta \geq 0$, then Π is complete and the conclusions of Theorem 5.3.3 hold.*

Proof. The set $\Gamma^{(L,U)}$ of multishapes determined by lower bound vector L and upper bound vector U is in product form with $\Gamma^{(L,U)} = \Gamma^{(L^1,U^1)} \times \cdots \times \Gamma^{(L^t,U^t)}$, where for $u = 1, \ldots, t$, $\Gamma^{(L^u,U^u)}$ is the set of nonnegative vectors that are bounded below by $L^u = (L_{u1}, \ldots, L_{up})$ and bounded above by $U^u = (U_{u1}, \ldots, U_{up})$. It now follows from Theorem 5.2.6 of Vol. I that each $\Gamma^{(L^u,U^u)}$ is shape-complete. The completeness of Π now follows from Theorem 5.3.6, which assures that Π satisfies the conclusion of Theorem 5.3.3. □

Corollary 5.3.8. *Let Γ be the set of all nonnegative integer tp-vectors $(n_{11},\ldots,n_{tp})$ satisfying $\sum_{j=1}^{p} n_{uj} = n_u$ for $u = 1,\ldots,t$ and let $\Pi \equiv \Pi^\Gamma$. If either $\theta \leq 0$ or $\theta \geq 0$, then Π is complete and the conclusions of Theorem 5.3.3 hold.*

Proof. The result is immediate from Corollary 5.3.7 with $L = (L_{11},\ldots,L_{tp}) = (0,\ldots,0)$ and $U = (U_{11},\ldots,U_{tp}) = (n,\ldots,n)$. □

Corollary 5.3.9. *Let $(n_{11},\ldots,n_{tp})$ be a nonnegative integer tp-vector and let Γ be the set of nonnegative integer tp-vectors obtained by coordinate-permutation of each of the t subvectors $(n_{11},\ldots,n_{1p})$, $(n_{21},\ldots,n_{2p}),\ldots,(n_{t1},\ldots,n_{tp})$ of $(n_{11},\ldots,n_{tp})$, independently. Then Γ is multishape-complete; further, if either $\theta \leq 0$ or $\theta \geq 0$, then $\Pi \equiv \Pi^\Gamma$ is complete and the conclusions of Theorem 5.3.3 hold.*

Proof. The set Γ is in product form with $\Gamma^1 \times \cdots \times \Gamma^t$, where for $u = 1,\ldots,t$, Γ^u is the set of nonnegative vectors obtained by coordinate-permutation of $(n_{u1},\ldots,n_{up})$. By Theorem 5.2.12 of Vol. I, for each $u = 1,\ldots,t$, Γ^u is shape-complete. Thus, by definition, Γ is multishape-complete. It now follows from Theorem 5.3.6, that if either $\theta \leq 0$ or $\theta \geq 0$, then $\Pi = \Pi^\Gamma$ is complete, and therefore the conclusions of Theorem 5.3.3 apply. □

5.4 Combinatorial Properties of Multipartitions

We have discussed the monotonicity property in the previous sections. Hwang, Wang and Lee [62] introduced two other properties and proved results about them that we shall report in the current section. Before we name these properties, we call attention to the fact that nonempty parts can have empty subparts in a particular type. Let $\pi_{uj} \geq \pi_{uk}$ denote the situation where $\theta^i \geq \theta^{i'}$ for all $i \in \pi_{uj}$ and $i' \in \pi_{uk}$ (which allows π_{uj} and/or π_{uk} to be empty). We shall also use the relation $\leq$, $>$ and $<$ with the obvious interpretation.

Monotone multipartitions were defined in Section 5.1; we next add three additional properties of multipartitions, and also restate the definition of monotone for easy comparisons.

- Consecutive (C): A multipartition π is consecutive if every type u, either $\pi_{uj} \geq \pi_{uk}$ or vice versa, i.e., π_u is a consecutive partition.
- Monotone (M): Let $\pi_j \geq \pi_k$ denote the situation where $\pi_{uj} \geq \pi_{uk}$ for

all $u = 1, \ldots, t$. Then a multipartition is monotone (M) if there exists a permutation $(\{j_1\}, \ldots, \{j_p\})$ of $\{1, \ldots, p\}$ such that $\pi_{j_u} \geq \pi_{j_w}$ if $j_u > j_w$. Note that $\pi_i \geq \pi_j$ and $\pi_j \geq \pi_k$ do not imply $\pi_i \geq \pi_k$ since π_j may have an empty subpart $\pi_j(u)$ while $\pi_i(u) < \pi_k(u)$.

- Index-monotone (I): A monotone multipartition is index-monotone if $\pi_1 \leq \pi_2 \leq \cdots \leq \pi_p$.
- Monopolistic (M_p): A monotone partition is monopolistic if only one part can be nonempty (defined only for size or open partitions).

Clearly, $I \Rightarrow M \Rightarrow C$, and $M_p \Rightarrow M$.

The indices of parts may be assigned according to some parameter of the problem, e.g., the importance of a part, the cost of a part, or the size of a part. In these cases, we may also use the terms *importance-monotone*, *cost-monotone* or *size-monotone*. But for our problem, the indices are considered fixed.

Let $\#_Q(\{n_{ui}\})$, $\#_Q(p,t)$ and $\#_Q(t)$ denote the number of labeled shape-t-partitions, size-t-partitions and open-t-partitions, respectively, that satisfy Q. Again, equivalent partitions are counted as one (thus it is advantageous to use the index-partition in counting since it automatically discount equivalent partitions).

Theorem 5.4.1. $\#_I(\{n_{ui}\}) = 1$,
$\#_I(p,t) = p! \prod_{u=1}^{t} \sum_{j=1}^{p} \binom{n_u - 1}{j-1}$,
$\#_M \leqslant (p!)\#_I$, $\#_C \leqslant (p!)^t \#_I$ *for either* $\{n_{ui}\}$ *or* (p,t), *and*
$\#_{M_p(p,t)} = p$.

Proof. $\#_I(\{n_{ui}\}) = 1$ is obvious since there is only one way of assigning the smallest n_{u1} elements of type u to part 1, the next smallest n_{u2} elements of type u to part 2, and so on for $u = 1, \ldots, t$. The $\#_I(p,t)$ equation is obtained by multiplying the number of consecutive partitions over all $u = 1, \cdots, t$, where the number for each u is obtained from Theorem 6.2.1(i) of Vol. I.

$\#_M \leqslant (p!)\#_I$ since any permutation of the p parts can serve as an index. $\#_C \leqslant (p!)^t \#_I$ since any permutation of the p parts in a type in an I-partition results in a C-partition. The inequalities are due to the fact that different permutations can contain the same partition when empty $\pi_i(u)$ exists. For example, the partition $\pi_1(1) = 1$, $\pi_2(1) = \pi_1(2) = \emptyset$, $\pi_2(2) = 2$ is counted in both permutations $(1,2)$ and $(2,1)$.

The number of monopolistic partitions is simply the number of ways of selecting a nonempty part. $\square$

The k-consistency and (ℓ, k, t)-sortability are defined in Vol. I, p. 261 and p. 265, respectively. We first study the consistency issue.

Theorem 5.4.2. *The minimum consistency index is* 2 *for consecutiveness, monopolicity and index-monotonicity, but* ∞ *for monotonicity.*

Proof. Since consecutiveness, monopolicity and index-monotonicity are defined through 2-subpartitions, they are 2-consistent by the definition of k-consistency. It is also easy to verify that they are not 1-consistent. Hence, their minimum consistency index is 2.

We next show that monotonicity is not k-consistent for all k. Consider $p = k+1$, and $\mathcal{N}_u = \{\theta^{u1}, \theta^{u2}\}$ for $1 \leqslant u \leqslant t$. Then $\pi_u = \{\theta^{u1}, \theta^{(u+1)2}\}$, $u = 1, \ldots, p-1$, and $\pi_p = \{\theta^{p1}, \theta^{12}\}$ do not satisfy monotonicity, but every k subset does. For example, for $\pi_1, \ldots, \pi_{p-1}$, the linear order is $\pi_1 > \pi_2 > \cdots > \pi_{p-1}$. □

We now study the sortability issue. First observe that Table 6.4.1 of Vol. I on sortability implications for partitions remains valid for multipartitions. We reproduce Table 6.4.1 here for easy references.

(st,k,op) $\Rightarrow^*$	(st,k,supp) $\Rightarrow^*$	(st,k,shape) $\Rightarrow$	(s-s,k,shape) $\Rightarrow$	(s-s,k,supp) $\Rightarrow$	(s-s,k,op)
$\Downarrow$	$\Downarrow$	$\Downarrow$	$\Downarrow$	$\Downarrow$	$\Downarrow$
(p-s,k,op) $\Rightarrow^*$	(p-s,k,supp) $\Rightarrow^*$	(p-s,k,shape) $\Rightarrow$	(w,k,shape) $\Rightarrow$	(w,k,supp) $\Rightarrow$	(w,k,op)

We also record a result which is a straightforward partition-to-multipartition extension of Lemma 6.4.12 in Vol. I. Note that the argument used in the proof of Lemma 6.4.12 in Vol. I does not depend on the assumption that elements are distinct.

Lemma 5.4.3. *Let Q and Q' be k-consistent, hereditary properties of multipartitions with $Q' \Rightarrow Q$. If Q' is (sort-specific, k, t)-sortable, then the same holds for Q.* □

We first give a result specific to certain properties.

Lemma 5.4.4. *If $Q \in \{C, I\}$ is not (ℓ, k, t)-sortable for $\ell \in$ {strongly, part-specific}, then Q is not so for $k' > k$.*

Proof. It suffices to prove for $k' = k+1$. First consider ℓ= "strongly". Let $\Pi = \{\pi^1, \ldots, \pi^m\}$ be a family of partitions of $\mathcal{N}$ not satisfying Q but being Q-(weakly, k, t)-invariant, i.e., for every $\pi^i \in F$ not satisfying Q, there exists a set of k parts not satisfying Q and a k-Q-sorting which turns π^i

into $\pi^j \in Q$. Let $\mathcal{N}^*$ be obtained from $\mathcal{N}$ by adding $|\mathcal{N}|$ new θ^{uj}'s for each u, all greater than θ^{un_u}, and let π^{i*} be obtained from π^i by adding a new part P consisting of these new θ^{uj}'s. We consider only sorting in which P remains invariant. Let K denote a k-part not satisfying Q in π^i. Then $K \cup P$ is a $(k+1)$-part not satisfying Q in $(\pi^i)'$, and a k-Q-sorting of π^i into π^j corresponds to a $(k+1)$-Q-sorting of π^{i*} to π^{j*}. Hence $\Pi' = \{\pi^{i*} : \pi \in F\}$ is Q-(weakly, k', t)-invariant, but not satisfying Q. Therefore, Q is not (strongly, k', t)-sortable. The proof for ℓ="part-specific" is similar. □

We now prove some results on sortability.

Theorem 5.4.5. *I is (strongly, k, shape)-sortable for all $k \geq 2$.*

Proof. Define $s(\pi) = \sum_{j=1}^{p} j \sum_{u=1}^{t} \sum_{i \in \pi_{uj}} \theta^i$. Suppose that π contains a set K of k parts not satisfying I. I-sort K to obtain π'. Since π' assigns the smaller θ^i to the part with the smaller index j, it achieves the maximum of $s(\pi)$ of K and is strictly greater than $s(\pi)$ on K since the latter has at least one inversion. Finally, since parts not in K remain unchanged, $s(\pi') < s(\pi)$. □

Corollary 5.4.6. *C is (sort-specific, k, t)-sortable for all $k \geq 2$.*

Proof. As I is (strongly, k, shape)-sortable (Theorem 5.4.5), Table 6.4.1 implies that I is (sort-specific, k, t)-sortable, which by Lemma 5.4.3 (noting $I \Rightarrow C$) yields the conclusions of the corollary. □

Note that we cannot use Corollary 5.4.6 to obtain a similar corollary for M since M is not k-consistent.

Theorem 5.4.7. *M_p is (strongly, k, t)-sortable for all $k \geq 2$ and $t \in$ {support, open}.*

Proof. Suppose K is a set of k parts not satisfying M_p. Then K has at least two nonempty parts. M_p-sort K into a k-partition K'. Noting that the number of nonempty parts is strictly decreasing in such sorting, the claimed sortability is proved for size. The same argument also works for open partitions. □

Next we show that I is not (strongly, k, support)-sortable for all $k \geq 2$. By Lemma 5.4.4, it suffices to prove for $k = 2$. Let $\Pi = \{\pi^1, \pi^2, \pi^3, \pi^4\}$, where $p = 3$, $t = 4$, $|N_u| = 5$ for $u = 1, 2, 3, 4$ and the π^j's are given by:

π^1			π^2			π^3			π^4		
13	4	25	123	4	5	12	34	5	1	34	25
1	24	35	13	24	5	123	4	5	123	4	5
12	34	5	1	34	25	13	4	25	123	4	5
123	4	5	12	4	35	1	24	35	13	24	5

Note that Π is I-(weakly, 2, support)-invariant: 2-I-sorting parts 1 and 3, can be used to convert π^1 into π^2 and π^3 into π^4, also 2-I-sorting parts 1 and 2, can be used to convert π^2 into π^3 and π^4 into π^1. But, Π does not satisfy Q. So, the extension of Theorem 6.4.8 in Vol. I to multipartitions assures that I is not (strongly, 2, support)-sortable.

We next show that I is not (part-specific, k, support)-sortable for $k \geq 3$. Again, it suffices to prove for $k = 3$. Let $\Pi = \{\pi^1, \pi^2, \pi^3, \pi^4\}$, where $p = 4$, $t = 3$ and $|N_u| = 6$ for $u = 1, 2, 3$ and the π^j's are given by

π^1				π^2				π^3			
1	23		456	1	23		456	145	23		6
123	45		6	123	45		6	1	45		236
1		23	456	145		23	6	1		23	456
123		45	6	1		45	236	123		45	6
145			236	123			456	123			456

Let Π^1 denote a family of 120 partitions by permuting the five types of π^1, and let Π^2 and Π^3 be obtained from π^2 and π^3 similarly. Define $\Pi = \Pi^1 \cup \Pi^2 \cup \Pi^3$. Note that no partition in Π satisfies I. For any partition π in Π^2, and any three parts of π not satisfying I, there exists a 3-I-sorting which turns π into π' where π' is in one of the other two Π^j families. The labels of the links in Figure 5.4.1 show the set of parts involved in the 3-I-sorting. It follows that Π is (sort-specific, 3, support)-invariant and therefore the extension of Theorem 6.4.8 in Vol. I to multipartitions assures that I is not (part-specific, 3, support)-sortable.

We next show that for $\ell \in$ {strongly, part-specific}, C is not (ℓ, k, t)-sortable for all $k \geq 2$. By Table 6.5.1 and Lemma 5.4.4, it suffices to establish the case where ℓ="part-specific", $k = 2$ and t="shape". Let $\Pi = \{\pi^1, \pi^2, \pi^3, \pi^4\}$, where $p = 3$, $t = 4$, $|N_u| = 4$ for $u = 1, 2, 3, 4$ and the π^j's are given by:

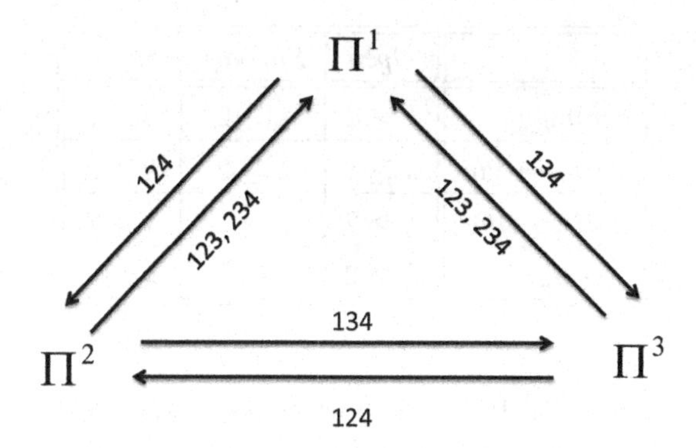

Figure 5.4.1

π^1			π^2			π^3			π^4		
24	3	1	23	4	1	12	4	3	12	3	4
12	4	3	24	1	3	23	1	4	23	4	1
3	4	12	4	3	12	1	3	24	1	4	23
4	1	23	1	4	23	3	4	12	3	1	24

Note that Π is C-(sort-specific, 2, shape)-invariant: 2-C-sorting parts 1 and 2 will convert π^1 into π^2, 2-C-sorting parts 1 and 3 will convert π^2 into π^3 and π^4 into π^1 and 2-C-sorting parts 2 and 3, will convert π^3 into π^4. Note that in each case, the pair of parts we sort is the only pair not satisfying C. Using the argument used earlier, we conclude that C is not (part-specific, k, shape)-sortable.

We next use 12-cell tables, in the format introduced in Section 6.5 of Vol. I, to summarize our finding of sortability of I, C and M_p with NA standing for not-applicable.

Table 5.4.1. Sortability of I

	Open	*Support*	*Shape*
strongly	NA	$\bar{k} \geq 2$	$k \geq 2$
part-specific	NA	$\bar{k} \geq 3$	$k \geq 2$
sort-specific	NA	$k \geq 2$	$k \geq 2$
weakly	NA	$k \geq 2$	$k \geq 2$

Table 5.4.2. Sortability of C

	Open	*Support*	*Shape*
strongly	$\bar{k} \geq 2$	$\bar{k} \geq 2$	$\bar{k} \geq 2$
part-specific	$\bar{k} \geq 2$	$\bar{k} \geq 2$	$\bar{k} \geq 2$
sort-specific	$k \geq 2$	$k \geq 2$	$k \geq 2$
weakly	$k \geq 2$	$k \geq 2$	$k \geq 2$

Table 5.4.3. Sortability of M_p

	Open	*Support*	*Shape*
strongly	$k \geq 2$	$k \geq 2$	NA
part-specific	$k \geq 2$	$k \geq 2$	NA
sort-specific	$k \geq 2$	$k \geq 2$	NA
weakly	$k \geq 2$	$k \geq 2$	NA

The lack of sortability results for M is perhaps due to its lack of k-consistency for any k. This situation can be improved by restricting our attention to a useful special case.

Theorem 5.4.8. *The minimum consistency index is 2 for M if the prescribed multishape is universal.*

Proof. Suppose every pair of types of π is monotone. Let type k be a universal type. Then every pair of parts is ordered in type k. Further, the parts in type u are also ordered according to the linear order established in type k, or the pair of types (u, k) would not be monotone. Since u is arbitrary, π is monotone. □

Theorem 5.4.9. *I is (part-specific, k, support)-sortable for all $k \geq 2$ if the prescribed multishape is universal.*

Proof. It is easily verified that if there exist k parts in π not satisfying I, then there exist k consecutive parts K not satisfying I. I-sort K to obtain π'. Define $s(\pi)$ to be the number of inversions in π, i.e., an inversion occurs if $x \in \pi_i$, $y \in \pi_j$ and $(x - y)(i - j) < 0$. Then the number of inversions is intact in $\pi \setminus K$ but decreases in K. Hence $s(\pi') < s(\pi)$. □

Corollary 5.4.10. *M is (sort-specific, k, support)-sortable for all $k \geq 2$ if the prescribed multishape is universal.*

Proof. From Theorem 5.4.9, I is (sort-specific, k, support)-sortable for all $k \geq 2$ if the prescribed multishape is universal. The proof then follows immediately from Theorem 5.4.8, Lemma 5.4.3 and the fact that $I \Rightarrow M$ is not affected by the restriction on multishape.. □

5.5 Constrained-Multishape Multipartition Problems with Asymmetric Schur Convex Objective: Optimization over Multipartition Polytopes

In this section we use results of the previous sections and the results of Chapter 3 of Vol. I about optimality of vertices to establish optimality of monotone multipartitions. The results extend those of Section 5.4 of Vol. I where partitions, rather than multipartitions, were considered. The main idea is to identify the class of objective functions which achieve optimality over the corresponding multipartition polytope, establish optimality at vertices of this polytope, and then argue that the vertices are associated with monotone partitions. This approach facilitates the solution of corresponding multipartition problems by enumerating either the vertices or the monotone multipartitions or by other methods over the polytope. Our development essentially parallels Section 5.4 of Vol. I.

Throughout this section we assume that positive integers n, p, t, $n_1, \ldots, n_t$ are given with $\sum_{u=1}^{t} n_u = n$ and the θ^{ui}'s are enumerated by lists $\theta^{u1}, \ldots, \theta^{u|\mathcal{N}_u|}$ for $u = 1, \ldots, t$, so that (5.1.2) is satisfied, that is, $\theta^{u1} \leq \cdots \leq \theta^{u|\mathcal{N}_u|}$ for $u = 1, \ldots, t$. Also, for a p-multipartition π, θ_π is given by (5.1.3). We consider a sum-multipartition problem where the value of the objective function F over p-multipartitions is given for p-multipartition π by

$$F(\pi) = f(\theta_\pi), \tag{5.5.1}$$

with f being a real-valued function on a subset of $\mathbb{R}^p$ which contains all the θ_π's.

Theorem 5.5.1. *Let $n_{11}, \ldots, n_{tp}$ be nonnegative integers with $n_u = \sum_{j=1}^{p} n_{uj}$ for $u = 1, \ldots, t$, and let $f : P^{(n_{11},\ldots,n_{tp})} \to \mathbb{R}$ be asymmetric Schur convex on $P^{(n_{11},\ldots,n_{tp})}$. Then f attains a maximum over $P^{(n_{11},\ldots,n_{tp})}$ at a vertex; further, if f is strictly asymmetric Schur convex on $P^{(n_{11},\ldots,n_{tp})}$, then every maximizer of f over $P^{(n_{11},\ldots,n_{tp})}$ is a vertex.*

Proof. By Theorem 5.2.7, the directions of the edges of $P^{(n_{11},\ldots,n_{tp})}$ are in $\{e^{ij} : (i,j) \in \Gamma\}$ and thus Theorem 3.4.5 of Vol. I implies that there exists a vertex of $P^{(n_{11},\ldots,n_{tp})}$ that maximizes f over this polytope. The conclusion that every maximizer of f over $P^{(n_{11},\ldots,n_{tp})}$ is a vertex when f is strictly asymmetric Schur convex follows from the second part of Theorem 3.4.5 in Vol. I. □

Corollary 5.5.2. *Let $n_{11},\ldots,n_{tp}$ be nonnegative integers, let $n_u = \sum_{j=1}^{p} n_{uj}$ for $u = 1,\ldots,t$ and let $f : P^{(n_{11},\ldots,n_{tp})} \to \mathbb{R}$ be asymmetric Schur convex on $P^{(n_{11},\ldots,n_{tp})}$. Then there exists an optimal multipartition over $\Pi^{(n_{11},\ldots,n_{tp})}$ which is monotone. Further, if f is strictly asymmetric Schur convex and $\theta^u i$'s are distinct for each $u = 1,\cdots,t$, then every optimal multipartition is monotone.*

Proof. The proof follows immediately from Theorems 5.5.1, 5.2.7 and Corollary 5.2.8. □

Solution of Single-Multishape Sum Multipartition Problems with f Asymmetric Schur Convex by Enumerating Monotone Multipartitions

Corollary 5.5.2 assures that when the objective function F over multipartitions has the form (5.5.1) with f asymmetric Schur convex, the single-multishape multipartition problem has a monotone optimal multipartition; consequently the problem can be solved by evaluating F for monotone multipartitions. Theorem 5.2.2 shows that a complete list of the monotone multipartitions with prescribed multishape can be determined from the permutations over $\{1,\ldots,p\}$. For each of the $p!$ permutations, the vector θ_π associated with the corresponding monotone multipartition can be computed with n additions and $F(\pi)$ can then be determined with a single evaluation of the function f. Thus, we get an algorithm for solving the single multishape multipartition problem with $np!$ additions and $p!$ evaluation of f and $p!$ comparisons (required to identify the highest $f(\theta_\pi)$ with π monotone).

The established optimality of monotone multipartitions in Corollary 5.5.2 depends on (5.1.2) holding. If a ranking of the θ^{ui}'s is not available, one can sort and renumber them with $O(n \lg n)$ comparisons (e.g., Knuth [63]). □

We next turn our attention to constrained-multishape multipartition problems. Using an argument similar to the one used in proving Lemma 1.2.1 of Vol. I, we extend results proved for single-multishapes to constrained-multishapes.

Theorem 5.5.3. *Let Π be a constrained-multishape set of p-multipartitions and let $f : P^{\Pi} \to \mathbb{R}$. If f is asymmetric Schur convex on P^{Π}, then a monotone optimal multipartition over P^{Π} exists.*

Proof. By the finiteness of the set of multipartitions, there is an optimal multipartition, say π. Let $(n_{11}, \ldots, n_{tp})$ be the multishape of π. Applying Corollary 5.5.2 with the nonnegative integers $n_{11}, \ldots, n_{tp}$, we conclude the existence of a monotone multipartition π' with multishape $(n_{11}, \ldots, n_{tp})$ and $F(\pi') \geq F(\pi)$. As Π is multishape-constrained and $\pi \in \Pi$, we have that $\pi' \in \Pi$ and the optimality of π implies that $F(\pi') = F(\pi)$. Thus, π' is a monotone optimal multipartition. □

We note from Theorem 5.2.7 that the vertices of constrained-multishape multipartition polytopes are associated with monotone multipartitions. With f asymmetric Schur convex, an example provided in Section 5.4 of Vol. I demonstrates that the f-optimality of vertices over single-shape partition polytopes cannot be extended to constrained-shape partition polytopes which may have edges with directions outside of $\{e^i - e^j : i, j = 1, \ldots, n\}$. Our proof of Theorem 5.5.3 did not rely on vertex optimality (cf., Theorem 3.4.5 of Vol. I) through an extension of the proof of Theorem 5.5.1, but relied on the special case of single-multishape multipartition problems examined in Corollary 5.5.2.

Solution of Constrained-Multishape Sum Multipartition Problems with f Asymmetric Schur Convex by Enumerating Monotone Multipartitions

Let Γ be a set of potential multishapes. Theorem 5.5.3 assures that when the objective function F over multipartitions has the form (5.5.1) with f asymmetric Schur convex, the constrained-multishape multipartition problem has a monotone optimal multipartition; consequently, the problem can be solved by evaluating F for all monotone multipartitions. Theorem 5.2.2 assures that the monotone multipartitions with prescribed multishape are determined from permutations over $\{1, \ldots, p\}$. As for single-multishape multipartition problems, $F(\pi)$ for a monotone multipartition π with a specified multishape in Γ that corresponds to any specified permutation over

$\{1,\ldots,p\}$ can be computed with n additions and one evaluation of f. Thus, we get an algorithm for solving the constrained-multishape multipartition problem with $n|\Gamma|p!$ additions and $|\Gamma|p!$ evaluations of f.

Similar to the single-multishape multipartition problem, if (5.1.2) is not available, sorting of the θ^{ui}'s can be executed with $O(n \lg n)$ comparisons. □

The above method for solving constrained-multishape multipartition problems is based on the restriction of the search of multipartitions to those that are monotone; when f is asymmetric Schur convex, Theorem 5.5.3 assures that the best multipartition found in the restrictive search will produce an optimal multipartition. An opposite approach is to expand the search rather than to restrict it, and maximize f over the whole multipartition polytope rather than over the relevant θ_π's. Such optimization will be geared towards the identification of a vertex which is optimal; once such a vertex is found, a (monotone) multipartition that corresponds to that vertex will be sought. When vertices are known to be optimal, as is the case for single-multishape multipartition problems (by Theorem 5.5.1), this two-stage approach will yield an optimal multipartition. It should be noted that as a multipartition polytope is defined as the convex hull of vectors associated with multipartitions, every vertex of the polytope is associated with multipartition, that is, has a representation as θ_π with π as a multipartition (but not necessarily monotone).

The above two-stage method depends on the capability of identifying a vertex of the multipartition polytope which is optimal (in the first stage). The method seems particularly useful when an explicit representation of the multipartition polytope is available as a system of linear inequalities; for example, this is the case for single-multishape multipartition polytopes as was demonstrated in Section 5.2.

Solution of Single-Multishape Sum Multipartition Problems with f Asymmetric Schur Convex by Optimization Over the Multipartition Polytope

Let $n_{11},\ldots,n_{tp}$ be nonnegative integers satisfying $\sum_{j=1}^{p} n_{uj} = n_u$ for $u = 1,\ldots,t$, and assume that f is asymmetric Schur convex. We will assume that (5.1.2) is in force, for otherwise, the θ^{ui}'s can be sorted with $O(n \lg n)$ comparisons. Theorem 5.2.4 combines with (5.2.2)–(5.2.1) to provide an explicit representation of the single-multishape multipartition polytope $P^{(n_{11},\ldots,n_{tp})}$ as the solution of the system of linear inequalities given by

$$\sum_{j\in I} x_j \geq \theta_*^{(n_{11},\ldots,n_{tp})}(I) = \sum_{u=1}^{t}\sum_{j=1}^{n_u(I)} \theta^{uj} \text{ for } \emptyset \neq I \subseteq \{1,\ldots,p\}, \quad (5.5.2)$$

and

$$\sum_{j=1}^{p} x_j = \sum_{i=1}^{n} \theta^i. \quad (5.5.3)$$

Further, we have seen (in the paragraph following (5.2.2)) that the $\theta_*^{(n_{11},\ldots,n_{tp})}(I)$'s are computable with $O(n+t2^p)$ additions. As f is asymmetric Schur convex, a vertex of $P^{(n_{11},\ldots,n_{tp})}$ that maximizes f over $P^{(n_{11},\ldots,n_{tp})}$ exists.

The next step is to identify a monotone multipartition π with θ_π as the optimal vertex, say vertex v. We next show that such recovery is available from part (a) of Theorem 5.2.7. Specifically, determine the set of subsets I with $v \in F_I$ and find a chain of length $p-1$ of such sets; Theorem 5.2.7 assures that such a chain exists and if $I_1,\ldots,I_{p-1}$ is such a chain then it defines a unique monotone multipartition π with $\theta_\pi = \cap_{t=1}^{p-1} F_{I_t} = v$. Now, $v \in F_I$ means that $v_I \equiv \sum_{j\in I} v_j$ equals $\theta_*^{(n_{11},\ldots,n_{tp})}(I)$. The partial sums $\{v_I : I \subseteq \{1,\ldots,p\}\}$ can be computed recursively by $v_I = v_{I\setminus\{j\}} + v_j$ for any selected j in I; the total effort is then $O(t2^p)$ arithmetic operations. To identify the desired chain, consider the directed graph with the subsets of $\{1,\ldots,p\}$ satisfying $v_I = \theta_*^{(n_{11},\ldots,n_{tp})}(I)$ as its vertices and with an edge from vertex I to vertex J if J is the result of augmenting I with a single element j. The search of a chain of length $p-1$ is a search of a path of length $p-1$ in this graph; the latter can be easily conducted with effort $O(t2^p)$, facilitating the recovery of a multipartition π with $\theta_\pi = v$ with effort $O(t2^p)$.

The total effort for solving linear single-multishape multipartition problems with the above method is $O(n+t2^p)$ and the maximization of a mathematical problem with p variables, 2^p constraints and asymmetric Schur convex objective function. Of course, the key issue is the availability of an (efficient) algorithm for finding a vertex of $P^{(n_{11},\ldots,n_{tp})}$ which maximizes f over $P^{(n_{11},\ldots,n_{tp})}$; this may be the case when f has special structure. The most natural case where such an algorithm exists is, of course, the case where f is linear; but, a much more efficient method for this special case is developed in the next section. We recall that the enumeration of monotone multipartitions required effort of $O(np!)$. □

The above solution method cannot be extended from single multishape to constrained-multishape multipartition problems. Let Π be a constrained-multishape set of multipartitions. Example 5.4.1 in Vol. I demonstrates that when f is asymmetric Schur convex, optimality over $P^{(\Pi)}$ needs not be realized at a vertex or at a point that is associated with a multipartition. Still, when f is quasi-convex, an optimal vertex does exist (see Theorem 3.3.4 of Vol. I); by Theorem 5.2.7, such a vertex is associated with a monotone multipartition. Thus, the above solution method extends to constrained-multishape partition polytopes with f quasi-convex. Still, the effectivity of the extended use of the approach depends on the availability of convenient representation of the corresponding constrained-multishape multipartition polytopes and availability of efficient methods to optimize f over P^{Π}; see Section 5.3 for instances where convenient representations of P^{Π} are available.

5.6 Sum Multipartition Problems: Explicit Solutions

In this section we study multipartition problems with objective functions belonging to classes other than asymmetric Schur convex to explore the possibility of obtaining optimal solutions in a more efficient or explicit way. Chapter 2 of Vol. I made such an attempt successfully and will serve as a guideline for this section. However, our efforts in transplanting the results from there to here are not totally successful as we will show that the multipartition structure imposes a severe limit on what we can do.

Define $\theta_{\pi_{uj}} = \sum_{\theta^{ui} \in \pi_{uj}} \theta^{ui}$.

Theorem 5.6.1. *Consider the single-multishape multipartition problems. Suppose*

$$f(\theta_\pi) = \sum_{j=1}^{p} \left[c_j \left(\sum_{u=1}^{t} \theta_{\pi_{uj}} \right) \right], \tag{5.6.1}$$

where c_j's are the coefficients in the linear function f. Relabel c_j's such that

$$c_1 \leq c_2 \leq \cdots \leq c_p.$$

Then there exists an index-monotone optimal multipartition (with the index-set of c_j's).

Proof. We can write

$$f(\theta_\pi) = \sum_{u=1}^{t}\sum_{j=1}^{p} c_j \theta_{\pi_{uj}}.$$

By Theorem 2.1.1 of Vol. I, each summand (under the sum on u) is maximized by an index-consecutive partition with the index set of c_j's. Theorem 5.6.1 follows immediately. □

This single-multishape result can be extended to bounded-multishape as was done in Theorem 2.1.3 of Vol. I.

Theorem 5.6.2. *Consider the bounded-multishape multipartition problems with linear objective functions as given in (5.6.1). Then there exists an index-monotone optimal multipartition.*

Proof. A bounded-multishape multipartition problem is separable as it is determined by a set Γ of multishapes having the representation $\Gamma^1 \times \cdots \times \Gamma^t$ and each Γ^u is a set of (partition) shapes that are determined by lower and upper bounds. It follows that the bounded-multishape sum-multipartition problem with f linear reduces to t bounded-shape sum-partition problems where $f(\theta_{\pi^u})$ is to be maximized over $\pi \in \Pi^{\Gamma^u}$. Efficient explicit solution methods for the latter class of problems was presented in Section 2.1 of Vol. I; the methods that were developed in Vol. I are particularly simple for single-shape problems and for problems where the θ^i's are one-sided (that is, either nonnegative or nonpositive).

Since for each of the t subproblems, there exists a consecutive optimal partition with the same order $1, 2, \cdots, p$, there exists an index-monotone optimal multipartition. □

We also observe that if the c_i's have to be ordered, their sorting needs to be executed only once (to be used for the t partitioning problems that are to be solved).

The above results can also be extended to constrained-multishape families of the product form type.

Theorem 5.6.3. *Consider the product form constrained-multishape multipartition problems with linear objective functions as given in (5.6.1). Then there exists an index-monotone optimal multipartition.*

Proof. Similar to the proof of Theorem 5.6.2; the proof is immediate by the separability of the t subproblems. □

Our next exploration is to see whether the above results for linear objective functions can be extend to Schur convex or Schur concave functions as done in Chapter 2 of Vol. I. The biggest obstacle under the new functions is that the t subpartitions are no longer separable; the contribution of a term $\theta_{\pi_{uj}}$ is no longer independent of the contribution of $\theta_{\pi_{u'j}}$ for $u \neq u'$, hence ruining the separability. Another obstacle is that while in Chapter 2 of Vol. I, the theory of Schur convexity and Schur concavity depend crucially on the notion of "majorization" among vectors, and the majorizations among shape-vectors correspond to majorizations among θ-vectors; but the last statement is no longer true for multipartitions (see the following example).

Example 5.6.1. Let $t = 2, p = 3, c_1 = 1, c_2 = 2, (n_{11}, n_{12}, n_{21}, n_{22}) = (1, 3, 2, 1)$, and

$$\Theta = \begin{pmatrix} \theta^{11} & \theta^{12} & \theta^{13} & \theta^{14} \\ \theta^{21} & \theta^{22} & \theta^{23} & \end{pmatrix} = \begin{pmatrix} 1 & 1 & 1 & 2 \\ 1 & 2 & 2 & \end{pmatrix}.$$

Then the shape vector is $(3, 4)$. If the larger part takes first, taking the three largest θ^i's from type 1 and the largest θ^i from type 2, the objective value is $(2+1+1)1 + (2)2 = 8$; while if we let part 1 gets the first take, then the objective value is $(2)1 + (2+2)2 = 10 > 8$, reversing the order of shape-vectors. □

As in Chapter 2 of Vol. I, we consider the special case that all $\theta^{ui} = 1$ for $u = 1, \cdots, t$ and $j = 1, \cdots, p$. Note that the θ-vector majorization is identical to shape-vector majorization. Then we can again use the analyses and algorithms in Sections 2.2 and 2.3 of Vol. I to obtain similar results as reported in these sections.

Chapter 6

Applications

In this chapter, we demonstrate applications of results obtained in the preceding chapters to various types of problems. The models we consider were introduced in Section 1.4 of Vol. I; in fact, the sections of the current chapter correspond to the subsections of Section 1.4 in Vol. I. Our approach is that for each topic we present a core problem and perhaps some other related problems for demonstrations. It is not our intension to do a complete and exhaustive coverage of the particular topic.

The examples of partition problems described in Vol. I Subsections 1.4.12–1.4.14—vehicle routing, graph partitions and the traveling salesman problem – are known to be NP-hard problems; we will explain why the methods we developed in previous sections do not offer efficient solutions for these problems.

6.1 Assembly of Systems

In this section we introduce a multi-type assembly problem, studied by Hwang and Rothblum [57,59], which generalizes models studied by Derman, Lieberman and Ross [25], El-Neweihi, Proschan and Sethuraman [33, 34], Du [29], Du and Hwang [30], Malon [67], Hwang, Sun and Yao [61] among others.

Consider a system having p modules as components and label these modules by $1, \ldots, p$. Each module can be either *operative* or *inoperative.* The *state* of the system is determined by the set of operative modules and is represented by a vector $s \in \{0, 1\}^p$, where $s_i = 0$ if module i is inoperative and $s_i = 1$ if module i is operative. The operativeness of the system is then determined by a *structure function* $\boldsymbol{J} : \{0, 1\}^p \to \{0, 1\}$, i.e., the system is *inoperative* if it is in a state s with $\boldsymbol{J}(s) = 0$ and the system is *operative* if it is in a state s with $\boldsymbol{J}(s) = 1$.

The structure function $\boldsymbol{J}$ is called *monotone* if

$$\boldsymbol{J}(s) \leq \boldsymbol{J}(s') \quad \text{for } s,\ s' \in \{0,1\}^p \text{ with } s \leq s'. \tag{6.1.1}$$

The monotonicity of the function $\boldsymbol{J}$, expressed in (6.1.1) asserts that an operative system does not become inoperative when the set of operative modules is increased. We call the system *coherent* if the structure function is monotone, and henceforth we assume that this is the case.

Parts of t different types which we denote $1, \ldots, t$ are used to construct the modules, with parts of the same type being functionally interchangeable. Specifications for the modules are given, and we assume that module $j \in \{1, \ldots, p\}$ requires exactly n_{uj} parts of type u for *each* $u \in \{1, \ldots, t\}$. By possibly ignoring modules that are empty and need no parts, we assume that $\sum_{u=1}^{t} n_{uj} > 0$ for each $j = 1, \ldots, p$. The modules are constructed in series, i.e., a module is operative if and only if each of its parts is operative.

For each $u \in \{1, \ldots, t\}$, all of the needed $n_u \equiv \sum_j n_{uj}$ parts of type u are assumed to be available. An *assembly* for the system is an assignment of parts to the modules in a way that matches the requirements of each module. Such an assignment is represented by a family of sets $\pi = \{\pi_{uj} : u = 1, \ldots, t \text{ and } j = 1, \ldots, p\}$ such that $\{\pi_{uj} : j = 1, \ldots, p\}$ is a partition of $\{1, \ldots, n_u\}$ for each $u = 1, \ldots, t$, and the number of elements in π_{uj} for each $u = 1, \ldots, t$ and $j = 1, \ldots, p$. Of course, an assembly is a multipartition as defined in Vol. I Chapter 1 and studied in Vol. II Chapter 5.

The *reliability* of a part, a module, and the system as a whole is the probability of being operative. We assume that the reliability of the parts are given and that the parts of each type are enumerated in a (weakly) decreasing order of their reliability. So, the reliability of the k-th part of type u is r_{uk} and

$$1 \geq r^{u1} \geq r^{u2} \geq \cdots \geq r^{un_u} > 0 \quad \text{for } u = 1, \ldots, t\,. \tag{6.1.2}$$

Operability of the parts is assumed to be stochastically independent. The goal is to determine an assembly which maximizes the system-reliability.

The formal data for the problem are positive integers t and p, nonnegative integers $\{n_{uj} : u = 1, \ldots, t \text{ and } j = 1, \ldots, p\}$, where $\sum_{u=1}^{t} n_{uj} > 0$ for every $j = 1, \ldots, p$, and t ordered lists of real numbers, where the u-th list has $n_u = \sum_{j=1}^{p} n_{uj}$ elements $r^{u1}, \ldots, r^{un_u}$ that satisfy (6.1.2). In addition, we have a monotone function $\boldsymbol{J} : \{0,1\}^p \to \{0,1\}$.

The reliability of a module depends on its composition. Given an assembly $\pi = \{\pi_{uj} : u = 1, \ldots, t \text{ and } j = 1, \ldots, p\}$, the series structure of the modules implies that the reliability of module j, denoted $r(\pi)_j$, is given by

$$r(\pi)_j = \prod_{u=1}^{t} \prod_{\{i: i \in \pi_{uj}\}} r^{ui} \quad j = 1, \ldots, p. \tag{6.1.3}$$

In particular, we let $r(\pi)$ be the vector in $\mathbb{R}^p$ whose coordinates are given by (6.1.3). Note that, under each assembly, the operabilities of the modules are stochastically independent.

The reliability of the system as a whole depends on the way it is constructed. Let r be a vector in $[0, 1]^p$ whose coordinates $r_1, \ldots, r_p$ are, respectively, the reliabilities of the modules. Then, the system's reliability is the expectation of $\boldsymbol{J}(s)$, where s is a random vector whose components have independent binomial distributions with coefficients $r_1, \ldots, r_p$, and is given by

$$f(r) = \sum_{s \in \{0,1\}^p} \boldsymbol{J}(s) \left[\prod_{\{j: s_j = 0\}} (1 - r_j) \right] \left(\prod_{\{j: s_j = 1\}} r_j \right). \tag{6.1.4}$$

In particular, if assembly $\pi = \{\pi_{uj} : u = 1, \ldots, t \text{ and } j = 1, \ldots, p\}$ is used, then the system's reliability is given by

$$R(\pi) \equiv f[r(\pi)].$$

We refer to the function R which maps each assembly π into its reliability $R(\pi)$, as the *reliability function of the given system.* Our objective is to find an assembly which maximizes the reliability function of the system, in particular, we refer to *optimal* assemblies as those which accomplish this task.

We say that an assembly π is *monotone* if there is a permutation τ of the integers $\{1, \ldots, p\}$ such that for $u = 1, \ldots, t$ and $j = 1, \ldots, p$

$$\pi_{u,\tau(j)} = \left\{ \sum_{s=1}^{j-1} n_{u,\tau(s)} + 1, \ldots, \sum_{s=1}^{j} n_{u,\tau(s)} \right\}.$$

Hwang and Rothblum [57] proved that the function in (6.1.4) is asymmetric Schur convex. Thus by Theorem 5.5.2, there exists a monotone optimal partition. We remark that the original proof by Hwang and Rothblum [57] of the existence of a monotone optimal partition uses a different approach by first proving the result for $p = 2$ and then extend it to general p by introducing a fiduciary part which is universal so that Corollary 5.4.10 can be used. Note that when all parts of the fiduciary type has reliability one, then the system reliability, hence the set of optimal partitions, is invariant to the addition of the fiduciary type.

Hwang and Rothblum [59] used the polytope approach to show that the above assembly problem can also be solved for bounded-multishape; while Theorem 5.5.3 shows a further extension to constrained-multishape problems.

6.2 Group Testing

Given a set $\mathcal{N}$ of n items each of which can be either good or defective, and the state of the items is assumed to be independent, let ρ denote the probability that an item is good and define $\overline{\rho} = 1 - \rho$ the probability that an item is defective. The problem is to identify all the defectives by using a small number of group tests. A *group test* takes an arbitrary subset S of $\mathcal{N}$ as input and produces either a positive or negative outcome as output. A negative outcome indicates that all items in S are good and a positive outcome indicates otherwise, namely, there exists at least one unspecified defective in S.

Dorfman [27] proposed a procedure which partitions the n items into disjoint groups where a group test applies to each such group. Items of all groups with positive outcomes will be tested individually (degenerate group tests). The expected number of group tests required for a group of size x is (clearly, no empty group needs to be tested)

$$g(x) = \begin{cases} 1 & \text{if } x = 1, \\ 1 + x(1 - \rho^x) & \text{if } x \geq 2 \end{cases} \tag{6.2.1}$$

and the total expected number of group tests for a procedure π with p groups $\pi_1, \ldots, \pi_p$ of size $n_1, \ldots, n_p$, respectively, is

$$F(\pi) = f(n_1, \ldots, n_p) = \sum_{j=1}^{p} g(n_j) . \tag{6.2.2}$$

The problem is to determine the optimal choice of p and $n_1, \ldots, n_p$ – an open-partition problem.

The solution of the above problem will be discussed below in a more general setting. We first turn to a variation of the Dorfman procedure. Sobel and Groll [89] observed that when a group of size x is tested to be positive but its first $x - 1$ items are found to be good by individual tests, then the last item of the group can be deduced to be defective without testing. Thus (6.2.1) becomes

$$\begin{aligned} g(x) &= 1 + x(1 - \rho^x) - \rho^{x-1}\overline{\rho} \\ &= 1 + x - \rho^{x-1} - (x - 1)\rho^x . \end{aligned} \tag{6.2.3}$$

Note that (6.2.3) also covers the case $x = 1$. Treating (6.2.3) as a continuous function of x, the second derivative of $g(x)$ is then

$$g''(x) = -\rho^{x-1}(\ln \rho)^2 - 2\rho^x \ln \rho - (x-1)\rho^x (\ln \rho)^2.$$

Let x^0 denote the root of $g''(x) = 0$, that is

$$x^0 = \frac{\rho \ln \rho - 2\rho - \ln \rho}{\rho \ln \rho}.$$

Then $x^0 \geq 1$ for $\rho \geq 0.422$ and $g''(x) \geq 0$ for $x \leq x^0$, or equivalently, $g(x)$ is strictly convex in the range $1 \leq x \leq x^0$. Gilstein [40] proved that any group of size $x > x^0$ can be broken down into smaller groups of size $x_1, \ldots, x_k$ for some $k \geq 2$ with $\sum_{j=1}^k x_j = x$ and

$$g(x) > \sum_{j=1}^k g(x_j).$$

Thus, we may assume that every group in an optimal procedure has size smaller than x^0. We interpret our problem as to maximize $-F(\pi)$. Since $\sum_{j=1}^p g(x_j)$ is Schur convex if g is convex, the conditions of Theorem 2.3.2 in Vol. I are satisfied and for each given p, the shape-minorizing partition is optimal. To find the optimal p, we want to minimize $g(x)/x$, the average number of tests per item. Gilstein proved that $g(x)/x$ has a unique minimum x^*. Set $p^* = n/x^*$. Then p^* is the optimal p, except for the fact that the group size and the numbers of group must both be integers. To accommodate for that, set

$$x^\circ \in \{\lfloor x^* \rfloor - 1, \lfloor x^* \rfloor, \lfloor x^* \rfloor + 1\},$$

where $\lfloor y \rfloor$ is the largest integer less than or equal to y, and

$$p^\circ = \lfloor n/x^\circ \rfloor.$$

Consider p°-partition with $n - p^\circ x^\circ$ groups of size $x^\circ + 1$ and $p^\circ - (n - p^\circ x^\circ)$ groups of size x°, then this partition is the shape-minorizing partition with $p = p^\circ$ and has group sizes x° and $x^\circ + 1$. Since the optimal integer x is either $\lfloor x^* \rfloor$ or $\lfloor x^* \rfloor + 1$ and g is convex, one of those three choices of x° must determine an optimal p (by the formula $p^\circ = \lfloor n/x^\circ \rfloor$).

Pfeifer and Enis [79] considered a different group testing model in which a test measures the amount of defective ingredients in the group. Thus, if a group is tested to be defective with amount w and then individual tests on the first k items identify defectives of amount $w_1, \ldots, w_z$, $z \leq k$, such that $\sum_{u=1}^z w_u = w$, then the remaining items in the group can be deduced to be all good and no more tests on them are needed. Also, note that the last item never needs a test. Thus, one test is needed if all x items in the group are good, $k+1$ tests are needed if item k is defective and all remaining items are good, and x tests are needed if at least one of the last two items is defective. The expected number of tests for a group of size x is

$$\begin{aligned} g(x) &= \rho^x + \sum_{k=1}^{n-2} (k+1)\,\overline{\rho}\rho^{n-k} + x(1-\rho^2) \\ &= x - (1-\rho^{x-1})\,\rho^2/\overline{\rho}\,. \end{aligned}$$

It is easily seen that $g(x)$ is strictly convex. Pfeifer and Enis showed that

$$h(x) \equiv g(x)/x = 1 - (1-\rho^{x-1})\rho^2/\overline{\rho}x$$

is also strictly convex. By Theorem 2.3.2 of Vol. I, for each given p, the shape-minorizing partition is optimal. Further, by an argument analogous to the previous case, the optimal value of p is restricted to three choices.

Hwang [49] considered the original Dorfman procedure but allowed ρ to vary from item to item. Let ρ^i denote the probability that item i is good and without loss of generality, assume $\rho^1 \le \rho^2 \le \cdots \le \rho^n$. We consider the shape-partition problem. Note that the expected number of tests for a group π_j of size $n_j \ge 2$ is

$$G(\pi_j) = 1 + n_j\left(1 - \prod_{i\in\pi_j} \rho^i\right).$$

It is easily verified that if an optimal Dorfman procedure contains a group of size k and another group consisting of a singleton item with probability ρ, then $\rho \le \frac{k}{k+1}$. Note that $\frac{k}{k+1}$ is increasing in k, hence $\frac{k}{k+1} \le \frac{n-1}{n}$. So, if every $\rho^i > 1 - 1/n$, which is a realistic assumption in practice, then an optimal Dorfman procedure cannot contain a singleton group. Assume this is the case. Then, for a Dorfman procedure π with groups $\pi_1, \ldots, \pi_p$ of sizes $n_1, \ldots, n_p$, the expected number of group tests is

$$F(\pi) = \sum_{j=1}^{p} G(\pi_j) = \sum_{j=1}^{p}\left[1 + n_j\left(1 - \prod_{i\in\pi_j}\rho^i\right)\right]. \tag{6.2.4}$$

Note that when ρ^i is a constant ρ, then (6.2.4) is reduced to (6.2.2).

Define $\theta^i = \log \rho^i$ for $i = 1, \ldots, n$. Then $F(\pi)$ can be expressed by

$$f(g_{n_1}(\sum \theta_{\pi_1}), \cdots, g_{n_p}(\sum \theta_{\pi_p})),$$

where f is additive and

$$g_y(x) = 1 + n_y(1 - e^x)$$

is concave in x for each y. Transforming our problem to a maximization problem by considering $-F(\pi) = f^*(g_{n_1}(\sum \theta_{\pi_1}), \cdots, g_{n_p}(\sum \theta_{\pi_p}))$, where $f^* = -f$. Then each g_j is still concave, but f^* is decreasing. By Theorem 7.2.1 of Vol. I, there exists a consecutive maximum partition of $-F(\pi)$, hence a consecutive minimum partition of $F(\pi)$.

The original proof of Hwang used a different approach: Let $\ell \in \pi_1$ and $k \in \pi_2$. Let π' be obtained from π by interchanging ℓ with k. Then

$$F(\pi) - F(\pi') = (\rho^\ell - \rho^k)\left(-n_1 \prod_{i \in \pi_1 \setminus \{\ell\}} \rho^i + n_2 \prod_{i \in \pi_2 \setminus \{k\}} \rho^i\right).$$

Since $H_j(\{\rho^i : i \in \pi_j\}) = -n_j \prod_{i \in \pi_j} \rho^i$ is nonincreasing in each ρ^i, by Lemma 7.5.1 of Vol. I, there exists a consecutive optimal partition as an optimal Dorfman procedure. However, unlike the uniform ρ case, no explicit solution is available for the optimal Dorfman procedure.

In some situations, a defective item can be contagious and placing it in a part ruins all other items in that part. Thus, one has to balance the cost of testing against the cost of losing items to defectiveness. Let c be the cost of losing an item. Since for fixed p the cost of testing is a constant, it suffices to consider the cost of a Dorfman procedure $\pi = (\pi_1, \ldots, \pi_p)$ which is

$$F(\pi) = \sum_{j=1}^{p} c\left(1 - \prod_{i \in \pi_j} \rho^i\right) n_j. \tag{6.2.5}$$

For fixed p, (6.2.5) differs from (6.2.4) only by subtracting and multiplying constants, which do not affect the applicability of either Theorem 7.2.1 or Lemma 7.5.1 in Vol. I. Therefore the same conclusion holds.

6.3 Circuit Card Library

Garey, Hwang and Johnson [38] considered the following circuit partition problem, generalizing the model considered by Kodes [64]; it has the following structure:

(i) A set of n circuits. Circuit i is characterized by a pair (c^i, r^i), where c^i is the cost and r^i the requirement, the latter interpreted as the number of such circuits required annually.

(ii) An upper bound $U \leq n$ on the number of circuits allowed per card.

(iii) A stocking cost s per type of card which is the annual fixed cost associated with the stocking and production of any given type of circuit card.

Since upon the request of a given circuit, the card with minimum cost among those which contains it will always be issued, one may assume that no circuit appears on more than one card. Thus, a library design corresponds to a partition of the n circuits into a set of disjoint cards $\pi = (\pi_1, \ldots, \pi_p)$. The annual cost of a card is simply the total cost of circuits in it multiplied by the number of annual requests, plus a constant cost s of stocking a card. Hence, the annual cost of a library π is

$$F(\pi) = \sum_{j=1}^{p} (\sum_{i \in \pi_j} c^i)(\sum_{i \in \pi_j} r^i) + ps. \tag{6.3.1}$$

In general, the circuit card library problem is a 2-dim open partition problem which is unsolved yet. Here we consider the special cases when c^i or r^i or both are constants. We solve (6.3.1) for each p and then select the best solution over p.

When both $c^i = c$ and $r^i = r$ for each i, (6.3.1) is reduced to

$$F(\pi) = \sum_{j=1}^{p} (cr|\pi_j|^2 + s) = \sum_{j=1}^{p} g(|\pi_j|),$$

where $g(\cdot)$ is the (quadratic) function with $g(x) = crx^2 + s$. Evidently, $g(x)$ is strictly convex; thus $\sum_{j=1}^{p} g(|\pi_j|)$ is strictly Schur convex. By Theorems 2.3.2 and 2.3.3 in Vol. I, for each p the shape-minorizing partition is an optimal partition. Further, suppose x^* minimizes $g(x)/x$, which is also strictly convex, and $x^* + 2 \leq U$. Set $x^\circ \in \{\lfloor x^* \rfloor - 1, \lfloor x^* \rfloor, \lfloor x^* \rfloor + 1\}$ and $p^\circ = \lfloor n/x^* \rfloor$. Then an optimal partition is from one of the three choices analogous to the Gilstein problem; see Section 6.2. (Garey, Hwang and Johnson gave $x^\circ = \lfloor x^* \rfloor$ as the only choice.)

Next, consider the case that $c^i = c$ for each i. Then

$$F(\pi) = \sum_{j=1}^{p} \left(c|\pi_i| \sum_{i \in \pi_j} r^i + s \right).$$

Define

$$F^*(\pi) = -F(\pi) = \sum_{j=1}^{p} \left[c|\pi_i| \sum_{i \in \pi_j} (-r^i) - s \right].$$

Then, for fixed p, minimizing $F(\pi)$ becomes maximizing $F^*(\pi)$.

Without loss of generality, assume

$$r^1 \le \cdots \le r^p .$$

Write $F(\pi) = f(r^1, \ldots, r^p)$. Since f is linear, there exists an optimal partition corresponding to a vertex of the partition polytope. By Corollary 5.2.7 of Vol. I, every such vertex corresponds to a reverse size-consecutive partition.

Finally, consider $r^i = r$ for each i. Due to the symmetry of $F(\pi)$ with respect to $\{c^i\}$'s and $\{r^i\}$'s, the analysis is the same as the c^i's constant case, leading to the same conclusion that there exists a reverse size-consecutive optimal p-partition.

Garey, Hwang and Johnson gave an $O(nm + n \lg n)$-time dynamic programming algorithm to find an optimal partition. They also gave an (nm/d)-time algorithm for the general model where there are d_r distinct values of r^i and d_c distinct values of c^i, with $d = \min\{d_r, d_c\}$.

6.4 Clustering

Given a set of n elements where element i is assigned a weight w^i and a numerical measure θ^i and given a positive integer $p < n$, consider the problem of finding a systematic and practical procedure for grouping the n elements into p mutually exclusive and exhaustive subsets $\pi_1, \ldots, \pi_p$ such that the weighted sum of squares

$$\sum_{j=1}^{p} \sum_{i \in \pi_j} w^i (\theta^i - \bar{\theta}^j)^2$$

is minimized; here, each $\bar{\theta}^j$ denotes the weighted mean of the θ^i's of the elements i assigned to subset π_j.

Fisher [35] proposed the above problem and proved that there exists a consecutive optimal p-partition for $d = 1$. This result also follows from Theorem 7.4.5 of Vol. I by setting

$$D_i(\theta^i, c_j) = w^i (\theta^i - \bar{\theta}^j)^2.$$

Fisher actually stated his wish to see this result extended to $\mathbb{R}^d$ with a weight attached to each dimension to reflect the difference in importance; so, each i is associated with a (nonnegative) vector $A^i \in \mathbb{R}^d$ and

$$F(\pi) = \sum_{j=1}^{p} \sum_{i \in \pi_j} \sum_{k=1}^{d} u_k \| A^i_k - \bar{A}_{kj} \|^2 ,$$

where each A^i_k represents the k coordinate of A^i, the u_k's are weight parameters, and $\bar{A}_{kj}$ represents the average of the A^i_k's assigned to π_j.

It took almost 40 years before Boros and Hwang [10] came up with an answer. In fact, they considered a slightly more general model by introducing weights w_j that depend on the parts such that the objective function $F(\pi)$ is expressed by

$$\begin{aligned} F(\pi) &= \sum_{j=1}^{p} w_j \sum_{i\in\pi_j} \sum_{k=1}^{d} u_k \|A^i_k - \bar{A}_{kj}\|^2 \\ &= \sum_{j=1}^{p} w_j \sum_{i\in\pi_j} \sum_{k=1}^{d} u_k \|A^i_k - \bar{A}_{kj}\|^2 \\ &= \sum_{j=1}^{p} w_j \sum_{i\in\pi_j} \|x^i - \bar{x}_j\|^2 \end{aligned}$$

where for $k = 1, \ldots, d$, $i = 1, \ldots, n$ and $j = 1, \ldots, p$, $x^i_k = \sqrt{u}_k A^i_k$, $x^i = (x^i_1, \ldots, x^i_d)^T \in \mathbb{R}^d$ and $\bar{x}_j$ is the average of the x^i's assigned to π_j. Assume that π is an optimal partition and π' is obtained from π by interchanging $A^u \in \pi_1$ and $A^v \in \pi_2$. Let $\overline{A}_j$ denote the mean of $\{A^i : i \in \pi_j\}$ and let $\overline{A}'_j$ denote the mean of $\{A^i : i \in \pi'_j\}$. Define $\Delta_F(A^u, A^v) = F(\pi') - F(\pi)$. Boros and Hwang showed that

$$\begin{aligned} 0 \le \Delta_F(A^u, A^v) &= w_1 \left(\sum_{A^i\in\pi'_1} \|A^i - \bar{A}'_1\|^2 - \sum_{A^i\in\pi_1} \|A^i - \bar{A}_1\|^2 \right) \\ &\quad + w_2 \left(\sum_{A^i\in\pi'_2} \|A^i - \bar{A}'_2\|^2 - \sum_{A^i\in\pi_2} \|A^i - \bar{A}_2\|^2 \right) \\ &= w_1 \left(\sum_{A^i\in\pi'_1} \|A^i\|^2 - \frac{\|\sum_{A^i\in\pi'_1} A^i\|^2}{n_1} - \sum_{A^i\in\pi_1} \|A^i\|^2 + \frac{\|\sum_{A^i\in\pi_1} A^i\|^2}{n_1} \right) \\ &\quad + w_2 \left(\sum_{A^i\in\pi'_2} \|A^i\|^2 - \frac{\|\sum_{A^i\in\pi'_2} A^i\|^2}{n_2} - \sum_{A^i\in\pi_2} \|A^i\|^2 + \frac{\|\sum_{A^i\in\pi_2} A^i\|^2}{n_2} \right) \\ &= w_1 \left[\|A^v\|^2 - \|A^u\|^2 - \frac{\|A^v\|^2 - \|A^u\|^2 + 2(A^v - A^u)^T(\sum_{A^i\in\pi_1} A^i - A^u)}{n_1} \right] \\ &\quad + w_2 \left[\|A^u\|^2 - \|A^v\|^2 - \frac{\|A^u\|^2 - \|A^v\|^2 + 2(A^u - A^v)^T(\sum_{A^i\in\pi_2} A^i - A^v)}{n_2} \right] \end{aligned}$$

$$= w_1 \left[\|A^v\|^2 - \|A^u\|^2 - \frac{\|A^v\|^2 + \|A^u\|^2 - 2[(A^v)^T A^u]}{n_1} - 2(A^v - A^u)^T(\bar{A}_1) \right]$$
$$+ w_2 \left[\|A^u\|^2 - \|A^v\|^2 - \frac{\|A^u\|^2 + \|A^v\|^2 - 2[(A^v)^T A^u]}{n_2} - 2(A^u - A^v)^T(\bar{A}_2) \right]$$
$$= (w_2 - w_1 - \frac{w_1}{n_1} - \frac{w_2}{n_2})\|A^u\|^2 + (w_1 - w_2 - \frac{w_1}{n_1} - \frac{w_2}{n_2})\|A^v\|^2$$
$$+ 2(\frac{w_1}{n_1} + \frac{w_2}{n_2})[(A^v)^T A^u] - 2(A^v - A^u)^T(\bar{A}_1 - \bar{A}_2).$$

Since the sum of coefficients of the $\|A^u\|^2$ and $\|A^v\|^2$ is $-\frac{2w_1}{n_1} - \frac{2w_2}{n_2} < 0$, at least one of the coefficients is negative. View $\Delta_F(A^u, A^v)$ either as a function $\Delta_F(A^u, \cdot)$ or as a function $\Delta_F(\cdot, A^v)$. It follows that at least one of the two functions is strictly concave. By Theorem 4.1.17 every optimal partition for a shape-problem is weakly convex separable. If $w_j = w$ for all j, then by Theorem 4.1.18, every optimal partition is weakly nonpenetrating. Golany, Hwang and Rothblum [41], using Theorem 4.1.3(i), concluded that every optimal partition is almost-sphere-separable. This strengthening is significant since the search of an optimal partition switches from an exponential class to a polynomial class. When all the w_j's are constant, this problem is also known as the k-mean problem in the literature (with $k = p$ in our case). Theorem 4.1.5(ii) allows one to further strengthen the property of optimal partitions to almost separability.

Suppose that, instead of the weighted sum of squares, the goal is to minimize the weighted sum of variances, i.e.,

$$F(\pi) = \sum_{j=1, n_j \neq 1}^{p} w_j \sum_{i \in \pi_j} \frac{\|x^i - \bar{x}_j\|^2}{n_j - 1}, \tag{6.4.1}$$

where w_j is the weight of part j. By setting $w'_j = \frac{w_j}{n_j - 1}$, we can apply Theorem 4.1.3(ii) to conclude that every optimal partition for a shape-problem must be almost-sphere-separable. Unlike the weighted sum of square case, setting $w_j = 1$ for all j does not bring a stronger conclusion since the term $1/(n_j - 1)$ then plays the role of w_j. Examples 7.4.2 and 7.4.3 in Vol. I, illustrating the existence of a unique optimal partition which is noncrossing but not consecutive, can also be used as examples here.

In the k-median problem ($k = p$ here),

$$F(\pi) = \sum_{j=1}^{p} \sum_{i \in \pi_j} \|A^i - x_j\|$$

where x_j is the median of π_j. Then Theorem 4.1.9 applies, with $h(x) = x$, to conclude the existence of a separable optimal partition.

Finally, in the k-center problem ($k = p$ here), a set $(t_1, \cdots, t_p)$ is given with $t_j \in \mathbb{R}^d$ for all j and

$$F(\pi) = \max_{j \in \{1,\cdots,p\}} \max_{i \in \pi_j} \|A^i - t_j\|.$$

Then Theorem 4.1.7 applies, with $f = \max$, to conclude the existence of a separable optimal partition.

6.5 Abstraction of Finite State Machines

Consider a finite state machine M with state-set Q, input-set X, output-set Y and two functions δ and λ which map $Q \times X$ into Q (the next-state) and Y, respectively. The problem is to construct an abstract finite state machine M_A by lumping each of Q, X and Y into classes, specified by partitions π^Q, π^X and π^Y of Q, X and Y, respectively; in particular, a part of π^Q (π^X, π^Y) represents a state (an input, an output) of M_A. Furthermore, the output function δ_A of M_A is defined by mapping $\delta_A(Q_i, X_j)$ to a set called an *admissible set* of outputs consisting of all Y_k such that there exist $q \in Q_i$, $x \in X_j$, $y \in Y_k$ with $\delta(q, x) = y$. The next-state function δ_A and its admissible set are similarly defined.

A λ-fault of M with domain $D \subseteq Q \times X$ is a mapping $\epsilon_D : D \to Y$ such that $\epsilon(q, x) \neq \lambda(q, x)$ for every pair in D; the multiplicity of such a λ-fault is $|D|$. In particular, a λ-fault with multiplicity 1 is called a single λ-fault. A δ-fault and its multiplicity can be similarly defined. Oikonomou [77] studied the abstraction problem to minimize the number of single undetected λ-faults, assuming no δ-fault can occur. In Section 1.4.5 of Vol. I, it was shown that the objective function can be written as

$$F(\pi) = \sum_{k=1}^{|\pi^Y|} |Y_k| \sum_{y \in Y_k} m(y), \tag{6.5.1}$$

where $m(y)$ for $y \in Y$ is the number of (q, x) pairs such that $\lambda(q, x) = y$. Note that (6.5.1) fits the format of (7.3.1) in Vol. I with f and h being the sum functions and $g(x, y) = xy$, except that here we want to minimize $F(\pi)$, rather than maximize it. To fit the "maximization" format, we change $F(\pi)$ to $-F(\pi)$ and $m(y)$ to $-m(y)$. Although $F(\pi)$ is in the right format, we find no results in Section 7.3 of Vol. I applicable to this $F(\pi)$.

Oikonomou attacked the above problem by first considering the shape version, namely: for a given shape $(n_1, ..., n_p)$ with $p = |\pi^Y|$ maximize

$$F^{(n_1,\ldots,n_p)}(\pi) = \sum_{k=1}^{p} n_k \sum_{y \in \pi_k} (-m(y)). \tag{6.5.2}$$

This expression fits the format (7.2.1) of Vol. I with $g_j(x, y) = g(x, y) = xy$ for all j and with f and h as the sum function. It is also easily verified that $g(x, y) = xy$ is strictly submodular and that it is convex and increasing in x. Further, the $-m(y)$'s are nonpositive for all y. Thus, Theorem 7.3.5(ii) of Vol. I applies and assures the existence of a reverse size-consecutive optimal partition. Note that despite all the strict conditions of the functions, we do not get the strict conclusion of Theorem 7.3.5(ii) in Vol. I because the $m(y)$'s may not be distinct.

To solve the original size-problem, we still need to compare the set of optimal shape partitions over all shapes of size p. By Theorem 6.2.1 of Vol. I, the number of unlabeled shapes with size p is $\binom{n-1}{p-1=O(n^{p-1})}$, while $p!O(n^{p-1}) = O(n^{p-1})$ is a bound on the number of labeled shapes. Thus we have a polynomial method for solving the problem (Oikonomou commented that no efficient algorithm exists to search the set of unlabeled shapes of size p). Oikonomou gave some method to reduce the computation burden of comparisons and relied on a branch and bound algorithm as the final step. This problem also serves as an example for the remarks after Lemma 7.3.1 in Vol. I, namely, the existence of an optimal reverse size-consecutive partition for the shape-problem does not imply the existence of an optimal reverse extremal partition for the size-problem.

In another paper, Oikonomou [78] shifted his attention to the number of single immediately detectable λ-faults. He showed that this number is

$$|Q||X||Y| - \sum_{i=1}^{|\pi^Q|} \sum_{j=1}^{|\pi^X|} |Q_i||X_j| \sum_{Y_k \in \delta_A(\theta_i X_j)} |Y_k|. \tag{6.5.3}$$

As (6.5.3) involves all three partitions, its maximization is hopelessly difficult. Oikonomou suggested to use the partition which minimizes (6.5.1) as π^Y, and then treating one of π^Q and π^X as a parameter to seek an optimal partition of the other. As π^Q and π^X are symmetric in (6.5.3), one could choose to optimize π^X without loss of generality. Thus, to maximize (6.5.3) becomes to minimize

$$F\left(\pi^X\right) = \sum_{j=1}^{|\pi^X|} |X_j| \sum_{i=1}^{|\pi^Q|} \sum_{Y_k \in \delta_A(Q_i, X_j)} |Q_i||Y_k|. \tag{6.5.4}$$

Unfortunately, even this trimmed version was shown to be an NP-complete problem. For each $x \in X_j$, define

$$S_x = \bigcup_{i=1}^{|\pi^Q|} \{|Q_i||Y_k| : Y_k \in \delta_A(Q_i\{x\})\}.$$

Then (6.5.4) can be written as

$$F(\pi^X) = \sum_{j=1}^{|\pi^X|} |X_j| W(X_j), \tag{6.5.5}$$

where $W(X_j)$ is the sum of the numbers in $\bigcup_{x \in X_j} S_x$. Equation (6.5.5) is now in the form of a subset-partition problem and Oikonomou proposed the heuristic reported in Vol. I Section 7.5 to solve it.

6.6 Multischeduling

Mehta, Chandraseckaran and Emmons [71] considered a problem of scheduling n jobs on two machines. The data for the problem consists of the time t_i required to execute job i on either machine, where $i \in \mathcal{N} = \{1, \dots, n\}$. The objective function is to minimize the *total flow time* – the sum of time the jobs spend in the system (from time 0 to the time the job is completed). A linear order of the n jobs is given such that job i must be worked on before job u if $i < u$. A 2-partition $\pi = (\pi_1, \pi_2)$ of $\mathcal{N}$ represents the assignment of the jobs to either of the two machines; given such a partition π, the *position* of job i, denoted $k_\pi(i)$, is the number of jobs to be worked on after job i plus one. As derived in Section 1.4.6 of Vol. I, the objective value associated with π equals

$$F(\pi) = \sum_{i=1}^{n} k_\pi(i) t_i.$$

Mehta, Chandraseckaran and Emmons proposed a dynamic programming solution. Let $f_i(k)$ denote the minimum total flow time required to do the first i jobs when job i has position k in the machine it is scheduled to be worked on. Suppose that in an optimal assignment of i jobs, job i is in position k. Evidently, job $i-1$ then either immediately precedes job i on the same machine i is assigned to, or is the last job on the other machine. The position of job $i-1$ is then, respectively, $k-1$ or $i-k$. As job i at position k contributes kt_i time to the total flow time, we have that

$$f_i(k) = kt_i + \min\{f_{i-1}(k-1), f_{i-1}(i-k)\} \text{ for } 2 \le k \le i$$

for $i = 1, \ldots, n$: $f_i(0) = \infty$, $f_i(1) = t_i$, and $f_i(k) = 0$ for $k > i$, since job $i-1$ either immediately precedes job i on the same machine or is the last job on the other machine, and since job i contributes kt_i time. Clearly, the minimal value of $F(\pi)$ equals

$$f_n = \min_k f_n(k).$$

Since we have to compute $f_u(k)$ for $1 \leq u \leq n$ and $1 \leq k \leq n$, $f_n = \min_k f_n(k)$ can be computed in $O(n^2)$ time.

Rothkopf [84] generalized the above model by assuming that t_{ij} time is required to do job i on machine j. A setup time c_{ij} is associated with doing job i on machine j and a cost $a_j t$ is associated with using t time of machine j. Let $f_i(k)$ be defined as before, except only for machine 1, and let $g_i(k)$ denote the corresponding cost for machine 2. Then, with the same boundary conditions as before,

$$f_i(k) = (a_1 + k)t_{i1} + c_{i1} + \min\{f_{i-1}(k-1), g_{i-1}(i-k)\},$$

and

$$g_i(k) = (a_2 + k)t_{i2} + c_{i2} + \min\{f_{i-1}(i-k), g_{i-1}(k-1)\}.$$

Again, $f_n = \min\{\min_k f_n(k), \min_k g_n(k)\}$ can be computed in $O(n^2)$ time.

Tanaev [91] extended the problem to p machines and defined

$$G_j^i[F(\pi_1, \ldots, \pi_{j-1}, \pi_j \setminus \{i\}, \pi_{j+1}, \ldots, \pi_p);\ |\pi_j| - 1] = (a_i + |\pi_j| - 1)t_{ij}$$

$$+ c_{ij} + \min_j G_j^{i-1}[F(\pi_1, \ldots, \pi_i \setminus \{j-1\}, \ldots, \pi_p); |\pi_j| - 1].$$

Since G is increasing in F, the problem can be solved by dynamic programming, as reported at the end of Section 7.5 in Vol. I.

Rothkopf also observed that if no prespecified ordering exists among the n jobs, then optimality consideration will order the jobs on machine j according to t_{ij}. Therefore, under the assumption $t_{ij} = t_i r_j$, the n jobs can be ordered according to t_i, say, $t_1 \leq t_2 \leq \cdots \leq t_n$, and there exists an optimal multischeduling such that this order is obeyed on each machine. Note that without the prespecified ordering of jobs, the multischeduling problem becomes a genuine partition problem. Also, note that the assumption $t_{ij} = t_i r_j$ is unnecessarily strict. All is needed is that the jobs can be labeled such that

$$t_{1j} \leq t_{2j} \leq \cdots \leq t_{nj} \text{ for all } j.$$

Furthermore, the number of machines can be extended to general p.

Graham [43] considered the problem of given p identical processing units and a set of n tasks which is to be processed by the processing units. Suppose that task i requires t_i units of processing time. The problem is to partition the tasks for assignment to the processing units to minimize the time that all tasks have been processed. This corresponds to a p-partition problem with the cost function

$$F(\pi) = \max\{t_{\pi_1}, \ldots, t_{\pi_p}\}. \tag{6.6.1}$$

This problem is NP-complete since it is a variation of the "number partition" problem. Graham proposed the heuristic, known as the "largest-first partition", by first ordering the tasks according to their processing time (from large to small) and assigning the currently largest task to the processor with the currently smallest processing time. Graham proved that the performance ratio of the largest-first partition is upper bounded by $4/3-1/3p$, while the bound can be achieved by the following set of $n = 2p+1$ processing time units:

$$(2p-1, 2p-1, 2p-2, 2p-2, \ldots, p+1, p+1, p, p, p)\,.$$

For example, for $p = 4$, the largest-first partition is: $\pi_1 = \{7,4,4\}$, $\pi_2 = \{7,4\}$, $\pi_3 = \pi_4 = \{6,5\}$, while the optimal partition is $\pi_1 = \pi_2 = \{7,5\}$, $\pi_3 = \{6,6\}$, $\pi_4 = \{4,4,4\}$. The performance ratio is $15/12$.

Cody and Coffman [23] considered the problem of partitioning a set of n equal-length records into p sectors of a drum, or a cylinder of a disk system, so as to minimize the expected cost of latency time. Let p^k denote the probability that a given request is for record k ($\sum_{k=1}^{n} p^k = 1$), let p_{π_j} denote the sum of the p^k in π_j and let d_{ij} denote the latency cost incurred in the motion of the drum from the end of sector i to the beginning of sector j. Then the cost of a partition $\pi = (\pi_1, \ldots, \pi_p)$ is

$$F(\pi) = \sum_{i=1}^{p} \sum_{j=1}^{p} p_{\pi_i} p_{\pi_j} d_{ij}. \tag{6.6.2}$$

More specifically, define

$$d_{ij} = \begin{cases} j-i-1 & \text{for } 1 \le i < j \le p, \\ p-2-d_{ji} & \text{for } 1 \le j < i \le p, \\ p-1 & \text{for } 1 \le i = j \le p. \end{cases} \tag{6.6.3}$$

Then (6.6.2) becomes

$$\begin{aligned} F(\pi) &= \frac{1}{2} \sum_{i=1}^{p} \sum_{j=1}^{p} p_{\pi_i} p_{\pi_j} (d_{ij} + d_{ji}) \\ &= \frac{p-2}{2} \sum_{i=1}^{p} \sum_{j \ne i} p_{\pi_i} p_{\pi_j} + (p-1) \sum_{j=1}^{p} p_{\pi_j}^2 \end{aligned}$$

by making use of (6.6.3). As $\frac{p-2}{2} \le p-1$ and

$$\sum_{i=1}^{p} \sum_{j \neq i} p_{\pi_i} p_{\pi_j} + \sum_{i=1}^{p} p_{\pi_i}^2 = 1,$$

the problem is converted to minimizing $\sum_{j=1}^{p} p_{\pi_j}^2$. Cody and Coffman noted that even for $p = 2$ this is an NP-hard problem since it can be reduced to the "number partition" problem. They proved that the performance ratio of the largest-first partition is upper bounded by the number $1 + 1/[16(p-1)]$. If each sector can contain at most u records, then the upper bound becomes $(p+1)/(p-1)$ (surprisingly, independent of u).

Chandra and Wong [20] improved the upper bound of the performance ratio of the largest-first partition to 25/24 for all p, and to 37/36 for p tending to infinity. For $p = 2$, they determined the exact value of the performance ratio to be 1.0285.

They also studied the cost function

$$F_\alpha(\pi) = \left(\sum_{j=1}^{p} p_{\pi_j}^{\alpha} \right)^{1/\alpha}, \tag{6.6.4}$$

where $\alpha > 1$. Note that for α tending to infinity, (6.6.4) is approaching (6.6.1). They proved that for general α, the performance ratio of the largest-first partition is upper bounded by

$$\frac{3^\alpha}{\alpha} \left(\frac{\alpha - 1}{2 \cdot 3^\alpha - 3 \cdot 2^\alpha} \right)^{\frac{\alpha-1}{\alpha}}.$$

Easton and Wong [31] studied the effect of the upper bound u. They proved that the performance ratio of the bounded case versus the unbounded case is at most 4/3.

6.7 Cache Assignment

A cache is an extremely fast memory. We assume that n items are stored in p memory pages $M_1, \ldots, M_p$ which, at a given time, are either in the cache or in the main memory, which is slow (a cache page may change its contents after each search). Suppose that the cache has L cache pages each with capacity C, i.e., only C items in each page are considered accessible. A requested item must be in the cache capacity to be quickly processed. (We call it a *cache hit* or a *cache miss* depending on whether the item is in the cache capacity.) If the item is in a memory page, we load the entire memory page to a cache page to replace its current content, and a performance penalty occurs each time this happens. Let p_i denote the probability that the item being requested is item i (which is positive in general or the item will not be considered), and the requests are independent. Without loss of generality, assume

$$0 < p_1 \le p_2 \le \cdots \le p_n.$$

The problem is to partition the n items to the p pages to maximize the expected number of hits. Gal, Holander and Itai [37] studied some special cases.

(i) $C = 1$ and $L = p$. Then a requested item is considered a hit only if it was the last item requested among all items in the same page. Suppose item $i \in M_j$. Then the probability that item i is the last requested item in M_j is

$$\frac{p_i}{\sum_{k \in M_j} p_k}.$$

Hence, the probability of a hit is

$$\sum_{j=1}^{p} \sum_{i \in M_j} \frac{p_i^2}{\sum_{k \in M_j} p_k}.$$

We interpret

$$\frac{\sum_{i \in M_j} p_i^2}{\sum_{k \in M_j} p_k}$$

as

$$\frac{\sum_{i \in \pi_j} w_i \theta_i}{\sum_{i \in \pi_j} w_i} = \bar{\theta}_j^{(w)},$$

where $w_i = \theta^i = p_i$ for $i = 1, \ldots, n$. Since $\theta^1/w_1 = \theta^2/w_2 = \cdots = \theta^n/w_n$, by Theorem 7.3.15 of Vol. I, the partition E^{-1}, i.e., each of the cache pages carries one of the $p - 1$ most frequent items and one cache page carries all other items, is optimal.

(ii) $L = 1$, the capacity covers the whole page. Then the requested item is a hit if it is in the cache page. The probability that M_j is the cache page is the probability that one of its items is the last request, which is $\sum_{i \in M_j} p_i$. The probability that M_j is the cache page, the current reference is a hit is also $\sum_{i \in M_j} p_i$. Therefore, the unconditional probability of a hit is

$$\sum_{j=1}^{p} \left(\sum_{i \in M_j} p_i \right)^2 .$$

Writing this in the format of

$$F(\pi) = f\left(\sum_{i \in \pi_1} \theta_i, \ldots, \sum_{i \in \pi_p} \theta_i \right),$$

where $\theta_i = p_i$ for $i = 1, \ldots, n$, then f is strictly Schur convex and θ^i's are positive. By Corollary 5.4.9 of Vol. I, every optimal partition is extremal.

(iii) Finally, we consider the general case $C \geq 1$ and $1 \leq L \leq p$. Suppose item $i \in M_j$. Define

$$P_j = \sum_{i \in M_j} p_i \quad \text{for } j = 1, \ldots, p.$$

The probability that M_j is in the cache is

$$\frac{\prod_{j=1}^{p} P_j C_{L-1}(R_{p \setminus \{j\}})}{\prod_{j=1}^{p} P_j C_L(R_p)}$$

where $R_p = \{P_1^{-1}, P_2^{-1}, \ldots, P_p^{-1}\}$ and $R_{p\setminus\{j\}} = R_p \setminus P_j^{-1}$.

Note that this probability is very messy and hard to handle. Gal, Holander and Itai [37] modified the model to replace this term by P_j.

Partition each M_j into sub-pages M_{jk} such that each sub-page can fit into the cache capacity C. In case M_j is in the cache but the requested item $I_i \in M_{jk}$ is not in the cache capacity, we load the entire M_{jk} into the cache capacity after the search. Then, analogous to case (ii), the probability that the requested item is a hit, given that M_j is in the cache, is

$$\left(\sum_{i \in M_{jk}} p_i \right)^2 / P_j^2 \equiv P(M_{jk})^2 / P_j^2$$

the difference being that the denominator is not 1. Hence, the unconditional probability of a hit is

$$\sum_{j=1}^{p} P_j \sum_{M_{jk}} P(M_{jk})^2 / P_j^2 = \sum_{j=1}^{p} \sum_{M_{jk}} P(M_{jk})^2 / P_j.$$

Set $\theta_{jk} = w_{jk} = P(M_{jk})$, then

$$\sum_{j=1}^{p}\sum_{M_{jk}} P(M_{jk})^2/P_j = \sum_{j=1}^{p}\sum_{k} w_{jk}\theta_{jk}/\sum_{k} w_{jk} = \sum_{j=1}^{p}\bar{\theta}_j.$$

So, Theorem 7.3.15 of Vol. I applies again and there exists a reverse extremal optimal partition. Note that each $\bar{\theta}_{jk}$ represents a set of C items. So, $p-1$ pages should each take a consecutive C-set of the $C(p-1)$ items with largest p_i, and one page gets all other items.

6.8 The Blood Analyzer Problem

Consider a set of $n = pm$ tests available on a blood analyzer (machine). A random blood specimen requires a subset of these tests for the detection of possible symptoms this specimen is suspected to possess. A probability, which may vary from test to test, is attached to each test for being included in the subset. The problem is to partition the n tests into p parts, each of size m, to minimize the expected number of parts intersecting the required set of tests, as explained in Section 1.4.8 of Vol. I.

For a random blood specimen i, define the random variable Z_i such that $Z_i = 1$ if the test of i is required and $Z_i = 0$ otherwise. As shown in Section 1.4.8 of Vol. I, the cost associated with a partition $\pi = (\pi_1, \ldots, \pi_p)$ of the n agents into the p groups is given by

$$F(\pi) = p - \sum_{j=1}^{p} Pr\{\sum_{i\in\pi_j} Z_i = 0\}$$

with $Pr\{\cdot\}$ denoting the probability of an event. Under the assumption that the requirement for tests is independent from the specimen, we have that the Z_i's are independent and

$$Pr\{\sum_{i\in\pi_j} Z_i = 0\} = \prod_{i\in\pi_j} Pr\{Z_i = 0\} = e^{\sum_{i\in\pi_j}\theta^i},$$

where for each i, $\theta^i \equiv \ln[Pr\{Z_i = 0\}]$ (in general, $P_r\{Z_i = 0\} < 1$, i.e., $\theta^i < 0$, or i would not be included in the blood analyzer). Thus,

$$F(\pi) = p - \sum_{j=1}^{p} e^{\sum_{i\in\pi_j}\theta^i} = f\left(\sum_{i\in\pi_1}\theta^i, \ldots, \sum_{i\in\pi_p}\theta^i\right),$$

with $f : \mathbb{R}^p \to \mathbb{R}$ mapping $x \in \mathbb{R}^p$ into

$$f(x) = p - \sum_{j=1}^{p} e^{x_j},$$

where $x_j = \sum_{i \in \pi_j} \theta^i$. Since p is a constant, we can ignore it in minimizing f. Then the problem becomes to maximize $f'(x) = \sum_{j=1}^{p} e^{x^j}$. As f' is strictly convex and the θ^i's are negative, Corollary 5.4.7 of Vol. I implies that every optimal partition for a shape-problem is reverse size-consecutive. Since shape is uniform in the current application, size-consecutive becomes simply consecutive.

In practice, the Z_i's are likely to be dependent on the specimen. Except for the trivial case, the problem becomes difficult and Nawijn formulated it as a linear integer programming problem, with a large number of inequalities.

We next consider a deterministic formulation of the blood analyzer problem. We may classify the blood specimen (actually, the patients) into types according to the set of tests required. Construct a bipartite graph $G(U, V; E)$, where the types are the U-vertices, the tests are the V-vertices, and an edge (u, v) for $u \in U$ and $v \in V$ means type u requires test v. Then the problem is to partition the V-vertices into p groups $\pi_1, \ldots, \pi_p$ so as to minimize

$$\sum_{u=1}^{n} |\{j : i \text{ has an edge to } j\}| Pr\{i\}.$$

This problem is a variation of the graph partition problem mentioned in Section 1.4.12 of Vol. I whose efficient solution is not known (see Section 6.12).

6.9 Joint Replenishment of Inventory

Consider an economic order quantity model involving n items, where the i-th item has (deterministic) demand rate D_i, a unit inventory holding cost h_i per unit time and a fixed cost K_i for placing an order. The problem is to partition the n items into p subgroups and choose order cycles for the groups out of a given set of allowable (joint) order cycles, so as to minimize the net average cost per unit time. This model was considered by Chakravarty, Orlin and Rothblum [19]; it generalizes the joint replenishment models considered by Goyal [42], Silver [88], Nocturne [76] and Chakravarty [15,16]. Specifically, the last references gave heuristic solutions to the problem in which a unit of order time λ is determined for the group of n items, and the order cycles for each item is an integer multiple of λ.

Let $a_i = 2^{-1}h_iD_i$ and $b_i = K_i$ where the items are labeled so that

$$\frac{a_1}{b_1} \le \frac{a_2}{b_2} \le \cdots \le \frac{a_n}{b_n}. \tag{6.9.1}$$

As in the ordinary *EOQ* model, if item i has order cycle τ, then its order quantity is τD_i, and the average net cost per unit time of a group S of items having the same order cycle τ is:

$$c(S,\tau) = \sum_{i\in S} 2^{-1}\tau D_i h_i + \sum_{i\in S} K_i\tau^{-1} = \tau\sum_{i\in S} a_i + \tau^{-1}\sum_{i\in S} b_i. \tag{6.9.2}$$

So, if the items are partitioned into groups $\pi_1,\ldots,\pi_p$ having order cycles $t_1,\ldots,t_p$, respectively, the total average cost per unit time is

$$c_p(\pi_1,t_1,\ldots,\ldots,\pi_p,t_p) = \sum_{j=1}^{p} c(\pi_j,t_j)\,. \tag{6.9.3}$$

The p-vector of order cycles is assumed to be taken out of a set T of allowable p-vectors of order cycles. This formulation allows one to impose joint restrictions on the order cycles, e.g., the requirement that all order cycles are integer multiples of the smallest one, or individual restrictions, e.g., that each order cycle is an integer in the set $\{1,7,30\}$ (representing, daily, weekly and monthly orders cycles). It follows that for a fixed partition $\pi_1,\ldots,\pi_p$ of the items, the minimum average cost per unit time is:

$$F_p(\pi_1,\ldots,\pi_p) = \inf_{t\in T} c_p(\pi_1,\ldots,\pi_p,t_1,\ldots,t_p) \tag{6.9.4}$$

$$= \inf_{t\in T}\sum_{j=1}^{p}[t_j\sum_{i\in\pi_j} a_i + t_j^{-1}\sum_{i\in\pi_j} b_i]. \tag{6.9.5}$$

Evidently, as $c_p(\pi_1,\ldots,\pi_p,t_1,\ldots,t_p) \ge 0$, the infimum defining F_p is finite for each partition. The problem is to find a partition of the items into subsets so as to minimize $F_p(\cdot)$. Now, as $c(S,t)$ is linear in $a_S \equiv \sum_{i\in S} a_i$ and $b_S \equiv \sum_{i\in S} b_i$ (for fixed t), $c_p(\pi_1,\ldots,\pi_p,t_1,\ldots,t_p)$ is linear in $a_{\pi_1},b_{\pi_1},\ldots,a_{\pi_p},b_{\pi_p}$ and therefore $F_p(\pi_1,\ldots,\pi_p)$, as the infimum of linear functions, is concave in $a_{\pi_1},b_{\pi_1},\ldots,a_{\pi_p},b_{\pi_p}$, i.e., the assumptions of Theorem 4.2.11 are satisfied. Hence, there exists a consecutive optimal partition.

We next consider several modification of the joint replenishment model considered by Chakravarty, Orlin and Rothblum. Our first modification has the terms $\tau\sum_{i\in\pi_j} a_i$ and $\tau^{-1}\sum_{i\in\pi_j} b_i$ in (6.9.4) replaced, respectively, by expressions $\eta_\tau(\tau a_S)$ and $\delta_\tau(\tau^{-1}b_S)$, where $\eta_\tau(\cdot)$ and $\delta_\tau(\cdot)$ are real valued functions. This formulation allows one to introduce discounting on the holding cost as well as the setup cost, where the value of the discount depends on the monetary volume. Under the above modification, (6.9.4) has to be replaced by

$$F_p(\pi_1, \ldots, \pi_p) = \inf_{t \in T} \sum_{j=1}^{p} \left(\eta_{t_j}(t_j a_{\pi_j}) + \delta_{t_j}(t_j^{-1} b_{\pi_j})\right) . \tag{6.9.6}$$

It is easily seen that if the function η_τ and δ_τ are concave, then the assumptions of Theorem 4.2.11 hold. Again, there exists an optimal grouping of the products where each set is consecutive. The concavity assumption of the functions η_τ and δ_τ is reasonable as it represents higher discounting as the volume increases.

Next, consider the case in which the cost of ordering in a period is determined by a concave function $\eta(\cdot)$ of the total quantity ordered in that period (the value of orders at a given period accounts for the reorders from different groups). Let q be the least common multiple of the order cycles. Then the total cost of ordering is:

$$\sum_{z=0}^{q-1} \eta \left(\sum_{t_j | z} t_j a_{\pi_j} \right) \tag{6.9.7}$$

where $t_j|z$ means that z is an integral divisor of t_j. As before, the infimum of concave functions is concave, and thus there is a consecutive optimal partition.

The next modification of the joint replenishment model has costs discounted based on monetary volume. Specifically, we consider various *discounting functions* that transform undiscounted costs into discounted costs. A reasonable property of such functions is concavity, which represents higher discounting (on marginal cost) as the monetary volume increases. We consider two main forms of discounting. In the first, costs of each group are discounted independently, based on the undiscounted costs associated with each particular group. In the second, discounting is based on total cost at each particular period. Throughout, as before, for each item i, we use the notation $a_i = 2^{-1} h_i D_i$ and $b_i = K_i$, where h_i is the undiscounted unit inventory holding cost per period, D_i is the (deterministic) demand rate and K_i is the undiscounted cost of placing an order. Also, for a set of items S, let $a_S = \sum_{i \in S} a_i$ and $b_S = \sum_{i \in S} b_i$.

We first consider the case in which costs associated with each particular group are discounted independently. Let $\gamma^j(\cdot)$ be the discounting function of the j-th group for the cost of placing an order. Then, if π_j is the j-th group, the corresponding discounted cost for placing an order is

$$\gamma^j\left(\sum_{i\in\pi_j} K_i\right) = \gamma^j(b_{\pi_j}).$$

We will examine two forms of discounting the holding costs: *contractual* and *periodic*. Under *contractual discounting*, the undiscounted holding cost for the entire order cycle is discounted. Let $\eta^j(\cdot)$ be the discounting function of the j-th group for the holding cost for the entire order cycle. Then, if π_j is the j-th group and τ is the order cycle, the corresponding discounted holding cost is

$$\eta^j\left\{\int_0^\tau \left[\sum_{i\in\pi_j}(\tau-\xi)D_ih_i\right]d\xi\right\} = \eta^j(\tau^2 a_{\pi_j}).$$

Under *periodic discounting*, the undiscounted holding cost at each period is discounted separately. Let $\eta^j(\cdot)$ be the corresponding discounting function of the j-th group. Then, if π_j is the j-th group and τ is its order cycle, the corresponding discounted holding cost is

$$\int_0^\tau \eta^j\left[\sum_{i\in\pi_j}(\tau-\xi)D_ih_i\right]d\xi = \int_0^\tau \eta^j\left[2(\tau-\xi)a_{\pi_j}\right]d\xi\ .$$

Consequently, if partition $\pi = (\pi_1,\ldots,\pi_p)$ is used with vector of order cycles $t = (t_1,\ldots,t_p)$, the total average cost per unit time under contractual discounting of holding cost is

$$c_p(\pi,t) = \sum_{j=1}^{p} t_j^{-1}[\eta^j(t_j^2 a_{\pi_j}) + \gamma^j(b_{\pi_j})]\ ,$$

and under periodic discounting of holding cost is

$$c_p(\pi,t) = \sum_{j=1}^{p} t_j^{-1}\left\{\int_0^\tau \eta^j[2(\tau-\xi)a_{\pi_j}]d\xi + \gamma^j(b_{\pi_j})\right\}\ .$$

The minimum average cost per unit time associated with partition Π is then given by (6.9.4) and the problem remains a 2-dim, size, sum-partition problem, with $f_p : \mathbb{R}^{2\times p} \to \mathbb{R}$ given by

$$f_p(X) = \inf_{t\in T}\sum_{j=1}^{p} t_j^{-1}\left\{\int_0^1 \eta^j[2(\tau-\xi)X_{1j}]d\xi + \gamma^j(X_{2j})\right\}$$

for each $X \in \mathbb{R}^{2\times p}$. It is natural to assume that the discounting functions η^j and γ^j are nonnegative and concave, in which case we have that f_p is concave. So, the assumptions of Theorem 4.2.11 are satisfied. Hence, there exists an optimal partition consisting of consecutive sets.

We next consider the case in which costs are discounted at each period, based on the total corresponding undiscounted costs at that period. In this case, we cannot compute discounted cost for each particular group. Alternatively, discounted costs must be calculated with respect to a given partition, say π, and a corresponding vector of ordering cycles, say t. For simplicity, we assume that $t_1, \ldots, t_p$ are integers. Let $q(t)$ be the least common multiple of $t_1, \ldots, t_p$. We next calculate the average cost per unit time over a *partition cycle* consisting of q consecutive time periods. Without loss of generality, we assume that the partition cycle starts in period zero. For $p = 0, \ldots, q(t) - 1$, let $J(p,t)$ be the set of all indices $j \in \{1, \ldots, p\}$ such that an order for the items in group j takes place in period p. Let $\gamma(\cdot)$ be the discounting function of the cost of placing an order. Then, the discounted costs of placing all orders that are due in period p is

$$\gamma\left[\sum_{j\in J(p,t)}\left(\sum_{i\in\pi_j} K_i\right)\right] = \gamma\left(\sum_{j\in J(p,t)} b_{\pi_j}\right) .$$

We next consider, as before, both contractual and periodic discounting of holding costs. Let η be the corresponding discounting function. Under contractual discounting of holding costs, the corresponding discounted holding cost in period p is

$$\eta\left[\sum_{j\in J(p,t)}\left(\sum_{i\in\pi_j} 2^{-1}t_j^2 D_i h_i\right)\right] = \eta\left(\sum_{j\in J(p,t)} t_j^2 a_{\pi_j}\right) .$$

Alternatively, under periodic discounting of holding costs, the corresponding discounted holding cost in period p is

$$\eta\left\{\sum_{j=1}^{p}\sum_{i\in\pi_j}[t_j - \xi_j(p)]D_i h_i\right\} = \eta\left\{\sum_{j=1}^{p} 2[t_j - \xi_j(p,t)]a_{\pi_j}\right\} ,$$

where $\xi_j(p)$ is last ordering period of items in the j-th group at or before period p. Therefore, the total average cost per unit time under contractual discounting of holding cost is

$$c_p(\pi,t) = q(t)^{-1}\left\{\sum_{p=0}^{q(t)-1}\left[\eta\left(\sum_{j\in J(p,t)} t_j^2 a_{\pi_j}\right) + \gamma\left(\sum_{j\in J(p,t)} b_{\pi_j}\right)\right]\right\} ,$$

and under periodic discounting of holding cost, it is

$$c_p(\pi, t) = q(t)^{-1} \left\{ \sum_{p=0}^{q(t)-1} \left\{ \eta \left\{ \sum_{j=1}^{p} 2[t_j - \xi_j(p,t)] a_{\pi_j} \right\} + \gamma \left(\sum_{j \in J(p)} b_{\pi_j} \right) \right\} \right\} .$$

In either case we have, again, a 2-dim, size, sum-partition problem at hand. If the functions $\eta(\cdot)$ and $\gamma(\cdot)$ are concave, the corresponding function f_p is concave. So, the assumptions of Theorem 4.2.11 are satisfied. Hence, there exists an optimal partition consisting of consecutive sets.

We note that there are further variants of the above models that preserve "the concave structure" and therefore, by Theorem 4.2.11, there exist optimal solutions that are consecutive. For example, the discounting functions can be allowed to be time-dependent over each cycle. Also, the holding costs and the ordering costs need not be discounted independently, i.e., the total cost need not be the sum of the discounted holding cost and ordering cost, but can be given by alternative functions of the undiscounted holding costs and ordering costs. Finally, the discounting function of ordering cost can be used to introduce a group-ordering cost that is independent of the individual items' ordering costs (and is imposed onto the top of them).

Chakravarty [17] considered an alternative model in which purchasing costs of the different items are discounted, and the holding cost per dollar value per unit time is the same for all items. Let V_i be the undiscounted unit price of the i-th item; let D_i be the (deterministic) demand rate of item i; let A_i be the ordering cost associated with item i, and let h be the holding cost per dollar value per unit time. Also, let $a_i = V_i D_i$ and $b_i = K_i$. Then, the net cost (consisting of purchasing costs, holding costs, and ordering costs) over an order cycle of a group S of items having the same order cycle τ is

$$c(S, \tau) = \sum_{i \in S} (V_i D_i \tau + 2^{-1} r V_i D_i \tau^2 + A_i) = \tau a_S + 2^{-1} h \tau^2 a_S + b_S .$$

Discounting the purchasing costs, holding costs and ordering costs can now be introduced correspondingly, resulting in each case a 2-dim, size, sum-partition problem.

6.10 Statistical Hypothesis Testing

The problem is to partition the set of experimental outcomes into two regions: acceptance and rejection (of the null hypothesis). Let a_i and b_i denote the probabilities that outcome i occurs under the null hypothesis and the alternative hypothesis, respectively. Without loss of generality assume that

$$\frac{a_1}{b_1} \leq \frac{a_2}{b_2} \leq \cdots \leq \frac{a_n}{b_n}. \tag{6.10.1}$$

Also, let q denote the a priori probability that the null hypothesis is true and let C_1 and C_2 be the costs of type-1 and type-2 error.

From Section 1.4.10 in Vol. I, the cost of a partition $\pi = (\pi_A, \pi_R)$ is

$$F(\pi) = C_1(1-q)\sum_{i\in\pi_A} b_i + C_2 q \sum_{i\in\pi_R} a_i.$$

With

$$A = \begin{pmatrix} a_1, \ldots, a_n \\ b_1, \ldots, b_n \end{pmatrix},$$

we have that $\pi = (\pi_A, \pi_R)$ and

$$A_\pi = \begin{pmatrix} \sum_{i\in\pi_A} a_i & \sum_{i\in\pi_R} a_i \\ \sum_{i\in\pi_A} b_i & \sum_{i\in\pi_R} b_i \end{pmatrix}.$$

We see that $F(\pi)$ is a linear function of A_π. Hence, by Theorem 4.2.11 there exists a consecutive optimal partition. This result is commonly known as the Neyman–Pearson Lemma in Hypothesis Testing, see Neyman and Pearson [75].

6.11 Nearest Neighbor Assignment

Consider the problem of assigning each of the n households in a city to one of the p post offices as its host office. It is practical to assume that the number of households office j can host is lower bounded by L_j, and upper bounded by some U_j. Let $d(H_i, O_j)$ denote the distance from household i to office j. The assignment problem can then be treated as a bounded-shape problem of partitioning the n households to the p offices. It is reasonable that the objective is to minimize the total distances from the n households to their hosts, i.e.,

$$F(\pi) = \sum_{j=1}^{p}\sum_{i\in\pi_j} d(H_i, O_j).$$

For all practical purposes, we may consider the city to be a piece of flat land. While a city may have many high-rise buildings such that the location of a household is really a 3-dimensional vector, we will ignore the height dimension since even for a household living on the 100th floor, the residents won't go to the post office by helicopter. In practice, it is almost certain that the residents would take the elevator to go down to the first floor. Thus we will treat the 2-dimensional location of the main entrance (or a designated entrance) of a building as the common location for all households in the building. Hence the set of household sites is actually a multiset.

The most natural distance to use in this problem is perhaps the Euclidean distance. However, there are urban areas (such as the Manhattan borough of New York City) where the rectilinear distance can be the more relevant one. Further, there are variations such as slanted or curved streets, 1-way streets and no-left-turn lights. Hence the actual distance function can be a complicated one. The big surprise is that regardless of how complicated the distance function is, Example 1.1.1 guarantees that an optimal partition can be found by the efficient linear programming method.

If the distance is sum-of-square, i.e., $d(A, B) = \|A - B\|^2$, then Theorem 4.1.5(iii) guarantees the existence of an almost-separable partition and every optimal partition is weakly separable. Further, if the sum of distance in the j-th office zone is assigned a weight w_j for $j = 1, \ldots, p$, perhaps reflecting the degree of traffic jam, then Theorem 4.1.3(iii) assures the existence of a sphere-separable optimal partition.

Sometimes, host offices can take almost all reasonable number of households assigned to it, or the capacity of the offices can be easily increased to accommodate the assignment, then the assignment problem can be treated as a size partition problem. If the objective is to minimize the distance between each office and its farthest customer, then Corollary 4.1.11 assures the existence of a separable optimal partition which achieves the minimum longest distance for each office.

6.12 Graph Partitions

Let G denote the given graph with n vertices and e edges. Let $N = \{1, \cdots, n\}$ denote the index-set of the vertices which is to be partitioned and let $A = \{A^{ik} : 1 \le i < k \le n\}$ where $A^{ik} = 1$ if (i, k) is an edge and $A^{ik} = 0$ otherwise. Then the objective function is to maximize

$$F(\pi) = \sum_{j=1}^{p} f(\pi_j) = \sum_{j=1}^{p} \sum_{i,k \in \pi_j} A^{ik}.$$

Note that this model differs from the usual models in that the parameter $\{A^{ik}\}$ is not associated with a single $i \in N$, but with a pair (i, j). Such a binary relation between N and A is not covered in our book.

6.13 The Traveling Salesman Problem

The traveling salesman problem is really a sequencing problem rather than a partition problem since the partition portion is indeed trivial. A part in a partition corresponds to a site, hence the set of partitions corresponds to the $n!$ permutations of the n sites. The difference between two partitions come from their sequences of labeling. Let π_σ denote the partition corresponding to a permutation sigma and let σ^j denote the element in π_j. The objective function is

$$F(\pi_\sigma) = \sum_{j=1}^{n} d(\sigma^j, \sigma^j + 1),$$

where the superscript runs in mod(n) and d is the distance function. Again d is defined on a binary relation on N, hence not covered in our book.

6.14 Vehicle Routing

Unlike the traveling salesman problem, the vehicle routing problem has a genuine partition portion which partitions the n students into p groups (buses) and then determine the best route for each such group (which is the traveling salesman problem). If we insist on an exact optimal solution, then we have to solve the traveling salesman problem first, which we know takes exponential time. If we settle for a heuristic, then some results in this book can help the partition portion. For example, it is reasonable to partition the n students into p groups to minimize the sum of radii (or diameters, or perimeters) over the p groups. Then Theorem 4.1.14 can be used to assure the existence of a separable optimal partition, which can be computed in polynomial time.

6.15 Division of Property

Consider the problem of partitioning n items among p players where player j places a value A_j^i on item i. Thus the set A to be partitioned is a $p \times n$

matrix where A_j^i is the entry in cell (i, j). The utility to player j from getting a bundle S of items is assumed to be $h_j(S) = \sum_{i \in S} A_j^i$. The welfare function of a p-partition π is given by

$$F(\pi) = f(h_1(\pi_1), \cdots, h_p(\pi_p)) = f(\sum_{i \in \pi_1} A_1^i, \cdots, \sum_{i \in \pi_p} A_p^i) \tag{6.15.1}$$

and the goal is to maximize F over a given family Π of partitions.

The objective function in (6.15.1) has several unique features. First, A^i is in $\mathbb{R}^d = \mathbb{R}^p$, i.e., $d = p$. Secondly, although (6.15.1) looks like a sum-partition problem, it is not. In a sum-partition problem, the sum of the p variables in f is $\sum_{j=1}^n A^i$, a constant independent of π. In (6.15.1), the sum is $\sum_{j=1}^p \sum_{i \in \pi_j} A_j^i$, a number depending on π. Thirdly, while A^i is in $\mathbb{R}^p$, f is a function defined on $\mathbb{R}^1$. These features have the following consequence. From feature two, the division of property problem does not fit the sum-partition problem format. Thus we have to look elsewhere for solution. From feature three, we can use the 1-dim methods to solve this p-dim problem. In particular, when f is linear, then it is easily verified that $F(\pi)$ has the monotone marginal structure, and the dynamic programming method introduced at the very end of Section 7.5 in Vol. I (the writing of that section contains some errors; but this section is self-sufficient) does apply.

First, consider the single-shape problem with shape $(n_1, \cdots, n_p)$. Let $g_k^i(n_1', \cdots, n_p')$ denote the welfare function of an optimal partition of the set $\{A^1, \cdots, A^i\}$ into p parts with shape $(n_1', \cdots, n_p')$ satisfying $\sum_{j=1}^p n_j' = i$ and $A^i \in \pi_k$, and let

$$g^i(n_1', \cdots, n_p') = \max_{\substack{k=1,\ldots,p \\ n_k'>0}} \{g_k^i(n_1', \cdots, n_p')\}. \tag{6.15.2}$$

Because f is linear, we have that

$$g_k^i(n_1', \cdots, n_p') = g^{i-1}(n_1', \cdots, n_k' - 1, \cdots, n_p') + h_k(A_k^i) \quad \text{when } n_k' > 0, \tag{6.15.3}$$

where h_k is a function depending on k. Combining (6.15.2) with (6.15.3), we obtain the recursive equation

$$g^i(n_1', \cdots, n_p') = \max_{\substack{k=1,\ldots,p \\ n_k'>0}} \{g^{i-1}(n_1', \cdots, n_k' - 1, \cdots, n_p') + h_k(A_k^i)\}. \tag{6.15.4}$$

An optimal partition is obtained by determining $g^n(n_1, \cdots, n_p)$ and tracing the path of $g^1, g^2, \cdots, g^{n-1}$ which leads to this g^n. So the question becomes how many of the g^i functions we have to compute. At first glance, it seems that for each i, g^i can be obtained by comparing up to p choices of parts (to contain i). Since i goes through $1, \cdots, n$, we may have to compute p^n g^i functions. But there is a telescopic way to compute the size of the set of g^i functions. For a given vector $(n_1, \cdots, n_p)$, define its down set $D = \{(n'_1, \cdots, n'_p) : 0 \le n'_j \le n_j\}$. Then $|D| = \prod_{j=1}^{p}(n_j + 1)$ since at position j, any number from 0 to n_j can be chosen. It is easily verified that $|D|$ is maximized when the n_j's are about equal and the maximum is about $(n/p+1)^p = O(n^p)$. For each $i = 1, \cdots, n$ with shape $(n'_1, \cdots, n'_p)$, we need to compute all members $(n'_1, \cdots, n'_k - 1, \cdots, n'_p)$ of D with $\sum_{j=1}^{p} n'_j = i$. Therefore the number of g^i functions to be computed is just $|D|$. Further, each g^i function can be computed by comparing at most p constants. So the time complexity is bounded by $pO(n^p) = O(n^p)$.

Example 6.15.1. Let $n = 4, p = 2$ and

$$A = (A_k^i) = \begin{pmatrix} 8\ 5\ 1\ 2 \\ 2\ 4\ 9\ 3 \end{pmatrix}.$$

Consider the single-shape problem with shape $(1, 3)$. Then the dynamic programming can work as the following table shows. The values of the g^i functions are computed from the upper left to the lower right.

$g^0(0,0) \equiv 0$	$\leftarrow$	$g^1(1,0) = g^0(0,0) + A_1^1 = 8$
		$\uparrow$
$g^1(0,1) = g^0(0,0) + A_2^1 = 2$		$g^2(1,1) = \max\{g^1(0,1) + A_1^2, g^1(1,0) + A_2^2\} = 12$
		$\uparrow$
$g^2(0,2) = g^1(0,1) + A_2^2 = 6$		$g^3(1,2) = \max\{g^2(0,2) + A_1^3, g^2(1,1) + A_2^3\} = 21$
		$\uparrow$
$g^3(0,3) = g^2(0,2) + A_2^3 = 15$		$g^4(1,3) = \max\{g^3(0,3) + A_1^4, g^3(1,2) + A_2^4\} = 24$

Thus, we obtain that the maximum welfare of this problem is $g^4(1, 3) = 24$ and the optimal partition is $(\{A^1\}, \{A^2, A^3, A^4\})$ by tracing back the g^i-path. □

The above dynamic programming solution also applies to bounded-shape problems. Let L_j and U_j denote the lower and upper bound of the number of items player j can get. Then the total number of p-vectors smaller than U_j but larger than L_j is $\prod_{j=1}^{p}(U_j - L_j + 1)$, which bounds the total number of recursive equations we need to solve. The time complexity of the dynamic programming solution is still polynomial because even in the single-size problem, the total number of shapes is at most $\binom{n+p-1}{p-1} = O(n^{p-1})$ and therefore the number of down shapes is bounded by $n\binom{n+p-1}{p-1} = O(n^p)$.

For $p = 2$, Granot and Rothblum [44] used the polytope approach to identify Pareto-optimal partitions. Let P^2 denote the convex hull of all 2-partitions. A 2-partition π is *Pareto-optimal* if, as a vector, it is not smaller than any other partition (also as a vector) in P^2. They proved that a Pareto-optimal partition is either a vertex of P^2 or a convex combination of two neighboring vertices whose corresponding partitions differ only in one item. Further, they proved that a Pareto-optimal partition always satisfies

$$A_1^i/A_2^i \geq A_1^k/A_2^k$$

for all $i \in \pi_1$ and $k \in \pi_2$ or vice versa. Note that A is a $2 \times n$ matrix. We can follow the text preceding Lemma 4.2.10 to reduce the problem to $d = 1$ by defining

$$A'^i = A_1^i/A_2^i.$$

Rearrange A'^i such that

$$A'^1 \leq A'^2 \leq \cdots \leq A'^n. \tag{6.15.5}$$

Then when $F(\pi)$ is as specified in Theorem 4.2.11, there exists an almost-separable optimal partition for any single-shape. Further, if the inequalities in (6.15.5) are strict, then almost-separable can be replaced by separable.

6.16 The Consolidation of Farmland

In some regions of Northern Bavaria, Germany, farmers often own scattered lots over a wide area. Thus it is desirable to redistribute the lots such that each farmer owns a contiguous set of lots to save time and cost in traveling from one lot to another, subject to the condition that the worth of his (her) property before and after the redistribution varies within an acceptable range.

Brieden and Gritzmann [11] first used partition models to study this problem. Let the given farmland be split into n lots $L_1, \cdots, L_n$ owned by f farmers, and let $w(L_i)$ denote the worth of L_i (depending on size, quality and possibly other factors). Let P_j denote the set of lots owned by farmer j before the redistribution and P'_j after. Then the before-worth of farmer j is

$$W_j = \sum_{i \in P_j} w(L_i).$$

Consider a redistribution $R = (r_{11}, r_{12}, \cdots, r_{fn})$ where $r_{ji} = 1$ if $L_i \in P'_j$ and $r_{ji} = 0$ otherwise, for $j = 1, \cdots, f$ and $i = 1, \cdots, n$. The constraints on R are:

$$(1-\epsilon_j)W_j \leq \sum_{i=1}^{n} w(L_i)r_{ji} \leq (1+\epsilon_j)W_j, j = 1, \cdots, f,$$

$$\sum_{j=1}^{f} r_{ji} = 1, \tag{6.16.1}$$

where ϵ_j is the tolerance index of farmer j for worth variation of his (her) property due to redistribution.

As it is difficult to set the "contiguity" requirement into an objective function, Brieden and Gritzmann replaced it by maximizing the sum (over all pairs) of distances between lands owned by every pair of different farmers. To further simplify, a lot L_i is associated with a pair of coordinates $g_i = (g_{i1}, g_{i2})^T$. The location of P'_j is represented by its gravity center

$$c_j = \frac{1}{W_j} \sum_{i \in P'_j} w(L_i) g_i r_{ji}.$$

Then the objective function is to minimize

$$\sum_{j \neq h} \|c_j - c_h\|^2.$$

Note that c_j is not normalized by the after-worth $\sum_{i=1}^{n} w(L_i)r_{ji}$, but by the before-worth W_j. Thus, even if P'_j consists of a single lot L_i, the definition of c_j would not yield the expected $c_j = g_i$. The reason of setting the definition in this unnatural way is to make the objective function linear, hence easier to attack.

Thus the farmland redistribution problem is reduced to a 0-1 integer linear programming which is still hard to solve. Brieden and Gritzmann gave a heuristic algorithm which was put to use in real situations and proved to be effective. Here we make an observation on a special case. When ϵ_j is small for all j and the lots are about equal worth, then the worth-variation constraints essentially limit the redistribution to become a single-shape partition, i.e., the number of lots owned by each farmer is preserved since adding or subtracting a lot would cause too great a worth variation. By Lemma 1.3.11, we can use any algorithm proposed in Section 1.1 to solve the 0-1 integer linear programming problem presented here.

Borgwardt, Brieden and Gritzmann [8] studied the same problem but proposed a different model in which each farmer j is allowed to name a subset of his lots $(L_{j1}, \cdots, L_{ju_j})$, called *center lots*, free from redistribution. The goal of redistribution then becomes to assign lots to farmer j which are close to at least one of his (her) center lots. Let $L_j(i)$ be the center lot of farmer j closest to L_i and $c_j(i)$ its location. Then the objective function is to minimize

$$\sum_{j=1}^{f}\sum_{i=1}^{n} r_{ji}\|g_i - c_j(i)\|^2.$$

The constraints are similar to those in (6.16.1) except there are upper-bound lower-bound inequalities for each subgroup of lots (with $L_j(i)$ as center) owned by farmer j, not just one such inequality for each P_j' (as if the partition were to distribute lots to $\sum_{j=1}^{f} u_j$ farmers instead of f).

Bibliography

[1] Ahuja, R.K., Magnanti, T.L. and Orlin, J.B. [1993]. *Network Flows: Theory, Algorithms, and Applications*, Prentice Hall, New Jersey.

[2] Ahuja, R.K., Orlin, J.B. and Tarjan, R.E. [1989]. Improved time bounds for the maximum flow problem, *SIAM Journal on Computing* **18**, 939–954.

[3] Alon, N. and Onn, S. [1999]. Separable partitions, *Discrete Applied Mathematics* **91**, 39–51.

[4] Asano, T., Bhattacharya, B., Keil, M. and Yao, F. [1988]. Clustering algorithms based on minimum and maximum spanning trees, in *Proc. 4th Ann. Sympos. Comput. Geom.*, 252–257.

[5] Aviran, S., Lev-Tov, N., Onn, S. and Rothblum, U.G. [2002]. Vertex characterization of partition polytopes of planar point sets, *Discrete Applied Mathematics* **125**, 1–15.

[6] Barnes, E.R. and Hoffman, A.J. [1984]. Partitioning, spectra and linear programming, in *Progress in Combinatorial Optimization*, W. Pulleybank (ed.), Academic Press, Toronto, 13–26.

[7] Barnes, E.R., Hoffman, A.J. and Rothblum, U.G. [1992]. Optimal partitions having disjoint convex and conic hulls, *Mathematical Programming* **54**, 69–86.

[8] Borgwardt S., Brieden, A. and Gritzmann, P. [2011]. Constrained minimum-k-star clustering and its application to the consolidation of farmland, *Operational Research*, **11**, 1–17.

[9] Boros, E. and Hammer, P.L. [1989]. On clustering problems with connected optima in Euclidean spaces, *Discrete Mathematics* **75**, 81–88.

[10] Boros, E. and Hwang, F.K. [1996]. Optimality of nested partitions and its application to cluster analysis, *SIAM Journal on Optimization* **6**, 1153–1162.

[11] Brieden, A. and Gritzmann, P. [2004]. A quadratic optimization model for the consolidation of farmland by means of lend-lease agreements, in *Operations Research Proceedings 2003: Selected papers of the Internatonal Conferenceon Operations Research (OR2003)*, D. Ahr, R. Fahrion, M. Oswarld and G. Reinelt (eds.), Springer Verlag, 324–331.

[12] Brieden, A. and Gritzmann, P. [2012]. On optimal weighted balanced clus-

terings: gravity bodies and power diagrams. SIAM J. Discrete Math. **26**, No. 2, 415–434.

[13] Brucker, R. [1977]. On the complexity of clustering problems, in *Optimization and Operations Research*, R. Henn, B. Korte and W. Oettli (eds.), Springer Verlag, 45–54.

[14] Capoyleas, V., Rote, G. and Woeginger, G. [1991]. Geometric clusterings, *Journal of Algorithms* **12**, 341–356.

[15] Chakravarty, A.K. [1982a]. Multi-item inventory into groups, *Journal of the Operations Research Society of the United Kingdom* **32**, 19–26.

[16] Chakravarty, A.K. [1982b]. Consecutiveness rule for inventory grouping with integer multiple constrained group-review periods (unpublished manuscript).

[17] Chakravarty, A.K. [1983]. Coordinated multi-product purchasing inventory decisions with group discounts (unpublished manuscript).

[18] Chakravarty, A.K., Orlin, J.B. and Rothblum, U.G. [1982]. A partitioning problem with additive objective with an application to optimal inventory grouping for joint replenishment, *Operations Research* **30**, 1018–1022.

[19] Chakravarty, A.K., Orlin, J.B. and Rothblum, U.G. [1985]. Consecutive optimizors for a partitioning problem with applications to optimal inventory groupings for joint replenishment, *Operations Research* **33**, 820–834.

[20] Chandra, A.K. and Wong, C.K. [1975]. Worst case analysis of a placement algorithm related to storage allocation, *SIAM Journal on Computing* **4**, 249–263.

[21] Chang, F.H., Hwang, F.K. and Rothblum, U.G. [2006]. The mean-partition problem, *Journal of Global Optimization* **36**, 21–31.

[22] Chang, H.L. and Guo, J.Y. [2006]. Strongly 2-shape-sortability of vector partitions, *Journal of Combinatorial Optimization* **11**, 407–410.

[23] Cody, R.A. and Coffman, E.G. Jr. [1976]. Record allocation for minimizing expected retrieval costs on drum-like storage devices, *Journal of the Association on Computing Machinery* **23**, 103–115.

[24] Denardo, E.V. [1982]. *Dynamic Programming: Models and Applications*, Prentice Hall, Englewood Cliffs, NJ.

[25] Derman, C., Lieberman, G.J. and Ross, S.M. [1972]. On optimal assembly of systems, *Naval Research Logistics Quarterly* **19**, 564–574.

[26] Dietrich, B. [1990]. Monge sequences, antimatroids, and the transportation problem with forbidden arcs, *Linear Algebra and Its Applications* **139**, 133–145.

[27] Dorfman, R. [1943]. The detection of defective members of large populations, *Annals of Mathematical Statistics* **14**, 436–440.

[28] Drezner, Z. [1984]. The p-center problem-heuristic and optimal algorithms, *Journal of Operational Research Socity* **35**, 741–748.

[29] Du, D.Z. [1987]. When is a monotonic grouping optimal?. *Reliability Theory and Applications*, S. Osaki and J. Cao (eds.), World Scientific, NJ, pp. 66–76.

[30] Du, D.Z. and Hwang, F.K. [1990]. Optimal assembly of an s-stage k-out-of-n system, *SIAM Journal on Discrete Mathematics* **3**, 349–354.

[31] Easton, M.C. and Wong, C.K. [1975]. The effect of a capacity constraint on the minimal cost of a partition, *Journal of the Association on Computing Machinery* **22**, 441–449.

[32] Edelsbrunner, H., Robison, A.D. and Shen,X.-J. [1990]. Covering convex sets with non-overlapping polygons *Discrete Mathematics* **81**, 153–164.

[33] El-Neweihi, E., Proschan, F. and Sethuraman, J. [1986]. Optimal allocation of components in parallel-series and series-parallel systems, *Journal of Applied Probability* **23**, 770–777.

[34] El-Neweihi, E., Proschan, F. and Sethuraman, J. [1987]. Optimal allocation assembly of systems using Schur functions and majorization, *Naval Research Logistics Quarterly* **34**, 705–712.

[35] Fisher, W.D. [1958]. On grouping for maximum homogeneity, *Journal of the American Statistics Association* **53**, 789–798.

[36] Fortune, S. [1992] Voronoi diagrams and Delaunay triangulations, *Computing in Euclidean Geometry*, Du, D.Z. and Hwang, F.K. (eds.), World Scientific, pp. 193–233.

[37] Gal, S., Hollander, Y. and Itai, A. [1994]. Optimal mapping in direct mapped cache environments, *Mathematical Programming* **63**, 371–387.

[38] Garey, M.R., Hwang, F.K. and Johnson, D.S. [1977]. Algorithms for a set partitioning problem arising in the design of multipurpose units, *IEEE Transactions on Computing* **26**, 321–328.

[39] Garey, M.R. and Johnson, D.S. [1979]. *Computers and Intractability: A Guide to the Theory of NP-Completeness*, W.H. Freeman and Company, San Francisco, CA.

[40] Gilstein, C.Z. [1985]. Optimal partitions of finite populations for Dorfman-type group testing, *Journal of Statistics, Planning and Inference* **12**, 385–394.

[41] Golany, B., Hwang, F.K. and Rothblum, U.G. [2008]. Sphere-separable partitions of multi-parameter elements, *Discrete Applied Mathematics*, **156**, 838–845.

[42] Goyal, S.K. [1974]. Determination of optimum packing frequency of items jointly replenished, *Management Science* **21**, 436–443.

[43] Graham, R.L. [1969]. Bounds on multiprocessing timing anomalies, *SIAM Journal on Applied Mathematics* **17**, 416–429.

[44] Granot, D. and Rothblum, U.G. [1991]. The Pareto set of the partition bargaining game, *Games and Economic Behavior* **3**, 163–182.

[45] Harding, E.F. [1967]. The number of partitions of a set of n points in k dimensions induced by hyperplanes, *Proceedings of the Edinburgh Mathematical Society* **15**, 285–289.

[46] Hershberger, J. and Suri, S. [1991]. Finding Tailored partitions, *Journal of Algorithms* **12**, 431–463.

[47] Hoffman, A.J. [1963]. On simple linear programming problems, in *Convexity: Proceedings of Symposia in Pure Mathematics* **7**, V. Klee (ed.), American Mathematical Society, 317–327.

[48] Hoffman, A.J. [1985]. On greedy algorithms that succeed, in *Surveys in Combinatorics*, I. Anderson (ed.), Cambridge University Press, 97–112.

[49] Hwang, F.K. [1975]. A generalized binomial group testing problem, *Journal of the American Statistics Association* **70**, 923–926.

[50] Hwang, F.K. [1989]. Optimal assignment of components to a two-stage k-out-of-n system, *Mathematics of Operations Research* **14**, 376–382.

[51] Hwang, F.K., Lee, J.S., Liu, Y.C. and Rothblum, U.G. [2003]. Sortability of vector partitions, *Discrete Mathematics* **263**, 129–142.

[52] Hwang, F.K., Onn, S. and Rothblum, U.G. [1998]. Representations and characterizations of vertices of bounded-shape partition polytopes, *Linear Algebra and Its Applications* **278**, 263–284.

[53] Hwang, F.K., Onn, S. and Rothblum, U.G. [1999]. A polynomial time algorithm for shaped partition problems, *SIAM Journal on Optimization* **10**, 70–81.

[54] Hwang, F.K., Onn, S. and Rothblum, U.G. [2000a]. Explicit solution of partitioning programs over a one-dimensional parameter space, *Naval Research Logistics* **47**, 531–540.

[55] Hwang, F.K., Onn S. and Rothblum, U.G. [2000b]. Linear shaped partition problems, *Operations Research Letters* **26**, 159–163.

[56] Hwang, F.K. and Rothblum, U.G. [1995]. Some comments on the optimal assembly problem, *Naval Research Logistics* **42**, 757–772.

[57] Hwang, F.K. and Rothblum, U.G. [1996]. Directional-quasi-convexity, asymmetric Schur-convexity and optimality of consecutive partitions, *Mathematics of Operations Research* **21**, 540–554.

[58] Hwang, F.K. and Rothblum, U.G. [2005]. Partition-optimization with Schur convex sum objective functions, *SIAM Journal on Discrete Mathematics* **18**, 512–524.

[59] Hwang, F.K. and Rothblum, U.G. [2006]. A polytope approach to the optimal assembly problem, *Journal of Global Optimization* **35**, 387–403.

[60] Hwang, F.K and Rothblum, U.G. [2011]. On the number of separable partitions, *Journal of Combinatorial Optimization* **21**, 423-433.

[61] Hwang, F.K., Sun, J. and Yao, E.Y. [1985]. Optimal set partitioning, *SIAM Journal on Algorithms and Discrete Methods* **6**, 163–170.

[62] Hwang, F.K., Wang, Y.M. and Lee, J.S. [2002]. Sortability of multi-partitions, *Journal of Global Optimization* **24**, 463–472.

[63] Knuth, D. [1981]. *The Art of Computer Programming*, 2nd ed., Addison-Wesley, Reading, MA.

[64] Kodes, U.R. [1972]. Partitioning and card selection, in *Design Automation of Digital Systems*, Vol. 1, Breuer, M.A. (ed.), Prentice Hall, Englewood Cliffs, NJ.

[65] Kreweras, G. [1972]. Sur le partitions non croisées d'un cycle, *Discrete Mathematics* **1**, 333–350.

[66] Liu, Y.C. and Pan, J.J. [2011]. On the vertex characterization of single-shape partition polytopes, *Journal of Combinatorial Optimization* **22**, 563–571.

[67] Malon, D.M. [1990]. When is greedy module assembly optimal, *Naval Research Logistics Quarterly* **37**, 847–854.

[68] McMullen, P. [1970]. The maximum number of faces of a convex polytope,

Mathematicka **17**, 179–184.

[69] Meggido, N. and Supowit, K.J. [1984]. On the complexity of some common geometric location problems, *SIAM Journal on Computing* **13**, 182–196.

[70] Meggido, N. and Tamir, A. [1982]. On the complexity of locating linear facilities in the plane, *Operations Research Letters* **1**, 194–197.

[71] Mehta, S., Chandraseckaran, R. and Emmons, H. [1974]. Order-preserving allocation of jobs to two machines, *Naval Research Logistics Quarterly* **21**, 361–374.

[72] Monma, C. and Suri, S. [1991]. Partitioning points and graphs to minimize the maximum or the sum of diameters, *Proc. 6th Int. Conf. Theory and Appl. of Graphs*, Wiley, New York.

[73] Nash. J. [1950]. The bargaining problem, *Econometrica* **28**, 155–162.

[74] Nawijn, W.M. [1988]. Optimizing the performance of a blood analyser: Application of the set partitioning problem, *European Journal of Operations Research* **36**, 167–173.

[75] Neyman, J. and Pearson, E. [1933]. On the problem of the most efficient tests of statistical hypotheses, *Phylosophical Transactions of the Royal Society of London*, Series A, 231, 289-337.

[76] Nocturne, D. [1973]. Economic order frequency for several items jointly replenished, *Management Science* **19**, 1093–1096.

[77] Oikonomou, K.N. [1987]. Abstractions of finite-state machines optimal with respect to single undetectable output faults, *IEEE Transactions on Computers* **36**, 185–200.

[78] Oikonomou, K.N. [1992]. Abstractions of finite-state machines and immediately-detectable output faults, *IEEE Transactions on Computers* **41**, 325–338.

[79] Pfeifer, C.G. and Enis, P. [1978]. Dorfman-type group testing for a modified binomial model, *Journal of the American Statistics Association* **73**, 588–592.

[80] Pfersky, U., Rudolf, R. and Woeginger, G. [1994]. Some geometric clustering problems, *Nordic Journal of Computing* **1**, 246–263.

[81] Preparata, F.P. and Shamos, M.I. [1985]. *Computaional Geometry: An Introduction.* Springer-Verlag, New York - Berlin - Heidelberg - Tokyo.

[82] Rockafellar, T. [1970]. *Convex Analysis*, Princeton University Press, Princeton, NJ.

[83] Rothblum U.G. and Tangir Y. [2008]. The partition bargaining problem, *Discrete Applied Mathematics* **156**, 428–443.

[84] Rothkopf, M.H. [1975]. A note on allocating jobs to two machines, *Naval Research Logistics Quarterly* **22**, 829–830.

[85] Schrijver, A. [1986]. *Theory of Linear and Integer Programming*, John Wiley and Sons, New York.

[86] Shamir, R. and Dietrich, B. [1990]. Characterization and algorithms for greedily solvable transportation problems, in *Proceedings of the First ACM_SIAM Symposium on Discrete Algorithms*, 358–366.

[87] Shamos, M. and Hoey, D. [1975]. Closest-point problems, in *Proceedings of the 16th Ann. Symp. Found. Comput. Sci.*, 151–162.

[88] Silver, E.A. [1976]. A simple method of determining order quantities in joint replenishments under deterministic demands, *Management Science* **22**, 1351–1361.

[89] Sobel, M. and Groll, P.A. [1959]. Group testing to eliminate efficient all detectives in a binomial sample, *Bell System Technical Journal* **38**, 1179–1252.

[90] Supowit, K.J. [1981]. Topics in Computational Geometry, Ph.D. Thesis, Department of Computer Science, University of Illinois at Urbana-Champaign, Report UIUCDCS-R81-1062.

[91] Tanaev, V.S. [1979]. Optimal subdivision of finite sets into subsets, *Akad Nauk BSSR Doklady* **23**, 26–28.

[92] Ziegler, G.M. [1995]. *Lecture Notes on Polytopes*, Graduate Texts in Mathematics, Springer Verlag, New York.

Index